Skeleton Keys

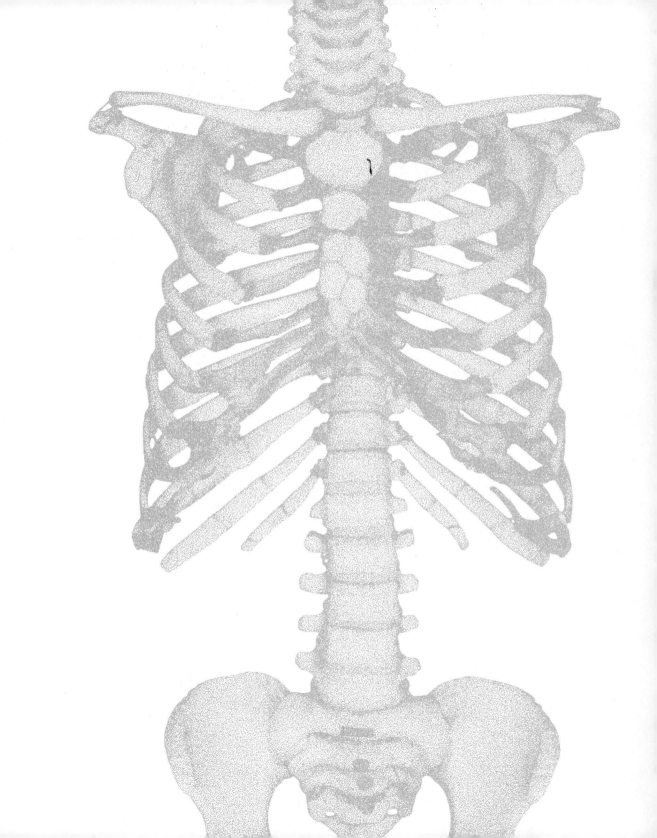

Skeleton Keys

An Introduction to Human Skeletal Morphology, Development, and Analysis

SECOND EDITION

Jeffrey H. Schwartz

University of Pittsburgh

New York   Oxford
OXFORD UNIVERSITY PRESS
2007

Oxford University Press, Inc., publishes works that further Oxford University's
objective of excellence in research, scholarship, and education.

Oxford New York
Auckland   Cape Town   Dar es Salaam   Hong Kong   Karachi
Kuala Lumpur   Madrid   Melbourne   Mexico City   Nairobi
New Delhi   Shanghai   Taipei   Toronto

With offices in
Argentina   Austria   Brazil   Chile   Czech Republic   France   Greece
Guatemala   Hungary   Italy   Japan   Poland   Portugal   Singapore
South Korea   Switzerland   Thailand   Turkey   Ukraine   Vietnam

Published by Oxford University Press, Inc.
198 Madison Avenue, New York, New York 10016
http://www.oup.com

Oxford is a registered trademark of Oxford University Press

Library of Congress Cataloging-in-Publication Data

Schwartz, Jeffrey H.
    Skeleton keys : an introduction to human skeletal morphology,
development, and analysis / Jeffrey H. Schwartz.—2nd ed.
    p. cm.
Includes bibliographical references and index.
ISBN-13: 978-0-19-518859-2 (alk. paper)
ISBN-10: 0-19-518859-4 (alk. paper)
    1. Human skeleton.   2. Forensic osteology.   I. Title.

QM101.S38 2007
611'.71—dc22

                                        2006043750

Printing number: 9 8 7 6 5 4 3 2 1
Printed in the United States of America
on acid-free paper

To Theya Molleson and in memory of Pru Napier.
Years ago they took me under their intellectual wings.

I have never forgotten.

Contents

Preface

This is not your usual textbook on how to analyze the human skeleton. First, even though I began my career as a traditional osteologist who analyzed human and animal bones from archaeological sites, soon thereafter I became engulfed in the method and theory of evolutionary systematics. That is, I became focused on how one thinks about organisms and their relationships to other organisms. The impact of this on how I thought about studying the human skeleton was irreversible. I could no longer deal with lists of characters that human osteologists had long used to identify an individual's sex, age, or supposed biological affinities. These concerns, especially with regard to affinity, are at base concerns of systematics and should be treated as such. Consequently, the approach I offer for thinking about these and related questions may seem unorthodox in the realm of human osteology, but it is commonplace among systematists.

That having been said, it is still the case that one must learn how to analyze skeletal morphology and how to put this knowledge into a meaningful comparative context. In this regard, my or anyone's theoretical and methodological bents are irrelevant. For one must first know the morphology of an organism in order to compare it to something else. And that is where we begin here, learning step by step how to appreciate the human skeleton as more than just a collection of elements of different sizes and shapes whose analysis serves merely to pigeonhole the past bearer of these bones and teeth into a particular category in terms of sex, age, and history of disease. Throughout I have sought to impart an understanding of development, including its theoretical considerations—something that has long been overlooked when doing not only human skeletal analysis but also comparative anatomy. With continuing insights into the regulation of development, this can no longer be ignored.

I also raise questions about the very nature of human skeletal analysis—from the basic "Why do it?" to the more hopeful "Why not do it better or more thoughtfully?"—that I hope will provoke the readers of this book to pursue new avenues of research. In this regard, I see this book as a primer for individuals to develop their own ideas as much as a textbook whose content is perceived as the last word.

The format of this edition has also changed. Not only are various chapters repositioned, but information that had originally been supplied in appendices has been included in the text in order to make it more easily accessible. Two appendices remain, however, one on osteometry (Appendix A) and the other a comparison of a human skeleton with those of animals whose isolated or fragmentary bones might be mistaken for a human's (Appendix B); the latter can also be used to identify these animals if their bones are encountered in the field. I have expanded the glossary in this edition, which I hope will prove useful.

This edition also includes a CD ROM of color images of virtually all of the black and white photographs that accompany the text as well as an array of specimens that represent (1) individuals that range in age from fetus to adult, (2) females and males, (3) degrees of individual variation, (4) different diseases, and (5) the effects of ante- versus postmortem insults. I have given clues to what lies in each folder but leave the depth of discovery to your inquisitiveness. There is as much and even more morphology to be plumbed from these images as you, the reader, care to learn. Let the book, with its descriptions and labeled illustrations, be your guide to the details each image contains.

Acknowledgments

This, the second edition of *Skeleton Keys*, owes its existence to the encouragement of my editor at Oxford University Press, Jan Beatty; to the generosity of Ian Tattersall, Gary Sawyer, and Kenneth Mowbray (all of the Department of Anthropology, American Museum of Natural History), who gave me unfettered access to vast collections of human skeletal material during my tenure as Kalbfleish Fellow and most recently in my quest to expand the photographic record of specimens presented in this edition; and to helpful comments (both obvious and by provocation) from reviewers of the proposal for, and especially from those who read with such care the manuscript of, this edition. I especially thank Annie Katzenberg for her support of this project and her openness in discussing parts that should be kept and those that should be changed in this edition. Throughout, with the first and now the second edition, I thank my students—undergraduate and graduate—who took seriously my request to comment on the material that would be included in this book and for their comments on how they would like to see the second edition improve upon the first.

Thanks again to Jon Anderton and Tim D. Smith for the drawings and to David Burr and Elizabeth Dumont for photomicrographs that are reproduced from the first edition. Thanks also to Douglas Ubelaker for allowing me to reproduce his dental-aging chart and to Owen Lovejoy for giving me his ilial auricular surface images to include in the accompanying CD ROM. David Stump organized and color-adjusted the color images included in the accompanying CD ROM and produced the new black and white images for this edition. My doctoral students Mary Elizabeth Kovacik and J. Christopher Reed helped me locate specimens whose morphologies and pathologies are presented in this edition and its accompanying CD ROM.

And for Lynn Emanuel, there is a level beyond thanks that exists for again enduring the years of effort that went into my "last" book.

Chapter 1

An Introduction to Skeletal Anatomy and the Development, Physiology, and Biochemistry of Bone

We begin with an overview of the skeleton and the terminology of anatomical direction, which allows the osteologist to orient a bone or bone fragment in its proper anatomical position and determine the side of the body from which it came. Since the philosophy behind this book is to provide multiple avenues to learning bone identification, I have described the positions of specific anatomical features and landmarks, as well as the positions of anatomical features relative to one another, to supplement the labeled drawings in the text. There is also an accompanying CD with hundreds of color photographs of the bones of fetuses, children, subadults, and adults in myriad positions, including, for the first time in any textbook, comparisons of bones from the right and left sides. These images should be used in conjunction with the labeled drawings to identify landmarks. From the broader perspective, we then focus on finer details: the anatomy of a bone, its organic and nonorganic constituents, its development at the molecular and cellular levels, and its subsequent growth, both normal and abnormal. Specific growth-related changes of a bone in shape and size are summarized in the relevant chapter.

The Terminology of Anatomical Position

It would be practical if one could refer to the sides of an animal's body as "front," "back," "up," "down," "right side," and "left side" and to relative positions of structures as "toward the head," "away from the head," "toward the middle of the body," and "away from the middle of the body." Since this intuitively obvious approach is considered unscientific, we have to use a set of standardized terms. But the situation is further complicated by the fact that while we humans, like dogs, for instance, are tetrapods (four-legged vertebrates), we walk upright and bipedally and our skull sits atop a vertebral column that, like our arms and legs, is vertically oriented. In contrast, the skull of dogs and most other mammals lies at the front of an essentially horizontal vertebral column that is suspended between vertical fore- and hindlimbs. Consequently, the sides of the

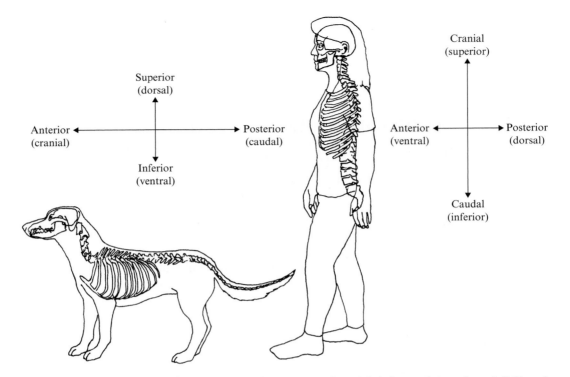

Figure 1–1   Terms used for defining anatomical position in the axial skeleton of a quadruped (*left*) and a biped (*right*).

same bone could be identified differently in a human and in a dog because of its different orientation.

The following terms and definitions are applicable to all vertebrates (see Figures 1–1 to 1–3): **anterior** (toward the front of the body, in the direction of the eyes), **posterior** (toward the rear of the body, in the direction away from the eyes), **cranial** (toward the head), **caudal** (toward the tail or tail region), **superior** (toward the highest or uppermost part of a bone), **inferior** (toward the lower or underside of a bone), **proximal** (toward the point of attachment or articulation of a bone), **distal** (away from the point of attachment or articulation of a bone), **dorsal** (toward the vertebral column), **ventral** (toward the belly), **medial** (toward the midline of the body), **lateral** (away from the midline of the body).

Because humans are upright, bipedal animals, some of these terms are interchangeable (Figures 1–1 to 1–3).

In humans, the equivalent pairs are "anterior" and "ventral" and "posterior" and "dorsal." As a rule, we will use "anterior" and "posterior" but will substitute "dorsal" and "ventral" in certain circumstances. As in referring to leg bones in quadrupeds, "anterior" and "posterior" are used to identify the front and back sides of a human's long bones of the arm or leg. This would be the terminology used for these bones in quadrupeds. For example, it is perhaps more meaningful to refer to the relevant sides of the human sternum (breastbone), the scapula (shoulder blade), or the os coxae (pelvis, innominate, or hipbone) as "dorsal" and "ventral" rather than "anterior" and "posterior."

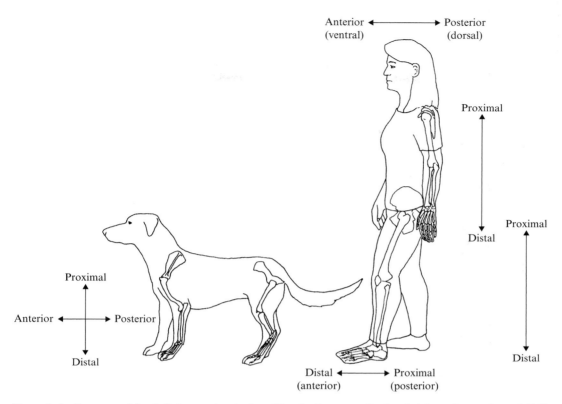

Figure 1–2   Terms used for defining anatomical position in the appendicular skeleton of a quadruped (*left*) and a biped (*right*).

In a quadruped, "cranial" and "caudal" are equivalent to the directions "anterior" and "posterior," respectively, in a human. In humans, the terms "cranial," "superior," and (in most cases) "proximal" are synonymous, as are "caudal," "inferior," and (in most cases) "distal." "Cranial" and "caudal" most appropriately refer, for example, to opposite sides of a vertebra; one side points toward the head and the other toward the tail, regardless of whether the animal is a biped or a quadruped. As defined above, "proximal" refers to the end of a bone that attaches to the body or the end of the bone that is closer to the appendage's attachment to the body. Thus, the upper end of the humerus is its proximal end because it attaches to the body via the shoulder joint,

and the upper end of the radius is its proximal end because it attaches to the body via the humerus and shoulder joint. The end of a bone that is farther away from the appendage's attachment to the body is its "distal" end.

Summaries of terms of direction and position and of terms reflecting the major planes of orientation are provided in Tables 1–1 and 1–2, respectively.

Units of the Skeleton and Their Articular Relations

The skeleton is often subdivided in two different ways. In one scheme, the head region is referred to as the **cranial skeleton** and the rest of the skeleton is the **postcranial skeleton**

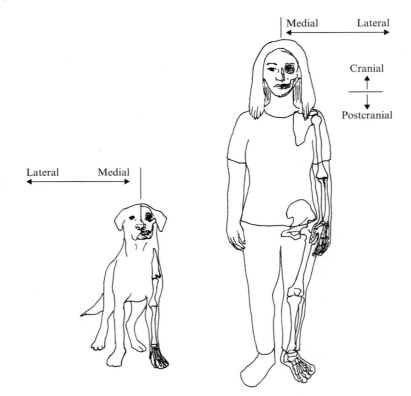

Figure 1–3   Terms used for defining anatomical position in the appendicular skeleton of a quadruped (*left*) and a biped (*right*).

(the latter being most literally applicable to quadrupedal animals in which the bulk of the skeleton, indeed, does lie posterior to the skull; in this regard, **infracranial** is sometimes used when discussing the human skeleton). The other scheme distinguishes **axial** (skull, vertebral column, sternum, ribs) and **appendicular skeletons** (limbs, scapula, clavicle, os coxae).

The region of contact between two or more bones is a **joint**. There are different categories of joints, each defined by certain ranges of motion and tissue components (see Figures 1–4 and 1–5). From the most to least mobile, the three major categories of joints are **synovial**, **cartilaginous**, and **fibrous**. A summary of terms of movement is provided in Table 1–3.

A **synovial** joint [obsolete terminology, *diarthrosis* (pl. *-es*)] is basically a freely movable joint; it is characterized by the presence of lubricated articular cartilage on the opposing bony surfaces and an articular cavity that is bound by a fibrous joint capsule lined with a fluid-secreting synovial membrane. Types of synovial joints are **plane**, **hinge**, **pivot**, **ellipsoidal**, **saddle**, and **ball-and-socket**. Plane, hinge, and pivot joints are *monaxial* joints (i.e., motion is confined to a single axis); ellipsoidal and saddle joints are *biaxial*; and ball-and-socket joints are *multiaxial*. Composite joints incorporate features of more than one joint type (e.g., although often identified as a hinge joint, the knee is actually a sliding hinge joint).

Table 1–1   Terms of Direction and Position Pertaining Especially to the Human Skeleton

Anatomical Position: In humans, lying with palms facing forward.

Anterior (also *frontal*, *ventral*): Toward the front; opposite of **posterior**.

Apical: Pertaining to the apex or highest part of a structure; opposite of **basilar**.

Basilar: Pertaining to the base or lowest part of a structure; opposite of **apical**.

Contralateral: Refers to a structure, feature (even motion), etc., on the side of the body opposite that on which a structure, etc., of interest occurs; opposite of **ipsilateral**.

Distal: (Limbs) Away from the trunk; opposite of **proximal**.

Dorsal (also *posterior*): Pertaining to (1) the back (i.e., posterior surface) of the hand or (2) the superior surface of the foot; opposite, respectively, of **palmar** and **plantar**.

External (also *exterior*; sometimes incorrectly referred to as *lateral*): Toward the outside; opposite of **internal**.

Inferior [also *caudal* (toward the tail)]: Toward the feet; opposite of **superior**.

Internal (also *interior*; sometimes incorrectly referred to as *medial*): Toward the inside; opposite of **external**.

Ipsilateral: Referring to a structure, feature (even motion), etc., occurring on the same side of the body as the structure, etc., of interest; opposite of **contralateral**.

Lateral: Away from the median plane; opposite of **medial**.

Medial: Toward the median plane; opposite of **lateral**.

Palmar (also *volar*): The anterior surface or palm of the hand; opposite of **dorsal**.

Plantar (also *volar*): The inferior surface or sole of the foot; opposite of **dorsal**.

Posterior (also *dorsal*): Toward the back; opposite of **anterior**.

Prone: Facing downward with the back up; opposite of **supine**.

Proximal: (Limbs) Toward the trunk; opposite of **distal**.

Superior (also *cephalic*, *cranial*, and *rostral*): Toward the head; opposite of **inferior**.

Supine: Facing upward with the back down; opposite of **prone**.

Table 1–2   Terms Related to Major Planes, Especially of the Human Skeleton

Anteroposterior (AP) plane: see **median plane**.

Coronal plane: A vertical plane that passes through the body (or structure, e.g., skull) from side to side (i.e., parallel to the coronal suture) and divides the body (or structure) into anterior and posterior portions; it lies perpendicular to the **median plane**.

Frankfort horizontal (FH) plane (also *plane of Virchow*): A horizontal plane on which the anthropometric landmarks porion (i.e., the midpoint of the superior margin of the external auditory meatus) and orbitale (i.e., the inferiormost point on the inferior margin of the orbit) are positioned in order to provide a standard orientation of the skull (e.g., for measuring, describing, illustrating).

Frontal plane: see **coronal plane**.

Horizontal plane: see **transverse plane**.

Median (sagittal) plane: A vertical plane that passes through the midline of the body (or the skull coincident with the sagittal suture) from front to back, dividing it into symmetrical left and right halves; it lies perpendicular to the **coronal plane**.

Midsagittal plane: see **median plane**.

Oblique plane: A plane not parallel to the **coronal**, **median**, or **transverse planes**.

Paramedian (or parasagittal) plane: A vertical plane parallel to the **median plane**.

Sagittal plane: see **paramedian plane**.

Transverse plane: A horizontal plane that passes through the body (or structure) at right angles to both the **median** and **coronal planes**, dividing it into superior and inferior portions and creating a cross section of the body (or structure).

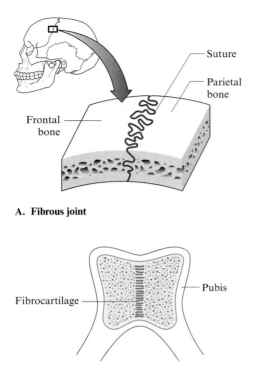

A. Fibrous joint

B. Cartilaginous joint

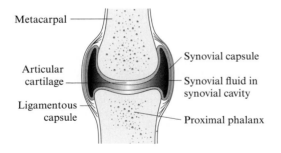

C. Synovial joint

Figure 1–4   Schematic representations of fibrous (*A*), cartilaginous (*B*), and synovial (*C*) joints (see text for detail) [reprinted with permission from Langdon (2005); copyright © Oxford University Press].

Definitions of terms of movement are presented in the Glossary.

In a **plane joint** (syn. *gliding* or *sliding joint*), the opposing bones can slide across one another because their articular surfaces are flat or nearly flat [the acromioclavicular (shoulder blade–collarbone), sacroiliacal (posterior pelvic), proximal tibiofibular, intercarpal (wrist), carpometacarpal (wrist–palm, except that of the pollex), and intermetacarpal (palm) joints and the corresponding joints of the foot]. Movement across a plane joint is limited. In a **hinge joint** (syn. *ginglymus*), the motions of flexion and extension of opposing bones rotate around an axis that is at a right angle to the long axis of these bones [e.g., tibiofemoral (knee), talocrural (ankle), humeroulnar (elbow), and interphalangeal joints]. In a **pivot joint** (syn. *trochoidal joint*) one bone rotates around the long axis of the opposing bone [e.g., the atlantoaxial joint (between the first and second cervical vertebrae) and proximal and distal radioulnar joints].

An **ellipsoidal joint** (syn. *condyloid joint*) allows movement in two axes that are at right angles to one another [e.g., metacarpophalangeal (palm–finger), metatarsophalangeal (midfoot–toe), radiocarpal (wrist) joints]. The shapes of the articular surfaces of opposing bones are ellipsoidal when viewed on end. One bone's articular surface is mildly convex in two dimensions, but the curvature is unequal along the two axes that define the ellipse; the opposing bone's articular surface is a concave mirror image of the former. The unequal curvature of these articular surfaces limits rotation. However, flexion, extension, abduction (movement away from the midline of the body), adduction (movement toward the midline of the body), as well as the composite movement of circumduction (pivotal rotation at the joint with the end that is away from the joint transcribing a circle) are all possible.

The carpometacarpal joint of the pollex (i.e., the joint at the base of the thumb) is a **saddle joint** (syn. *sellar joint*). The articular surfaces of opposing bones are reciprocally

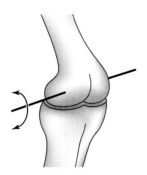
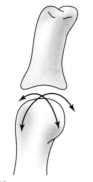
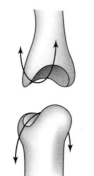
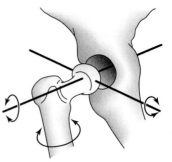

(a)
One degree–a hinge joint such as a knee or elbow (specifically between the ulna and humerus)

(b)
Two degrees–a condylar joint, such as a metacarpophalangeal joint

A sellar joint at the base of the thumb (first carpometacarpal, between the metacarpal and the trapezium)

(c)
Three degrees–a ball and socket joint such as the shoulder or hip

Figure 1–5   Schematic representation of degrees of movement of various synovial joints [reprinted with permission from Langdon (2005); copyright © Oxford University Press].

concavoconvex (resembling two horses' saddles) and fit snugly together because one surface is rotated 90° relative to the other. A saddle joint is capable of all the motions of an ellipsoidal joint as well as of some rotation.

A **ball-and-socket joint** [syn. *enarthrosis* (pl. *-es*) or *spheroidal joint*] has the widest range of possible motions: flexion, extension, abduction, adduction, circumduction, and medial and lateral rotation. The articular surface of one bone is ball-shaped and fits into a variably cup-shaped articular surface of the opposing bone [e.g., the glenohumeral (shoulder), coxofemoral (hip), and incudostapedial (inner ear bone) joints].

A **cartilaginous joint** [obsolete terminology, *amphiarthrosis* (pl. *-es*)] is formed when bones are connected by either hyaline cartilage or fibrocartilage. In a **synchondrosis** (pl. *-es*; syn. *primary* or *hyaline cartilaginous joint*), the bones are connected by a plate of hyaline cartilage; there is little or no movement between contiguous bones. A synchondrosis may be temporary [e.g., the epiphyseal plate between the diaphysis (shaft) and the epiphysis (end) of a growing bone] or permanent

(e.g., the costal cartilage that connects the first rib to the manubrium of the sternum or in the juncture of the sphenoid and occipital bones). In a **symphysis** (pl. **-es**; syn. *secondary cartilaginous* or *fibrocartilaginous joint*), a thin layer of hyaline cartilage covers the articular surfaces of opposing bones that are connected by a plate of motion-limiting fibrocartilage. Symphyseal joints occur between vertebrae (intervertebral joints), the manubrium and body of the sternum (manubriosternal joint), the xiphoid process and the body of the sternum (xiphisternal joint), and right and left pubic bones (pubic symphysis).

Sutures, **syndesmoses**, and **gomphoses** are **fibrous joints** [obsolete terminology, *synarthrosis* (pl. *-es*)] that are held together by bands of fibrous tissue. With the least amount of fibrous tissue of any fibrous joint, sutures (the joints between the flat bones of the skull) have the greatest amount of movement. **Sutures** may fuse or close over completely (e.g., with increasing age) and consequently transform a fibrous joint into a bony joint called a *synostosis*. In a **syndesmosis**, the abundance of fibrous connective tissue may

Table 1–3   Terms of Movement

Abduction: A laterally directed movement in the coronal plane away from the median sagittal plane; opposite of **adduction.**

Adduction: A medially directed movement in the coronal plane toward the median sagittal plane; opposite of **abduction.**

Circumduction: A circular movement created by the sequential combination of **abduction, flexion, adduction,** and **extension.**

Depression: An inferiorly or inwardly directed movement; opposite of **elevation.**

Dorsiflexion: A bending of the foot that decreases the angle between it and the anterior surface of the leg (e.g., squatting); opposite of **plantar flexion.**

Elevation: A superiorly or outwardly directed movement; opposite of **depression.**

Eversion: A movement of the foot that causes its plantar surface to face laterally; opposite of **inversion.**

Extension: A movement in the sagittal plane around a transverse axis that separates two structures (e.g., straightening the leg or arm); opposite of **flexion.**

Flexion: A bending movement in the sagittal plane and around a transverse axis that draws two structures toward each other; opposite of **extension.**

Inversion: A movement of the foot that causes its plantar surface to face medially; opposite of **eversion.**

Lateral rotation: Movement of a structure around its longitudinal axis, causing its anterior surface to face laterally; opposite of **medial rotation.**

Medial rotation: Movement of a structure around its longitudinal axis, causing its anterior surface to face medially; opposite of **lateral rotation.**

Opposition: A movement of the thumb across the palm such that its "pad" contacts the "pad" of another digit; this movement involves **abduction** with **flexion** and **medial rotation** at the carpometacarpal joint of the thumb.

Plantar flexion: A bending of the foot in the direction of its plantar surface such that the top of the foot lies in the same plane as the anterior surface of the leg (e.g., standing on tiptoe); opposite of **dorsiflexion.**

Pronation: A medial rotation of the forearm, which causes the palm of the hand to face posteriorly and the radius to cross over the ulna: opposite of **supination.**

Protraction (also *protrusion*): An anteriorly directed movement, usually used to describe the forward movement of the mandible at the temporomandibular joint; opposite of **retraction.**

Retraction: A posteriorly directed movement, usually used to describe the backward movement of the mandible at the temporomandibular joint; opposite of **protraction.**

Supination: A lateral rotation of the forearm that causes the palm of the hand to face anteriorly and the ulna and radius to lie parallel to one another; opposite of **pronation.**

form a ligament or an interosseous membrane. A syndesmosis permits slight-to-moderate movement between bones (e.g., as in the distal tibiofibular joint and the radioulnar and tibiofibular interosseous membranes). The fibrous connection between the root of a tooth and its alveolus is a **gomphosis**, in which movement is limited.

Gross Anatomy of Bone

A bone is not uniform in its characteristics. In **adult** or **mature bone**, the outer layer (variably called the **cortex, cortical bone,** or **compact bone**) is hard and denser than the bone it surrounds (Figure 1–6). The compact bone of an articular region is typically smoother and more finely granular in quality than compact bone elsewhere on the same bone. In its dried state, bone that is encased by compact bone is characterized by a network of thin bony plates (**trabecula,** pl. -ae) that surround variably small air cells; in living bone, these "air" cells contain the soft, fibrous connective tissue collagen. In postcranial bones, the air-celled or *pneumaticized* bone is

Figure 1–6 Examples of postcranial cortical (*large arrow*) and cancellous (*small arrow*) bone: (*left*) shaft of tibia also illustrating medullary cavity and nutrient foramen (*asterisk*); (*right*) proximal tibia.

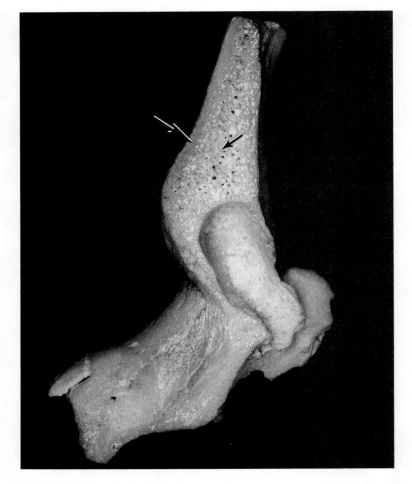

Figure 1–7 Example of cranial cortical bone (*large arrow*) and diploë (*small arrow*): parasagittal section of frontal bone (with frontal sinus and crista galli behind it), left nasal bone, and frontal process of left maxilla.

spongy or **cancellous bone** (Figure 1–6); in cranial bones, it is referred to as **diploë** and is sandwiched between outer as well as inner layers or tables of compact bone (Figure 1–7).

Technically, postcranial spongy bone is **secondary spongiosa**, which overlies and is continuous with **primary spongiosa**, which, in turn, is a continuation of the cartilaginous matrix that precedes the deposition of bone. In general, the degree to which spongy bone is densely or loosely "packed" depends on the region of a bone itself under study and the age of the individual. With age, compact bone thins and spongy bone becomes less densely concentrated (e.g., in long bones, it "retreats" to the ends of the shaft). Many postcranial bones develop a **medullary cavity** (Figure 1–6) within the core, which enlarges as bone **marrow** is laid down. **Nutrient arteries** branch or arborize within the medullary cavity and send finer capillary branches into the walls of the bone itself. With age, the

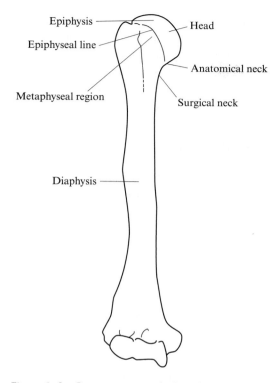

Figure 1–8   Gross anatomy of a long bone.

medullary cavity becomes even more pervasive throughout the diaphysis as spongy bone decreases.

Many postcranial bones are longer than wide and have two identifiable ends. This describes long bones as well as bones of the fingers and toes (phalanges), the palm (metacarpals), and the sole (metatarsals). The somewhat tubular portion of these bones is the shaft or **diaphysis**, and one or both ends may bear an **epiphysis** (pl. -es) (Figure 1–8). Developmentally, the diaphysis and its epiphyses are separate elements that eventually fuse, typically leaving as evidence of their former separateness a thin, indented **epiphyseal line** around the perimeter. The diaphyseal region with which an epiphysis merges is the **metaphysis**. Juvenile and young adult age can be estimated from the sequence of

epiphyseal–diaphyseal fusion (see Chapters 5, 6, and 8). In skeletons of young individuals, unfused epiphyses will be separate from diaphyses, and care should be taken during archaeological excavation or at a potential crime scene to collect them.

The upper or proximal epiphysis of a long bone is the **head** (which, as in the femur and humerus, may be rounded). However, the round head of a metatarsal or metacarpal lies at the distal end of the bone, and only in the second to fifth metatarsals and metacarpals is the head a true epiphysis. A rib's head articulates with the vertebral column (the other end is its sternal end); the right and left heads of the mandible are the projections that articulate with the skull; the head of the scapula articulates with the head of the humerus. With the exception of the first metatarsals and metacarpals, all other osteological heads are true epiphyses. A bone may have two ends, but each may not be an epiphysis. In bones for which a head can be defined, the region immediately adjacent to it is often identified as the **anatomical neck** (which may not be in the same place as the **surgical neck**, where the bone tends to break).

A bone is ensheathed in a connective tissue called **periosteum**. Tendons connect muscles to the bone via their attachment to the periosteum. Periosteum also carries blood cells, lymphatics, and nerves. **Endosteum** lines bone marrow cavities, trabeculae of spongy bone, and vascular canals of compact bone. The inner layer of periosteal and endosteal cells is capable of producing bone cells that participate in growth as well as in fracture repair. Endosteum, like the marrow it surrounds, can also contribute to the production of blood cells (i.e., it has hemopoietic properties).

Blood is supplied to a bone via vascular systems that correspond to different functional or developmental units of that bone. **Periosteal arteries** (within the periosteum) encase the bone in a continuous vascular network; in long bones, this network is least dense around the center of the diaphysis and increases markedly (with the proliferation of hundreds of small vessels) toward the metaphyseal ends. Each epiphysis is supplied its own set of epiphyseal arteries (usually one to three). A nutrient artery supplies the interior of a bone. In a long bone, the single nutrient artery penetrates the diaphysis obliquely, often within the midregion of the shaft; in the long bones of the arm, the direction of penetration is toward the elbow joint, while in the long bones of the leg, it is away from the knee. Within the bone, the nutrient artery arborizes into ascending and descending branches, with branches serving the endosteum as well as the inner walls of the cortical bone. Near the metaphyseal region of the diaphysis, periosteal arterial branches course through the compact bone and connect, or *anastamose*, with nutrient arterial branches inside. The metaphyseal regions of a dried long bone are riddled with variably small canals and holes (*foramina*, s. *foramen*) that attest to this anastamosis. In young individuals who are still growing and in whom a plate of cartilage separates the epiphysis from the diaphysis, this plate is also supplied by perforating metaphyseal as well as by superficial epiphyseal arteries.

Osteogenesis: The Formation of Bone

Bone is a highly specialized type of **connective tissue** that is distinguished by its being mineralized and thus hard. Bone not only serves a structural function but also stores inorganic minerals for use when needed by the body. A bone contains many cell types: mesenchyme, cartilage, bone, macrophages (cells that engulf particulate matter, e.g., leukocytes, or white blood cells), monocytes (large, agranular leukocytes), and endothelial cells.

Bone formation (**bone morphogenesis** or **osteogenesis**) refers broadly to the creation of both "new" and "mature" bone. New bone is laid down either at the beginning of **skeleto-genesis** (the formation of the skeleton) or in the process of healing a break or wound. Mature bone replaces new bone. The organlike property of renewal distinguishes bone from another mineralized tissue, tooth enamel. Although the details of formation of various hard tissues may differ somewhat, there are certain elements that all hard tissues share (Ten Cate and Osborn, 1976).

A primary feature of all hard tissues is the modification of an **organic matrix**. Its transformation requires a specialized type of cell, the differentiation of which is correlated with areas of concentrated vascularization. This kind of cell is characterized by having higher concentrations of ribonucleic acid (RNA) and higher levels of enzyme activity involved in oxygen and water uptake. It both synthesizes and secretes material, including proteins that form a fibrous, extracellular component of the organic matrix (i.e., **extracellular matrix**) as well as a substance (*ground substance*) that accumulates between these fibers. Conversion of the extracellular organic matrix into a hard substance is achieved by the deposition or accretion primarily of **calcium** and **phosphate ions** in localized areas of **hydroxyapatite crystals** (vesicles that bud off from the extracellular matrix) in concentrations higher than those normally occurring throughout the body.

Specialized bone cells are **osteoblasts** and **osteocytes**. The extracellular matrix contains **glycosaminoglycans** (mucopolysaccharides, which are hydrophilic macromolecules), and the fibrous protein of the extracellular matrix is **collagen**, which contributes to bone's structure and tensile strength (Marieb, 1989; Parfitt, 1983). Type I collagen occurs in bone as well as in teeth, skin, ligaments, and tendons. Unmineralized extracellular organic matrix (**osteoid**) becomes hard via the deposition of calcium phosphate. Coincident with the onset of mineralization is the appearance of **osteonectin** and **osteocalcin**, which belong to a group of molecules called **phosphoproteins** that are specific to bone.

Bone arises via two induction systems that begin in the early fetus and continue post-natally: in **membrane** (producing intramem-branous or membrane bone) or in **cartilage**, through the interaction of degenerating carti-lage cells and perivascular connective tissue [producing cartilage or intracartilaginous, enchondral, endochondral, or replacement bone (the root "chond-" refers to cartilage)]. Regardless of the pathway, the cells involved are the same. With the exception of the clav-icle, membrane bone develops primarily in the head region, giving rise to the flat bones of the cranial vault (the frontals, parietals, squamous portions of the temporals, occipi-tals) as well as to the nasal bones, the vomer, and some parts of the sphenoid, palatines, and mandible. Cartilage bone forms the postcranial and some of the cranial skeleton [the petrosal portions of the temporals, basi-lar portions of the occipitals, and some part(s) of the sphenoid, palatines, and mandible]. During the "life" of a diaphysis, the ongoing processes of bone growth and resorption—rebuilding and remodeling—of the diaphysis also result from intramembranous induction (i.e., bone formation immediately beneath the periosteum is direct and not the result of replacement).

Bone deposition occurs in similar ways in membrane and cartilage bone, whose cells have experimentally been demonstrated to belong to the same **osteogenic cell lineage** (Hall, 1978; Parfitt, 1983). The transformation of cartilage into bone-forming cells may, however, occur in several possible ways. (1) As seen in the ribs, mandible, and articular areas of the mouse's pelvis, cartilage cells (**chondroblasts** and **chondrocytes**) directly transform into osteoblasts and osteocytes. (2) The cellular membrane surrounding the cartilaginous precursor (**perichondrium**) switches from producing cartilage cells to generating bone cells, which then replace the cells of the cartilaginous precursor as the latter deteriorate; the transformed, bone-producing cellular membrane is now identified as the periosteum. And (3) bone cells are trans-ported into the cartilaginous precursor via the vascular system as it spreads throughout the territory of the cartilaginous precursor (Hall, 1988). The latter is consistent with Caplan's (1988) work suggesting the presence of osteogenic progenitor cells prior to the differentiation of cartilage cells. The pathway of cartilage bone formation also occurs during the healing of fractures.

The postcranial skeleton ultimately derives from mesoderm. But whether membrane- or cartilage-derived, cranial bone arises from **neural crest cells** that migrate into the embryonically presumptive head region. Neural cells differentiate along the edges of the infolding **neural plate** and are "released" as the neural plate closes up to form the **neural tube** (Figure 1–9). The neural tube, which will develop into the presumptive spinal cord and brain, becomes covered with epithelium that will develop into skin (Gans and Northcutt, 1983; Hall, 1988). Once neural crest cells become situated in their intended regions, they are referred to as **mesenchymal** or **ectomesenchymal cells**. Molecular sig-naling pathways initiated in epithelial cells induce mesenchyme to differentiate into cells of specific structural identity, such as osteoblasts or bone-forming cells.

In terms of some aspects of the molecular underpinnings of bone formation, **trans-forming growth factor beta (TGF-β)** induces mesenchymal cell proliferation. **Transcription factors** belonging to the TGF-β superfamily (specifically **bone-modifying proteins**, or **BMPs**) target mesenchymal cells and trans-form them into osteoblasts. Expression of the transcription factor **Cbfa 1** in mesenchymal cells signals the initiation of osteogenesis. Cbfa 1 is transcribed in cells that will become osteoblasts (or chondroblasts). Osteoblast differentiation itself appears to be under the control of two "master control genes," *OSE 1* and *OSE 2* (Ducy et al., 2000). Replication of osteoblasts is induced by BMPs and various other proteins, such as **epidermal growth**

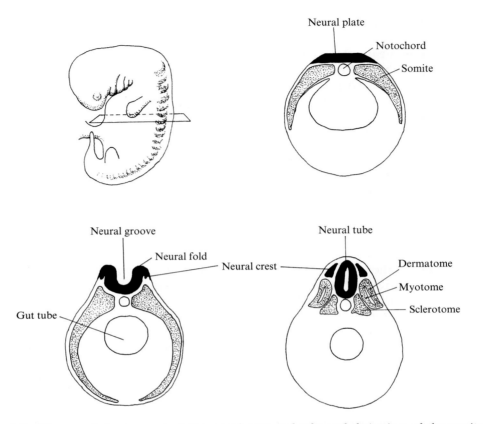

Figure 1–9 Diagram of development of the neural crest and tube and derivatives of the somite in a mammalian embryo. The plane of the section is indicated in the diagram of the embryo (*upper left*). Neural crest cells originating in the head region migrate into the presumptive jaws and face to contribute to the formation of teeth and bone.

factor (EGF), platelet-derived growth factor (PDGF), fibroblast growth factor (FGF), insulin, and insulin-like growth factor (IGF), (Tsukahara and Hall, 1994). [For a general overview of the molecular regulation of development see Carroll et al. (2005) and Wolpert et al. (2002).]

In membrane bone formation, there is an increase in vascularization of the mesenchyme, which proliferates in regions where ossification will begin and creates a fibrous intracellular matrix. Other mesenchymal cells transform into osteoblasts that produce the ground substance and extracellular matrix [including presumptive collagen (procollagen), osteocalcin, and other proteins] that, ultimately, give rise to the osteoid that surrounds the cells in which mineralization occurs (Sandberg, 1991). Mineralization begins within budlike extensions the osteoblast sends into the osteoid. Further mineralization via crystal growth eventually encases the osteoblast, at which time the cell is identified as an osteocyte. In addition to being captured in a bony space or lacuna, an osteocyte can be recognized histologically by its extensions or dendritic processes, which are contained within bony tubes or canaliculi. Osteocytes must keep from becoming mineralized by maintaining communication with the "outside." Osteoclasts enlarge the "space" around osteocytes by removing mineral and matrix (the process is called **osteocytic osteolysis**).

While in membrane bone, the expanding mesenchyme is the direct precursor of bone; in cartilage bone, mesenchymal cells precede the differentiation of **precartilage cells** that become the cartilaginous "model" of the bone that will mineralize (Ogden, 1980). The onset of ossification is evidenced by the transformation of the perichondrium that surrounds the cartilaginous "model" to periosteum, which produces a **bony collar** around the central region of the presumptive bone. The nutrient artery develops from fibrovascular tissue that penetrates the bony collar. As ossification via osteoblasts proceeds in the central region, the domain of the periosteum and bony collar expands toward the ends of the cartilaginous "model." Ossification begins in a diaphysis before its epiphysis.

New bone (**embryonic** or **coarse-fibered woven bone**) is laid down rapidly as plates or trabeculae that eventually come to surround collagenous connective tissue and blood vessels. Bone is then deposited as fine-fibered sheets or lamellae on the inner walls of the trabeculae. **Lamellae** (s. **lamella**) form series of concentric rings that eventually replace the soft connective tissue and come to surround a core region through which the blood vessels course. (In polarized light, these lamellae vaguely resemble tree rings.) A "unit" consisting of concentric lamellae encircling a core of soft connective tissue and vasculature is identified as a **primary osteon** (Figure 1–10). Primary osteons are also formed during circumferential bone growth—when the girth of a bone enlarges—as longitudinal blood vessels within the periosteum become encased in lamellae.

Mature bone results from the natural process of bone remodeling, that is, when **osteoclasts** begin resorbing primary osteons

and embryonic bone (Parfitt, 1983). An osteoclast is identified histologically as a giant cell with a variably large number of nuclei and some number of extending processes. It appears that osteoblasts play a role in activating the process of bone resorption, which is therefore a process of "give and take" between bone-forming and bone-dissolving cells (Chambers, 1988), collectively referred to as the **basic multicellular unit (BMU)** (Robling and Stout, 2000). As an osteoclast resorbs bone, either along a surface or through the cortex, it creates a tunnel or depression that becomes lined with dense, collagen-free connective tissue. The coordinated action of multiple osteoclasts, which create a "cutting cone" (Stout, 1989), gives a scalloped or irregular contour to the margin of the region being resorbed. The "tunnel" created by osteoclasts is generally 250–300 μm in diameter at it largest size (just behind the osteoclasts). Free osteoblasts (including osteocytes that are released as the bone surrounding them is broken down) as well as osteoblasts newly derived from mesenchyme follow in the path of the osteoclasts, laying down bone as a series of lamellae that fills in the "cutting cone" until they produce a (type I) **secondary osteon** of the same diameter as the maximum of the cutting cone. A distinct line of demarcation (**cement line**) delineates the outermost lamella from the unresorbed bone around it; the innermost lamella captures blood vessels, lymphatics, nerves, and connective tissue. When, as often happens, the process of secondary osteon formation is interrupted, a cement line (**arrest line**) develops between lamellae; since they are between lamellae, arrest lines have a smooth perimeter. One can thus distinguish a secondary from a primary osteon by a series of smooth-bordered, concentric layers of lamellae within a clearly defined but scallop- and irregularly shaped perimeter (Figure 1–8). Osteons that develop marked cement lines are referred to as "double zonal" (Robling and Stout, 2000).

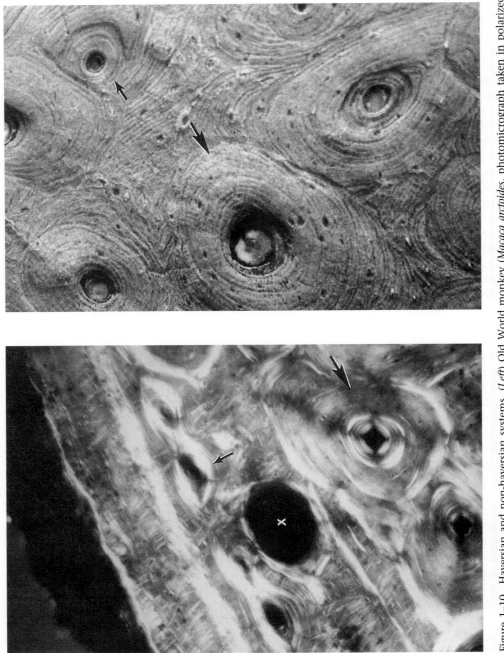

Figure 1–10   Haversian and non-haversian systems. (*Left*) Old World monkey (*Macaca arctoides*, photomicrograph taken in polarized light). (*From bottom right to top left*) Secondary osteons (harversian systems; e.g., see *large arrow*), region of remodeling (*x*), primary osteons (non-haversian systems; e.g., see *small arrow*), and layer of primary lamellar bone being apposed to surface of periosteum. (*Right*) Fossil hominid [Kabwe ("Broken Hill") skull, photomicrograph taken using reflected light with Nomarski optics): secondary osteons (e.g., see *large arrow*), osteon fragments (e.g., see *small arrow*), and interstitial lamellae (between whole and fragmentary osteons) (photomicrographs courtesy of D. Burr).

The canal of a secondary osteon is referred to as a **haversian canal**; it and its contents constitute a **haversian system**. A primary osteon is a **non-haversian system**. Osteoclasts may arise in haversian systems or in the walls of transverse canals (**Volkmann's canals**) that connect haversian canals to one another, permitting vascularization laterally. As osteoclasts burrow to create a tunnel of resorption, they destroy previously formed osteons. Thus, with increasing age, the number of primary osteons decreases and the number of osteon fragments increases. Secondary osteons that form in remodeled haversian canals are type II or embedded osteons, which are generally smaller than type I osteons (Robling and Stout, 2000).

The initial site of mineralization is an **ossification center**. In the development of membrane as well as cartilage bones, one or more centers of ossification may contribute to the formation of what, in the adult skeleton, is a single bone. The general rule with regard to bones with epiphyses is that each diaphysis and each of its epiphyses arise from a single center of ossification. **Primary centers of ossification** give rise to diaphyses. **Secondary centers of ossification** give rise to epiphyses. Other structures that, like epiphyses, begin development as independent elements that eventually fuse to a diaphysis are also secondary centers of ossification (e.g., the greater and lesser tubercles of the proximal humerus, or the iliac crest). The typical as well as variant numbers of ossification centers are discussed for each bone in its appropriate section.

Whereas the platelike membrane bones develop as ever-expanding sheets of crystallizing minerals, the girth of a bone increases by the surficial accretion of successive generations of lamellae encased in a compact bony sheath with concomitant removal of bone internally, thus contributing to the formation of the medullary cavity. Longitudinal bone growth is somewhat more complicated.

In mammals, longitudinal growth of epiphysis-bearing long bones proceeds as long as a diaphysis and its epiphysis remain separated by a **cartilaginous disc** (**epiphyseal disc** or **plate**) (McLean and Urist, 1964). Typically, a layer of bone covers the epiphyseal surface of this disc, while the diaphyseal surface is totally cartilaginous. Elongation of the diaphysis occurs via the proliferation of cartilage from the epiphyseal surface of the disc and the replacement of cartilage by bone at the metaphyseal region of the diaphysis, which also keeps the disc's thickness relatively constant. The cartilage cells in the zone of proliferation (which are separated by transverse slips of matrix) are arranged in columns that are in turn separated by longitudinal bands of extracellular matrix. As an epiphysis and its associated diaphyseal region increase in girth, so too does the epiphyseal plate. When reduction in growth hormone secretion signals the cessation of growth, the epiphyseal plate stops generating cartilage. Ossification continues, however, in the metaphyseal region until the perimeter of the epiphyseal plate becomes mineralized and begins to coalesce with the perimeter of the epiphysis, "capturing" a much reduced cartilaginous epiphyseal plate in between. Mineralization of the longitudinal and transverse bands of extracellular matrix produces a dense region of calcified tissue that persists as the epiphyseal line or scar, which delineates the region of the epiphysis from the diaphysis in adults.

Prior to the onset of union between epiphysis and diaphysis, the growth process may be interrupted or delayed. During such times, mineralization of the longitudinal and transverse bands of extracellular matrix at the more rapidly growing ends of long bones may occur. This produces radio-opaque bands of mineralization variably referred to as **growth-arrest** (Harris, 1926, 1931), **transverse** (e.g., Hunt and Hatch, 1981), and **Harris lines**. [The epiphyseal line represents the ultimate growth arrest line (Harris, 1926, 1931).] The most commonly cited causes of growth-arrest lines are illness/infection and malnutrition.

For this reason, their radiographic assessment is often included in paleoepidemiological and paleopathological studies; and attempts (e.g., McHenry and Schulz, 1976) are sometimes made to correlate these skeletal recordings of bodily insult with those that may be found reflected dentally, in linear enamel hypoplasias (see Chapter 7). The presence of several or more growth-arrest lines in the same bone has been interpreted as representing recurrent or seasonal conditions (e.g., nutritionally difficult times of year).

The bone most frequently scrutinized for growth-arrest lines is the tibia, especially its distal end, where the zone of proliferating cartilage is particularly well defined. This zone is also well delineated in the proximal ends of tibia and femur and the distal ends of humerus, radius, and ulna but less so in the other ends of these bones. The absence of detectable growth-arrest lines does not, however, necessarily testify to the absence of growth-interrupting phenomena. The regions of growth-arrest lines are as subject to osteoclastic activity and remodeling as any other part of a bone. Thus, even though the most tenacious growth-arrest lines appear to be those formed early in childhood (Garn et al., 1968), they begin to disappear relatively soon after they form.

Regulation of and Abnormalities in Bone Growth

Interference with the growth of cartilage results in **dyschondroplasia** (McLean and Urist, 1964), which may be hereditarily based (**Ollier's disease**), induced by chemical or metabolic alteration (e.g., **rickets**, which results from vitamin D and phosphate deficiencies), or due to interference with vascularization. Since the epiphyseal plate is vascularized by various epiphyseal and metaphyseal arteries, interruption of its blood supply will curtail cartilage and bone formation and, ultimately, growth. If metaphyseal arterial supply is cut off, the metaphyseal portion of the epiphyseal plate will not mineralize and growth will cease.

General body size and weight, as well as longitudinal bone growth, are maintained by **growth (somatotrophic) hormone**, which is produced in the **anterior lobe of the hypophysis** or **pituitary gland**. With the normal cycle of growth hormone production, bone growth continues until proliferation of cartilage in the epiphyseal plate ceases and mineralization leads to coalescence of the epiphysis and diaphysis around a much reduced epiphyseal plate (Ortner and Putschar, 1981).

If the hypophysis is obliterated prematurely (often by the development of a tumor), proliferation of cartilage in the epiphyseal plate ceases, as does bone growth, and fusion of epiphysis to diaphysis occurs; although smaller in size, the coalesced epiphysis and diaphysis will encase a reduced epiphyseal plate, thus producing a miniature version of the normal adult configuration. When growth is curtailed and the result is the proportioned miniaturization of an individual, the condition is **midgetism** (also **hypopituitarism**). Reduction in, or lack of, sufficient growth hormone production results in **pituitary dwarfism**, in which the body proportions remain those of an infant, the appearance of secondary centers of ossification is retarded, and closure of sutures and/or fusion of epiphyses is delayed until late in life.

If growth hormone is produced beyond the age at which bone growth normally diminishes or a new surge of growth hormone occurs at that time, bone formation and growth will continue, resulting in **gigantism** (**hyperpituitarism**). Activation of growth hormone production after epiphyses have joined diaphyses and longitudinal bone growth has ceased results in **acromegaly**. Since longitudinal growth is no longer possible in bones with epiphyses, bones thicken or expand. Often, hand and foot bones are affected first. Bone enlargement is also common in the region of the middle and upper face (e.g., broadening of the cheekbone regions) and mandible (creating the so-called lantern jaw). Acromegalic features may be superimposed on those of gigantism. When fetal monitoring is available, it is possible to detect imbalances in growth and other hormones that can be corrected in utero.

Thyroid hormone (thyroxin) secretion also affects proper bone growth, primarily via a feedback effect on the hypophysis' ability to secrete growth hormone (McLean and Urist, 1964). Diminished or curtailed thyroxin secretion leads to overall stunting of bone size and retention (nonreplacement) of cartilage and primary spongiosa. **Cretinism** is a type of hypothyroidism that is characterized by dwarfism and mental deficiency; its effects emerge in utero because the fetus' thyroid gland either does not develop or is inhibited because the mother suffers from hypothyroidism. With regard to the maternal/fetal case (**endemic cretinism**), the degree to which the fetus is impacted depends on the severity of the mother's **hypothyroidism**; dwarfing is much more dramatic with glandular hypertrophy (**sporadic cretinism**).

Experimentally induced **hyperthyroidism** can cause early fusion of epiphyses and thus cessation of growth. Hyperthyroidism in young individuals is often expressed as retardation in the growth of long bones. The effect of hyperthyroidism on adults may be mild, perhaps accelerating the rate of resorption that will yield a more friable and porous (**osteoporotic** or **porotic**) condition.

Gonad-stimulating and **gonad-produced hormones** also have an effect on proper bone growth. Gonad-stimulating hormones (**chorionic gonadotropins**) are produced primarily by the placenta and have an effect on the fetus similar to that of growth hormone produced by the anterior lobe of the hypophysis. An insufficiency in gonad-produced hormones (**estrogen** in females and **androgen** in males; in the latter, the most extreme case results from castration) can retard the appearance of secondary centers of ossification, truncate cortical bone deposition, and delay epiphyseal closure—all of which leads, as in eunuchs, to thin-boned, typically longer-legged and longer lower-jawed individuals. Incomplete development or absence of the ovaries can lead to dwarfism with characteristics similar to those associated with hypopituitarism. Cessation of estrogen (e.g., in menopause) or androgen production after an individual reaches adulthood leads to general porosity (**osteoporosis**) of the postcranial axial skeleton.

Of the various steroidal hormones produced by the cortex of the **adrenal gland** (**adrenal corticosteroids**), some—most notably, the **glucocorticoids hydrocortisone** and **cortisone**—have demonstrable, although nonspecific, effects on skeletal growth (McLean and Urist, 1964). The action of these steroids is ultimately controlled via a feedback mechanism with the anterior lobe of the hypophysis through its secretion of **corticotropin** (formerly, adrenocorticotropic hormone, or ACTH); ablation of the hypophysis leads to reduced adrenal corticosteroid secretion. Overproduction of corticotropin or one of the glucocorticoids suppresses osteoblast activity, which retards growth in young individuals and the mending of fractures. Hypersecretion of cortisone (**hypercortisonism** or **Cushing's syndrome**) may be caused by more than one agency (e.g., basophil adenoma of the hypophysis, primary hyperplasia, a tumor of the adrenal cortex). Hypercortisonism results in severe overall osteoporosis of the ribs and vertebrae (leading to their potential collapse), as well as to internal osteoporosis of the long bones.

Vitamin A is critical for the proper growth and regulation of bone (McLean and Urist, 1964). It is necessary for growth of endochondrally and periosteally derived bone, as well as for the ongoing process of bone remodeling. In cases of **vitamin A deficiency**

(**avitaminosis A**), bones become misshaped and foramina fail to keep pace with the increasing size of the nerves or neurovascular bundles that course through them, which leads, ultimately, to the death (**necrosis**) of these neurovascular structures. Too much vitamin A (**hypervitaminosis A**) results in increased osteoclastic activity, which causes an increase in bone destruction (**osteolysis**) and also an increase in bone cell turnover. This can lead to inconsistencies both in mineralization and in the development of compact cortical bone as well as to overall bone fragility.

Vitamin C (**ascorbic acid**) plays a major role in the proper formation and maintenance of the collagenous extracellular matrices of bone, cartilage, and other connective tissue, as well as in the formation and maintenance of the dentine of teeth (Ortner and Putschar, 1981). With **vitamin C deficiency (scurvy)**, collagen formation is either suppressed or the collagen that is produced cannot mineralize. The effects of scurvy are most deleterious in infants and children but decrease in severity as bone growth itself slows down. **Infantile scurvy (Möller-Barlow disease)** is characterized by subperiosteal hemorrhaging with induced reactive bone formation and fractures or separations of fast-growing metaphyseal regions. Not uncommon in infantile scurvy is a sinking inward of the sternum (breastbone), which creates the so-called **scorbutic rosary** (which, in the living, causes the sternal rib ends to poke up into the skin). Infantile scurvy is typically expressed by the end of the first year of life. In adults, scurvy most often affects the rib cage and leads to separation within, or fracture of, the regions where bone and cartilage interface. Since vitamin C is water-soluble and is not stored in the body, resuming its intake can correct the effects of scurvy, especially in adults.

Vitamin D is also crucial for proper bone growth since it plays a role in the absorption of calcium and thus in the mineralization of bone; it can influence the demineralization of bone as well (Ortner and Putschar, 1981). Although vitamin D is vital for overall systemic calcium metabolism and mineralization, it does not act alone in this process. The efficacy of vitamin D is mediated by **parathormone**, which is a hormone secreted by the **parathyroid glands**; it acts to maintain proper calcium levels in the blood plasma. Although in the process of regulating calcium ion concentration parathormone can induce the release of calcium from the skeleton, it apparently does not play a role in the process of mineralization. Parathormone also acts on the renal tubules; its increased secretion results in increased excretion of phosphate into the urine.

In contrast to vitamins C and A, little vitamin D comes from the diet. Instead, vitamin D is synthesized in the skin under the influence of ultraviolet light. Although populations indigenous to tropical and subtropical regions tend to have a higher incidence of vitamin D deficiency than those in other locales, this is more likely due to dietary deficiencies in calcium than to a deleteriously high rate of ultraviolet light absorbed by the skin's melanin.

Vitamin D deficiency, which is not life-threatening, can affect infants and children typically 4 years of age and younger (**rickets**) as well as adults (**osteomalacia**). Since the most important stages of bone formation and growth occur during infancy and childhood, vitamin D deficiency during this period has the most profound consequences; in individuals who have stopped growing, the effects will be somewhat different. In both afflictions, the collagenous extracellular matrix does not become mineralized.

Like scurvy, rickets most notably affects areas of rapid bone growth (e.g., junctures of ribs, metaphyseal regions). Unmineralized matrix then accumulates next to growth plates (e.g., in ribs and long bones) and/or replaces bone during membrane bone growth. In cranial bone, this can lead to the bone(s) involved becoming thicker, with the smooth, compact-boned outer table acquiring an incised appearance. The diaphyses of postcranial bones become thicker-walled and the metaphyseal regions expanded. Ribs and their attendant cartilage may swell at their sternal ends, creating, in the living, an arrangement of bumps under the skin (**rachitic rosary**). Extreme porosity and fragility of bone would indicate that the individual had not only been rachitic but also suffered from malnutrition. In these cases, bones of the cranial vault appear porotic (in individuals with prolonged vitamin D deficiency, portions of cranial bone may become continuously and completely porous from the inner to outer surfaces); postcranial bones become thin-walled with expanded medullary cavities and thus become increasingly susceptible to stress fractures and bending. Rickets alone (i.e., uncomplicated by malnutrition) is described as **hyperplastic** rather than porotic.

Since osteomalacia affects mature bone that is maintaining stasis in resorption and reconstitution, the result is a decrease in ossified material; the effects are most notable in bones that are comprised predominantly of spongy bone and in which bone turnover is rapid (e.g., ribs, vertebrae). Over time, increased unmineralized tissue creates vacuities (making desiccated bone feel light and papery); the surface of the bone may also become porous. In severe cases—when bone becomes excessively fragile and malleable—a bone may become deformed or may fracture. Loss of calcium during pregnancy or lactation can put a vitamin D–deficient female at further risk of osteomalacia; indeed, the incidence of osteomalacia is greater in women between the ages of 20 and 40 years.

Large doses of vitamin D (**hypervitaminosis D**) can intensify the rate of bone resorption. This causes an increase in the amount of unmineralized extracellular matrix, which in turn expands the field of poorly calcified spongy bone and produces porotic bone.

Since parathyroid hormone (**parathormone**) is involved in maintaining calcium levels in the blood by stimulating osteoclastic activity, which in turn liberates calcium ions, **hypoparathyroidism** lowers and **hyperparathyroidism** elevates calcium levels. Increased parathormone secretion also appears to inactivate vitamin D, which is then eliminated in the bile.

Hypoparathyroidism, which results from the absence of, or interference with, the parathyroid glands, occurs infrequently in human populations. Because hypoparathyroidism decreases osteoclastic activity, compact and spongy bone builds up, creating thickened compact bone and densely packed trabecular bone. Lowered levels of calcium also often produce painful muscle spasms (**tetany**).

Oversecretion of parathormone (leading to **hyperparathyroidism**) may result directly from overactivity of the parathyroid glands (**primary hyperparathyroidism**) or as a consequence of the parathyroid glands responding to kidney disease (**secondary hyperparathyroidism**), which can interfere with proper phosphate levels in blood plasma and cause loss of calcium via the kidneys.

In general, since hyperparathyroidism increases osteoclast activity, which in turn releases calcium ions into the blood, bone becomes porous. Particularly affected in primary hyperparathyroidism are the surfaces of phalanges, alveoli (root sockets for teeth), and cranial vault bones; vertebral bodies may become depressed as a result of trabecular resorption and overall porosity. The continually demineralizing skeleton becomes increasingly susceptible to fracture and hemorrhage. Bones —especially the mandible, long bones, metatarsals, metacarpals, phalanges, and os coxae —can become brittle through the development of pockets of unmineralized matrix that are encased in thin bony walls; in dried bone, these "pockets" are air-filled. The accumulation of unmineralized matrix may also cause bone to expand. The effects of secondary

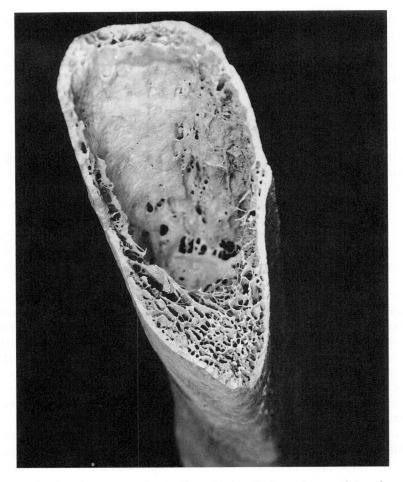

Figure 1–11   Example of senile osteoporosis: note loss of trabecular bone, increased size of medullary cavity, and thinned cortical bone (left tibia, prehistoric, Pennsylvania).

hyperparathyroidism in children are compounded by symptoms similar to those of rickets and hypopituitarism; in fact, hyperparathyroidism can affect joints in a manner similar to rickets. In adults, hyperparathyroidism may be accompanied by symptoms of osteomalacia. In young and old, hyperparathyroidism can also induce the buildup of calcium around joints. Diagnosing hyperparathyroidism can be difficult because porosity, which can result from any number of causes, may be the only clue. Hyper-

thyroidism may also be confused with hyperparathyroidism.

Other Pathological Aspects of Bone Growth and Physiology

Although the term "osteoporotic" has been used to describe various effects of abnormal bone growth, the *process* of **osteoporosis**, which is a systemic rather than localized increase in soft tissue at the expense of mineralized tissue, is an abnormality in its own right. In general, osteoporosis is an affliction of the aged (although it can occur in younger individuals) that affects the spongy bone, first and especially, of the ribs, vertebrae, os coxae, and femoral neck. With advancing age, the bones of the extremities may become involved. Cortical bone thickness and the density of trabeculae and transverse plates are reduced (Figure 1–11). Affected bone continues to lose calcium and phosphate in spite of the fact that diet, as well as vitamin and mineral intake and serum levels, may be normal. Hypercortisonism is known to cause osteoporosis. Other specific causes are less clearly delineated, although diets deficient in calcium, nutrition- or disease-related weight loss, castration, and hyperthyroidism appear to exacerbate active osteoporosis.

A pathological condition that begins as localized phenomena (perhaps because of disruptions in capillary circulation) but can eventually affect much of the skeleton is **Paget's disease (osteitis deformans)** (Ortner and Putschar, 1981). It manifests itself as an acceleration and distortion of osteoclastic and osteoblastic changes that spread along the bone from the point of origin, thickening cortical bone and creating a coarser and more interwoven trabecular pattern along the way. As the disease spreads, large numbers of osteoclasts resorb osteons, leaving behind vacuous and distorted haversian systems. Redeposited cortical bone is characterized by a mosaic pattern of deformed haversian systems amid mineralized lamellae that are bounded by thick and irregularly configured cement lines. As the disease progresses, bone thickens and hardens (i.e., becomes **sclerotic**) periosteally and endosteally; the domains of trabecular bone and medullary cavities of long bones diminish in size. A general descriptor of the bone formed as a result of Paget's disease is **pumice bone**. **Secondary kyphosis** (kyphosis being an exaggeration of the outward curvature of the thoracic region of the vertebral column), in which vertebrae collapse and become wedge-shaped, can be a consequence not only of Paget's disease but also of hyperparathyroidism, osteomalacia, and osteoporosis.

Anemias, which are either hereditarily **hemolytic** (causing the destruction of red blood cells) or due to **iron deficiency**, may lead to conditions in which cranial cortical bone presents itself as being at the same time both thickened (**hyperostotic**) and perforated by a series of quite visible pinholes (**porotic**) (e.g., Stuart-Macadam, 1989). It appears that this condition—which Angel (1966) was the first to refer to as **porotic hyperostosis**—is a response to anemia that entails expansion of the bone marrow (which produces red blood cells), ultimately resulting in the replacement of compact with spongy bone (e.g., Larsen, 1997). Iron-deficiency anemia can result from excessive blood loss (e.g., due to parasitic and microbial infection via contaminated water and/or congested living situations or to menstruation in the absence of adequate iron in the diet) and/or dietary factors (e.g., prolonged reliance on iron-poor foods or intake of substances that can decrease the availability or utility of iron); the latter are thought to be relevant to prehistoric populations (Klepinger, 1992), although cultivation (such as of maize in the New World) also brings with it increased population density (Larsen, 1997). When cranial vault bones are affected, porotic hyperostosis is referred to as **cribra crania** (although some osteologists simply identify it as **porotic hyperostosis**) (Figure 1–12). When localized in the orbital roofs, the lesions are often identified as **cribra orbitalia** (Figure 1–13). Cribra orbitalia typically occurs in younger individuals.

Bone Biochemistry and Diet

Techniques for analyzing the chemistry of bone make it possible to infer aspects of an individual's diet. Although of limited utility because they form early in life, teeth are otherwise useful for these tests because their hard outer covering of enamel protects the inside of the tooth from contamination that

Figure 1–12   Extensive cribra crania (porotic hyperostosis) on both parietal bones as well as the frontal bone of an adult (prehistoric, Pennsylvania).

might affect test results. Consequently, even if a specimen has been handled a great deal, a usable sample can be extracted through a small drill hole below the surface of a tooth's enamel or a bone's compact outer layer. The most commonly used tests are stable carbon and nitrogen isotope analyses. Isotopes are chemical elements (such as carbon and nitrogen) that have the same atomic number but different atomic weights. Sometimes strontium ($^{87}S/^{86}S$ ratios) and oxygen ($^{18}O/^{16}O$ ratios) isotopes are also analyzed (e.g., Katzenberg, 2000; Katzenberg and Harrison, 1987).

Carbon stable isotope analysis has focused on the ratio of the amount of the carbon 13 isotope (^{13}C) compared with the carbon 12 isotope (^{12}C) (in a sample under study divided by that same ratio in a standard substance,

minus 1, times 1000, or the $\partial^{13}C$ value (Katzenberg, 1992)). $\partial^{13}C$ values can be determined for collagen, which is the predominant protein found in bones and tooth dentine, and for amelogenin, which occurs in tooth enamel (Lee-Thorpe and van der Merwe, 1993; Sponheimer and Lee-Thorpe, 1999). During photosynthesis, plants (which are at the base of the food chain) discriminate against taking up ^{13}C, which is the heavier of the two isotopes, preferring instead to take up ^{12}C from the environment. Plants discriminate more against ^{13}C (and thus have proportionately higher levels of ^{12}C) when they follow what is called the C_3 pathway of photosynthesis than when they follow the C_4 pathway. Consequently, classes or types of plants can be distinguished by differences in

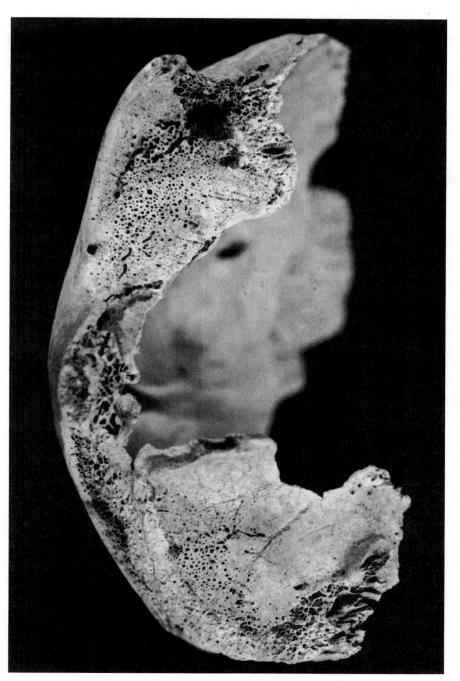

Figure 1–13   Cribra orbitalia in a 12- to 14-year-old (frontal fragment, Vandal or Byzantine period, Tunisia).

their $^{13}C/^{12}C$ values ("less negative" or higher in C_4 than in C_3 plants). In warm climates, grasses follow a C_4 pathway while other plants (trees, shrubs, forbs, corms, and tubers) follow a C_3 pathway. In temperate climates, maize, sorghum, millet, and sugar follow a C_4 pathway (Katzenberg, 2000). Levels of ^{13}C in C_4 plant-biased animals are also reflected in the collagen as well as tooth enamel of humans who consume these animals (Katzenberg, 1989). Because aquatic plants have more varied avenues of access to carbon than terrestrial plants (whose primary source of CO_2 is atmospheric), their $\partial^{13}C$ values typically lie between those of C_3 and C_4 plants (Schoeninger and Moore, 1992), which means that aquatic animals that consume these plants will have higher $\partial^{13}C$ levels than animals that favor C_3 plants but lower levels than animals that select C_4 plants.

Clearly, stable carbon isotope analysis has the potential to provide insight into aspects of an individual's diet, as well as differences in diets between individuals of different ages, sex, or population, and thus into subsistence patterns and shifts therein. But the utility of analyzing tooth versus bone is not the same.

Since bone is constantly reconstituting itself, $\partial^{13}C$ values will reflect the plants an individual consumes over a period of time. Teeth, however, do not represent an ongoing tape recording of an individual's life history because, once they erupt into the jaws, their development is essentially complete. This applies especially to the crown of the tooth, whose size and shape are set once its enamel covering is deposited. Thus, a $\partial^{13}C$ value obtained from tooth enamel will reflect diet only as long as enamel is being deposited. Since an individual's teeth develop and erupt at different times over a period of years, it is possible that $\partial^{13}C$ values could vary from tooth to tooth and reflect shifts in diet corresponding to life changes from childhood to young adulthood (when the last teeth in the jaws—third molars or "wisdom teeth" —erupt). Thereafter, teeth cannot provide

complete evidence of an individual's dietary habits, but it is probably likely that the components of an individual's diet would not change significantly after the first (deciduous or milk) set of teeth erupted into the jaws and the individual could chew with some efficacy.

Study of the ratio of nitrogen isotopes ($\partial^{15}N$) in bone collagen also provides clues to diet. While $\partial^{13}C$ analysis reveals the type and relative amount of plant foods that an individual ingests either directly or as a result of consuming herbivores, $\partial^{15}N$ analysis attempts to identify the relative amount of flesh (protein) consumed. Nearly 100% of earth's ^{15}N is either atmospheric or aquatic, with $\partial^{15}N$ values being slightly lower in terrestrial compared to aquatic (marine and freshwater) plants (Schoeninger and Moore, 1992). In turn, differences in $\partial^{15}N$ values between terrestrial and aquatic plants are incorporated into organisms that feed on them. Nitrogen stable isotope values therefore reflect the (trophic) level of an organism in the food chain. The food chains (i.e., number of organisms) in aquatic systems are typically longer than in terrestrial ones (the latter often consisting only of the sequence plant–herbivore–carnivore). Thus, a fish or a carnivorous aquatic bird or mammal will be expected to have higher $\partial^{15}N$ levels than an herbivore. If an individual consumes the former organisms, their $\partial^{15}N$ levels will be reflected in the consumer's bones.

One can also supplement interpretations of diet and subsistence based on stable isotope analysis with studies on teeth. For example, Lalueza et al. (1996) demonstrated via microscopic analysis of tooth-wear striations that the teeth of nonagricultural, relatively carnivorous humans (e.g., the Inuit) bear fewer striae (wear striations or small abrasive grooves) than those of more omnivorous humans (e.g., the San). And Larsen (1997) pointed out that changes in frequency of caries (dental lesions, reflecting changes in maize consumption) paralleled those in $\partial^{13}C$ levels from Terminal Classic to post-Classic Mayan populations.

Strontium values in bone and tooth enamel do not directly reflect diet, but they are related to the age of the rocks of a region. Consequently, because ingested strontium reflects the level of that element in a particular area, analysis of $^{87}S/^{86}S$ values could provide insight into permanent residence versus migratory behaviors of individuals or of groups of individuals (e.g., Larsen, 1997). In addition, analysis of $\partial^{18}O$ (oxygen) could illuminate aspects of climate and weather (e.g., relative temperature, humidity), which could be related to interpretations of dietary preferences determined on the basis of other studies (Larsen, 1997).

Chapter 2

The Skull

The **skull** comprises the **braincase** or **cranium** and the **lower jaw** or **mandible**; often, the **hyoid bone** is included. Since sutures are rather inflexible joints, the only significant points of articulation are between the mandible and the cranium.

It is convenient to think of the skull in terms of its major functions. Since the cranium protects the brain, it is basically a container. The bones that contribute to it on three sides are platelike; the basilar region, in particular, is pervaded by holes (**foramina**) or passageways for the exit of various nerves emanating from the brain (**cranial nerves**) as well as for the entrance and exit of blood vessels that serve the brain and its attendant structures. The bones of the facial skeleton accommodate the sense organs and other anatomies associated with sensation. Right and left sockets (**orbits**) protect the eyes. Between and below the orbits lies the **nasal aperture**, which opens into the **nasal cavity**. The upper jaw bears teeth. Although teeth are used to gather and process food, they were probably derived evolutionarily from sensory dermal scales, as in fish. Indeed, teeth are innervated by sensory nerves: mandibular teeth are innervated by a branch of the same sensory nerve (derived from the fifth cranial or trigeminal nerve) that innervates the upper teeth.

There are considerably fewer bones in an adult than in a neonate, each developing from separate centers of ossification. Thus, although it may be traditional to learn cranial anatomy by studying the adult first, one must not lose sight of the fact that this is the last stage of development. Because of this, the ontogeny of individual bones is reviewed in Chapter 3 along with the description of their adult features.

Certain analyses require the skull to be oriented in a standard position. To do so, one aligns the bottom, or inferior, edge of the orbit with the top, or superior, margin of the opening of the bony "ear" (the *external auditory* or *acoustic meatus*, which is the lateral aperture of the *tubular ectotympanic* or *ectotympanic tube*) and aligns this plane parallel to the ground. The vertical axis of the skull should be at a right angle to this horizontal plane. The skull is now in the **Frankfort plane** or **horizontal** (see Table 2–1) and can be studied properly from the front (in *norma facialis* or *frontalis*), from behind (*norma occipitalis*), from the side (*norma*

Table 2–1   Glossary of Osteometric Landmarks and Anatomical Regions

Midsagittal landmarks of the skull

Acanthion: The point at the base of the anterior nasal spine.

Alveolare (abbreviation *ids*; also *infradentale superius* or *hypoprosthion*): The inferiormost point of the alveolar margin between the maxillary central incisors; often confused with **prosthion**. Compare with **infradentale** and **prosthion**.

Alveolon (*alv*): The point on the hard palate at which a line drawn through the distal margin of the maxillary alveolar processes intersects the median sagittal plane.

Apex (*ap*): The highest point on the skull, which also lies in the same vertical plane as **porion**; the skull must be in the Frankfort horizontal.

Basion (*ba*): 1. (General) A point at the midline of the anterior border of the foramen magnum. 2. A point at the position pointed to by the apex of the triangular surface at the base of either condyle (i.e., the average position from the crests bordering this area), about halfway between the inner border directly facing the posterior border (the **opisthion**) and the lowermost point on the border (i.e., between **endobasion** and **hypobasion**); it is the same as **endobasion** if the anterior border of the foramen magnum is thin and sharp (Howells, 1973).

Bregma (*b*): The juncture of the coronal and sagittal sutures in the median sagittal plane; should an ossicle be present, the landmark can be located by drawing in pencil a continuation of the sutures until these lines intersect.

Endobasion (*endoba*): The most posterior point on the midline of the anterior margin of the foramen magnum; it is usually behind and below **basion** but coincident with it when the margin is thin and sharp.

Glabella (*g*; also *metopion*): The anteriormost region (not a point) of the frontal bone above the frontonasal suture and between the superciliary arches. This region generally protrudes anteriorly, but in certain skulls this region may be flat ("hyperfeminine" skulls, infants, and juveniles) or even depressed. When it is not protrusive, glabella may be identified by a change in the direction of the frontal bone.

Gnathion (*gn*; also *menton*): The "lowest" (anterior- and inferiormost) point on the chin.

Incision (*inc*): The point of contact [identified on the buccal (labial) side] between the mesial occlusal edges of the upper central incisors.

Infradentale (*idi*; also *infradentale inferius* and *symphysion*): The superiormost point of the alveolar margin (identified on the anterior surface) between the mandibular central incisors.

Inion (*i*): A point at the midline of the superior nuchal lines, which often coincides but should not be identified with the base of the external occipital protuberance. Measurement should not be taken from the protuberance.

Lambda [*l*; sometimes abbreviated as *la* (but see lacrimale)]: The juncture in the median sagittal plane of the lambdoid and sagittal sutures; should an ossicle be present, the landmark is located by drawing in pencil a continuation of the sutures until these lines intersect.

Menton: see **gnathion**.

Metopion: see **glabella**.

Nasion (*n*): The juncture in the median (sagittal) plane of the frontonasal and (inter)nasal sutures.

Nasospinale (*ns*): A midline point on a line drawn through the inferiormost margins of the nasal aperture; it may coincide but is not synonymous with **acanthion**.

Obelion (obsolete): The point of intersection between a line drawn through the two parietal foramina and the sagittal suture.

Ophryon (obsolete): A midline point on a line drawn across the forehead and which lies on a plane that passes through right and left **frontotemporale**.

Opisthion (*o*): A point in the midline of the posterior margin of the foramen magnum.

Opisthocranion (*op*): The most posterior point on midline of the skull when the skull is oriented in the Frankfort horizontal.

Orale (*ol*): The point of intersection of a line drawn through the lingualmost surfaces of the upper central incisors and the midline of the premaxillary region.

Pogonion (*pg*): The anteriormost midline point of the chin; it lies anterior to and above **gnathion**.

Prosthion (*pr*; also *exoprosthion*): The anteriormost midline point on the premaxillary alveolar process in the median (sagittal) plane; it is located slightly above **alveolare**, with which it is often confused.

Rhinion (*rhi*): The inferiormost point on the (inter)nasal suture.

Staphylion (*sta*): The midline point of a line drawn through the anteriormost invaginations of the posterior margins of the right and left palatine bones; it typically falls anterior to the posterior nasal spine.

Subspinale (*ss*): The deepest midline point in the concavity between the anterior nasal spine and the premaxillary alveolar; it is located on the crest of the midline suture.

Table 2–1   (*continued*)

Supradentale: see **prosthion.**

Symphysion: see **infradentale.**

Vertex (*v*): The highest midline point on the skull when the skull is in the Frankfort horizontal; it may coincide with **apex.**

Bilateral (paired) landmarks of the skull

Alare (*al*): The lateralmost point on the margin of the nasal aperture.

Asterion (*as*): The juncture of the temporal, parietal, and occipital bones (also of the lambdoid, temporoparietal, and occipitomastoid sutures).

Condylion laterale (*cdl*): The most lateral point on the mandibular condyle.

Coronale (obsolete): The most lateral point on the coronal suture.

Dacryon (*d*; may also be abbreviated as *dk*): 1. (Traditional; considered obsolete by Vallois, 1965) The juncture of the frontolacrimal, frontomaxillary, and lacrimomaxillary sutures; lies between **lacrimale** and **maxillofrontale.** 2. "The apex of the lacrimal fossa, as it impinges on the frontal bone" (Howells, 1973, p. 167); considered to be more easily located and useful in anteroposterior measurements.

Ectoconchion (*ec*; may also be abbreviated *ek*): The anteriormost point on the lateral rim of the orbit that contributes to the determination of maximum orbital breadth (as measured from **dacryon** or **maxillofrontale**); this point is also used in the measurement of facial flatness.

Ectomolare (*ecm*): The most lateral extent of the alveolar process in the middle of the second maxillary molar on its buccal side.

Endomolare (*enm*; may also be abbreviated as *endo*): The most medial extent of the alveolar process in the middle of the second maxillary molar on its lingual side.

Euryon (*eu*): The point that, on a parietal, contributes to the determination of maximum (biparietal) cranial breadth.

Frontomalare anterior (*fm:a*): The most anteriorly prominent point on the zygomaticofrontal suture.

Frontotemporale (*ft*): The point at which the temporal ridge is most inwardly or medially curved; also, the point which contributes to the determination of minimum frontal breadth.

Gonion (*go*; may also be abbreviated as *g* or *gn*): The intersection of a line tangent to the posterior margin of the ascending ramus and a line tangent to the inferior margin of the body of the mandible; also, the most marked point of transition in the upward curvature of the posteroinferior margin of the mandible.

Jugale (*ju*): The juncture on the posterior margin of the zygomatic bone of the horizontal and vertical portions (i.e., temporal and frontal processes, respectively) of this bone.

Krotaphion (*k*): The juncture of the sphenoparietal, sphenotemporal, and squamosal sutures; the location of this landmark must be estimated on those individuals in whom there is frontotemporal articulation.

Lacrimale (*la*): The juncture of the posterior lacrimal crest and the frontolacrimal suture; lies posterior to **dacryon.**

Mastoidale (*ms*): The inferiormost point on the mastoid process when the skull is in the Frankfort horizontal.

Maxillofrontale (*mf*): The juncture of the anterior lacrimal crest and the frontomaxillary suture; lies anterior to **dacryon.**

Mentale (*z*): The anteriormost point on the margin of the mental foramen (mandible).

Nariale: The most inferior extent of the anterior margin of the nasal aperture on one side or the other of the anterior nasal spine.

Orbitale (*or*; may also be abbreviated as *orb*): The most inferior point on the orbital rim; used with **porion** to orient the skull in the Frankfort horizontal.

Porion (*po*; may also be abbreviated as *p*): The most lateral point on the superior margin of the external auditory (acoustic) meatus; it is located vertically over the center and thus at the middle of the superior margin of the meatus.

Pterion (*pt*): A region (not a point) delineated by the frontal bone anteriorly, the squamous portion of the temporal bone posteriorly, and the area along the sphenoparietal suture; the landmark would not be recognized in those individuals with frontotemporal articulation (see **sphenion**).

Sphenion (*sphn*): The juncture of the frontal bone with the parietal bone, the greater wing of the sphenoid, and the squamous portion of the temporal bone.

Stephanion (*st*): The point at which the inferior temporal line crosses the coronal suture.

Zygion (*zy*): The most lateral point on the zygomatic arch.

Zygomaxillare anterior (*zm:a*): The point on the facial (anterior) surface where the zygomaticomaxillary suture intersects the attachment scar of the masseter muscle; considered more appropriate than **zygomaxillare (inferior)** for measuring flatness/protrusion of the face.

Zygomaxillare (inferior) (*zm*): The most inferior point on the zygomaticomaxillary suture.

Zygoorbitale (*zo*): The juncture of the zygomaticomaxillary suture with the inferior orbital rim.

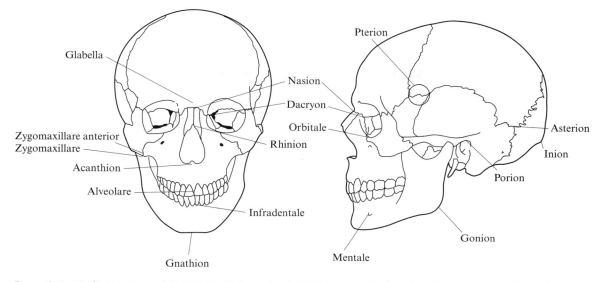

Figure 2–1 (*Left*) Anterior and (*right*) lateral views of a skull, illustrating the location of some commonly used anthropometric landmarks.

lateralis), from below (*norma basilaris*), or from above (*norma verticalis*). This standardized position also allows osteologists to make repeatable measurements from agreed-upon anthropometric landmarks (see Figures 2–1 and 2–2).

External Morphology

Cranium

The cranium (Figure 2–3) can be subdivided into two units: the **calvaria** (sometimes the word "calvarium" is used, but this is based on the misconception that "calvaria" is, in Latin, the neuter plural and not the feminine singular) and the **facial skeleton** or face; "calvaria" and "braincase" are interchangeable.

The calvaria is composed of eight major bones: the **ethmoid, frontal, occipital, parietal, sphenoid**, and **temporal**. In the adult, the ethmoid, frontal, occipital, and sphenoid are single bones whereas the parietal and temporal bones are paired (right and left). The frontal and occipital and especially the parietals are essentially platelike. Since the temporal

bones contain the bony contributions to the **ear region**, each houses the minuscule **inner ear bones**: the **incus, malleus**, and **stapes**.

The **frontal bone** lies at the front of the braincase, primarily constituting the forehead and anterior region of the top of the skull. It also contributes to the orbital roof, superior orbital rim, and posterior wall of the orbit and somewhat to the sides of the cranium superiorly. The frontal bone contacts the paired nasal bones, which lie between the orbits. One of the major muscles of mastication (*temporal muscle*) courses from the mandible superiorly, spreading over much of the lateral wall of the cranium. As it does, it rides up behind the frontal bone's contribution to the posterior wall of the orbit and arcs superiorly and posteriorly over the side of that bone. Features on the frontal bone reflect the presence of the temporal muscle: the portion of bone immediately behind the lateral and superior "corner" of the orbit is thickened into the *temporal ridge*, which bounds the temporal muscle anteriorly. This ridge, whose degree of development and rugosity differs

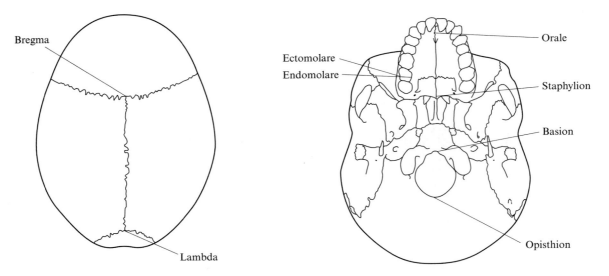

Figure 2–2   (*Left*) Superior and (*right*) basal views of a skull, illustrating the location of some commonly used anthropometric landmarks.

among individuals, eventually bifurcates into two low-lying, somewhat parallel *temporal lines*, which, because they represent the attachment scars left by the temporal muscle, reflect its course over the cranium.

At approximately the same height as, or a little below, the superiormost extent of the temporal lines, the frontal bone may swell variably into two rounded, perhaps bossed *frontal eminences*, one above each orbit. These swellings are more pronounced in young individuals. In adults, they are often more distinct in females than in males.

The anthropometric landmark *frontotemporale* is located at the most medial point of the temporal ridge. (Anthropometric landmarks are useful for taking standardized measurements and provide a shorthand reference for a point on a bone; see Table 2–1 for landmarks of the skull and mandible.) The landmark *glabella*, which is sometimes the most anterior point on the frontal bone, is defined on the frontal bone at the midline of the skull, above the nasal region and between

the orbits, paralleling the highest extent of the superior rims of the orbits.

Right and left **parietal bones**, which form the largest portion of the sides and top of the skull, abut along the midsagittal plane of the skull and articulate posteriorly with the frontal bone. In many individuals, particularly females, each parietal bone bears a somewhat swollen and bossed region (*parietal eminence*) that sits approximately halfway along the bone's length, lateral to the midline of the skull. In some individuals, this region is quite prominent and swollen, giving the skull the appearance of having a blunt corner. Right and left parietal eminences may coincide with the points of greatest cranial width, which is measured bilaterally at *euryon*.

The **occipital bone**, which houses the opening (*foramen magnum*) through which the spinal cord exits the skull, articulates with the parietal bones posteriorly; it also extends inferiorly and basally. Viewed from the rear, this bone presents two morphologically distinguishable areas: a relatively smooth-boned

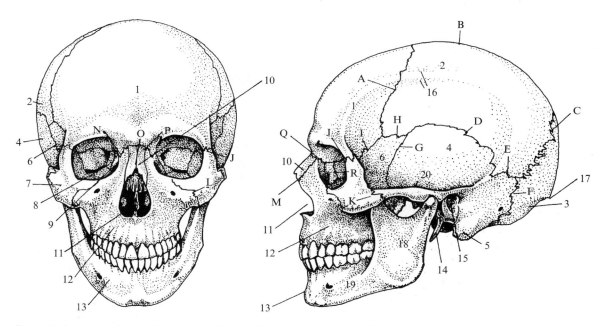

Figure 2–3   Articulated skull and mandible: (*left*) anterior and (*right*) lateral views. Major features/bones: (*1*) frontal, (*2*) parietal, (*3*) occipital, (*4*) temporal, (*5*) mastoid process (temporal bone), (*6*) greater wing of sphenoid, (*7*) zygoma, (*8*) ethmoid, (*9*) lacrimal, (*10*) nasal, (*11*) nasal aperture, (*12*) maxilla, (*13*) mandible, (*14*) styloid process, (*15*) external acoustic meatus, (*16*) temporal lines, (*17*) external occipital protuberance, (*18*) ascending ramus, (*19*) body of mandible, (*20*) temporal fossa. Sutures: (*A*) coronal, (*B*) sagittal, (*C*) lambdoid, (*D*) squamosal, (*E*) parietomastoid, (*F*) occipitomastoid, (*G*) sphenotemporal, (*H*) sphenoparietal, (*I*) sphenofrontal, (*J*) zygomaticofrontal, (*K*) zygomaticotemporal, (*L*) zygomaticomaxillary, (*M*) nasomaxillary, (*N*) frontomaxillary, (*O*) nasal, (*P*) frontonasal, (*Q*) frontolacrimal, (*R*) ethmolacrimal.

superior component (*planum occipitale*) and a rougher, muscle-scarred inferior portion (*planum nuchale*), to which the neck or nuchal muscles attach. These planes are typically delineated from each other by a roughly undulating *superior nuchal line*, which increases in bulk and thickness toward the midline of the skull, where it is often variably elevated into a blunt to peaked and distended *external occipital protuberance*. Sometimes, the external occipital protuberance is coincident with the anthropometric landmark *inion*, which lies at the midline of the superior nuchal line. The most posterior point on the skull, which is usually on the planum occipitale, is identified by the anthropometric

landmark *opisthocranion*. In some primates, including various fossil hominids, inion and opisthocranion coincide. Two landmarks are defined at the midline margin of the foramen magnum: *basion* anteriorly and *opisthion* posteriorly.

The **temporal bone**, which lies below the parietal and in front of the occipital, contributes to the lower portion of the lateral wall of the cranium. It appears platelike from the side, but the part that forms part of the cranial base is quite thick and bulky. The platelike portion of the temporal bone is the *squama* or *squamous portion of the temporal* ("squamous" means "flat" or "scalelike"). The bulky basicranial part is the *petrous portion of* *the temporal bone* or just *petrosal bone*. A variably gracile bony strut—*zygomatic process of the temporal bone*—emanates from the side of the squamous portion of the temporal bone, just in front of the external opening (*external auditory meatus*) of the bony tube of the ear region and projects anteriorly to contribute to the cheekbone (*zygomatic arch*). The "space" above the horizontal plane of the zygomatic arch is the *temporal fossa*; the *infratemporal fossa* lies below.

Behind the external auditory meatus, the *parietomastoid region* of the temporal bone is bulky, extending posteriorly along its contact with the parietal bone (*parietomastoid suture*) as a rather horizontal ledge and swelling inferiorly into a variably pointed and distended projection (*mastoid process*), the tip of which is the landmark *mastoidale*. The region at which the parietomastoid suture leaves the squamous portion is the variably angular *parietal notch*, into which a "corner" of the parietal bone nestles firmly. The midpoint of the superior margin of the external auditory meatus is the landmark *porion*. A vertical line projected upward from an imaginary line connecting right and left porions (and at a right angle to the Frankfort horizontal) terminates superiorly at the landmark *apex,* which frequently is the highest point on the skull. (Apex can be measured directly from a coronal radiograph. When measured on the cranium, the end of the arm of the osteometer, which is essentially a large sliding caliper, is placed on porion, the stem of the osteometer is held vertically and at a right angle to the Frankfort horizontal, and the other arm of the osteometer is extended over the top of the skull and brought down on the skull. The point of contact will be apex. See Appendix A for more measurements.)

The *greater wing of the sphenoid* intervenes between the temporal bone anteriorly and the frontal bone. Right and left greater wings spread out from the body of the sphenoid, the bulk of which lies in the midregion of the cranium, where it contributes to the base of the skull and the walls of the orbits. The *greater* and the much smaller *lesser wings of the sphenoid* are visible within the orbits. The greater wing forms much of the inferior and posterior portion of the lateral orbital wall; the lesser wing constitutes the most posterior and most minor contribution to the medial orbital wall as well as to the orbit's roof. The greater wing is separated inferiorly from the orbital floor by the deep *inferior orbital fissure*, which communicates primarily with the infratemporal fossa but also somewhat with the *pterygopalatine fossa* (see Facial Skeleton, later). The inferior orbital fissure originates laterally from the posterior edge of the zygomatic bone and courses medially and posteriorly to the back of the funnel-shaped orbit. At its terminus, the *superior orbital fissure* takes origin, coursing upward and arcing somewhat laterally to intervene between the lesser and greater wings of the sphenoid. (The oculomotor, trochlear, and abducent nerves and the terminal branches of the ophthalmic nerve and veins course through the superior orbital fissure.) The superior orbital fissure communicates with the middle cranial fossa in the interior of the cranium. The deepest part, or *apex*, of the orbit lies just medial to the superior orbital fissure. Just medial to the superior orbital fissure and the apex of the orbit lies the *optic foramen*, through which the optic nerve and ophthalmic artery course as they proceed from the anterior cranial fossa, through the sphenoid bone, and into the orbit.

The inferior orbital fissure transmits the maxillary nerve, which continues as the infraorbital nerve and artery through the maxilla in the floor of the orbit. As the infraorbital neurovascular bundle enters the orbital floor, it courses first through a groove of variable length (*infraorbital groove*, which is a groove because its "roof" is unossified) that becomes the *infraorbital canal* (whose "roof" is ossified), which opens onto the face just below the inferior orbital rim as the *infraorbital foramen*. Sometimes an infraorbital foramen may be bisected by a thin bony plate (ossified tissue of the neurovascular bundle), or there may be one or even, but more rarely, two smaller accessory foramina (representing branches of the neurovascular bundle).

The sphenoid bone also extends inferiorly from the greater wing into the basicranium, where it comes to lie adjacent to the anterior border of the petrous portion of the temporal. The "tip" of the petrosal, which is directed anteriorly and medially, in conjunction with the body of the sphenoid, subtends a large, jagged-edged *foramen lacerum*, which, in life, contains cartilage. The tip of the petrosal opens upon the foramen lacerum as the *carotid canal.* From the general region of each foramen lacerum, the sphenoid descends to the level of the palate as two platelike wings (*medial* and *lateral pterygoid plates* or *laminae*) that are confluent anteriorly and together subtend a shallow but tall depression (*pterygoid fossa*).

The **ethmoid** lies in front of the sphenoid and is contained largely within the nasal cavity. Its paper-thin lateral sides are exposed in, and therefore contribute to, the medial orbital walls; an ethmoid exposure in the medial orbital wall is sometimes referred to as the *os planum.* Humans and the two African apes—the chimpanzee and gorilla—are apparently the only extant primates in which the ethmoid is internally subdivided into air cells or chambers. The African apes develop only a few large air cells, whereas humans are distinguished by an abundance of ethmoidal air cells. In many human skulls, the orbitally exposed ethmoid is sufficiently translucent that one can see the pattern of air cells within.

All of the aforementioned bones—the ethmoid included—contribute to the interior surface of the braincase and bear evidence of their association with soft tissue structures associated with the brain.

The joint between the frontal and parietal bones is the *coronal suture*, which courses from side to side across the top of the skull. It is formed by an edge-to-edge contact of the bones involved and presents itself exter-nally as three segments, of which the middle segment is the most denticulate and severely undulating. The coronal suture terminates inferiorly at the superior extent of the greater wing of the sphenoid, which is roughly at the same level as the highest point of the superior margin of the orbit. A plane passing vertically through the body via the coronal suture is the *coronal plane.*

The *sagittal suture* courses posteriorly from the coronal suture (the juncture is the landmark *bregma*) along the midline of the cranium between the two parietal bones; posteriorly, it terminates at the occipital bone (the landmark *lambda*). A plane passing vertically through the sagittal suture is the *sagittal plane.* The sagittal suture is variably oscillating along its length: proceeding posteriorly from bregma, it is shallowly and mildly undulating for approximately one-third of its length, then deeply peaked for a short burst, very mildly undulating for a short distance, and then quite peaked for the remainder. The contact between the parietal bone and the greater wing of the sphenoid is the *sphenoparietal suture*, which marks the general area (not specific point) of the anthropometric landmark *pterion*. The contact between the frontal and the greater wing of the sphenoid is the *sphenofrontal suture.*

Sometimes small islands of bone form within segments of the sagittal suture. Such intrusive, supernumerary bony elements are referred to interchangeably as *ossicles* or *intrasutural, accessory,* or *wormian bones.* In the sagittal suture, they can be identified as, for example, *sagittal ossicles* or *wormian bones in the sagittal suture.* A wormian bone located at bregma is a *bregmatic bone* or *ossicle* or an *ossicle at bregma.* An ossicle at pterion is an *epipteric bone.* One in the region of the parietal notch is a *parietal notch bone.* An extremely large ossicle at lambda—created by a suture that courses essentially between the occipital and nuchal planes—is an *Inca* or *interparietal bone.*

The *lambdoid suture*, between the parietal and occipital bones, courses down to the

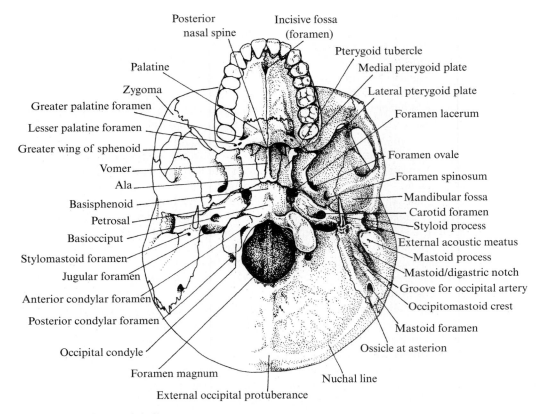

Figure 2–4   Basal view of skull.

lateral part of the suture between the occipital and the mastoid region. The contact between the occipital and the mastoid region is the *occipitomastoid suture*. Often, the *mastoid foramen* lies in the occipitomastoid suture, just behind the mastoid process; but sometimes it lies outside the suture (is *exsutural*), penetrating the mastoid process. There may also be smaller, accessory mastoid foramina, which are usually exsutural. The tripartite juncture of the occipital, parietal, and temporal bones (i.e., the parietomastoid, lambdoid, and occipitomastoid sutures) is the vaguely star-shaped landmark *asterion*.

Basicranially (Figures 2–4 and 2–5), the spit of occipital bone that proceeds anteriorly

from the foramen magnum—the *basilar portion of the occipital bone* or *basiocciput*—meets the *basilar portion of the sphenoid* or the *basisphenoid* at the *spheno-occipital synchondrosis* or *basilar suture*. Eventually, basiocciput and basisphenoid coalesce. The basiocciput and adjacent thick, craggy petrosal are separated throughout life by the *petro-occipital fissure*. Together, these bones contribute to the formation of a moderately large *jugular foramen*, through which the jugular vein exits the braincase.

The platelike squamous portion of the temporal bone forms an arcuate *squamosal suture* primarily with the parietal bone, which it overlaps rather than abuts. Thin striations

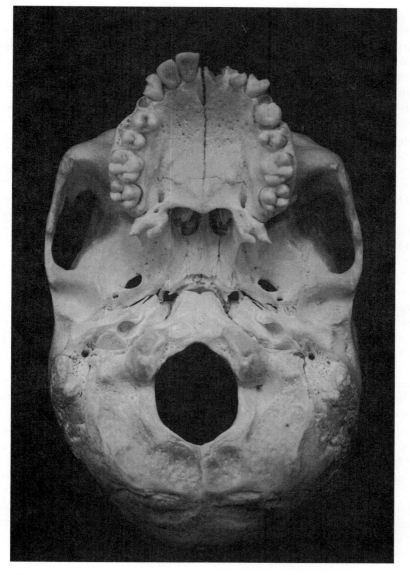

Figure 2–5   Basal view of skull.

often radiate out from the squamosal suture over the external surface of the parietal bone. The contact between the temporal bone and greater wing of the sphenoid is the *sphenotemporal suture*. The parietal and mastoid region of the temporal articulate along the *parietomastoid suture*.

Facial Skeleton

The facial skeleton proper (Figure 2–3) is typically described as being composed of 14 major bones: the **maxillary, nasal, zygomatic** or **malar, lacrimal,** and **palatine bones;** the **inferior nasal conchae;** and the

vomer, all of which are paired bones. If one includes the mandible, the number increases to 15. The **hyoid bone** is also often included in discussions of the skull because it derives embryologically from the second and third branchial arches and thus belongs to the cartilaginous branchial arch skeleton, as do two of the inner ear bones (the **incus** and **malleus**) and (part of) the mandible. An adult's hyoid is a single, "U"-shaped bone and the only bone in the body that does not contact another bone; it is suspended below the mandible by ligaments that attach to processes on the cranial base.

The typically short, inferiorly broadening **nasal bones** lie in the midsagittal plane of the face. They articulate with the frontal bone above and form the superior margin of the nasal aperture. On each side, the *frontal process* of the maxilla (actually, of the premaxilla) courses superiorly alongside a nasal bone to contact the frontal bone. Only in humans and various other species of the genus *Homo* do the nasal bones project away outward from the facial plane.

Lateral and inferior to the frontal process, the **maxilla** contributes to the *inferior border (rim, margin) of the orbit*, where it is folded or creased. A posteriorly tapering tongue of maxilla extends from this flexure into the orbital floor and maintains contact medially with the inferior border of the ethmoid exposed in the medial orbital wall. The *inferior orbital fissure* separates this maxillary "tongue" from the greater wing of the sphenoid and, thus, from the lateral orbital wall. The bulk of the maxilla lies below the infraorbital rim and expands to contribute to the *cheekbone*, or *zygomatic* or *malar region*, below which it becomes "waisted" as it approaches the *alveolar border* (in which the teeth are anchored).

The *nasal aperture*, which is taller than wide and typically broader inferiorly than superiorly, is subtended primarily by the maxilla. Its shape is technically trapezoidal but has also been described as broadly piriform or pear-shaped. At the midline of its inferior margin, the abutting right and left maxillae may be distended anteriorly into a small/blunt to large/spikelike, sometimes bifid projection, *anterior nasal spine*. The reason this spine may be bifid is that it derives from the coalescence of what were ontogenetically two separate spikelike projections; the suture between right and left anterior nasal spines often persists into the adult. The maxillary contribution to the floor of the nasal cavity is typically smooth and planar. The teeth sit in a raised peripheral portion of the maxilla—the *alveolar process, margin*, or *border*—medial to which the palatal region is variably vaulted. A single, relatively large *incisive foramen* or *fossa* perforates the midline of the palate approximately 1 cm posterior to the alveolar margin of the incisors.

The small **lacrimal bone** in the medial orbital wall is sandwiched between the frontal process and the exposed ethmoid. (In humans, the lacrimal lies entirely within the medial orbital wall, while in other mammals, including other primates, it may extend onto the face.) In humans, the lacrimal is dominated by a large *lacrimal fossa (groove, sulcus)* that is bordered posteriorly by the *posterior lacrimal crest*, which lies on the lacrimal bone. This crest may be crestlike but may also be a minimal swelling or even a posteriorly distended sheet of bone. The *anterior lacrimal crest* lies not on the lacrimal but on the frontal process of the premaxilla (maxilla), at the edge of the orbit. This crest typically presents as a blunt rise. The anterior and posterior lacrimal crests bound the broad and variably deep lacrimal fossa.

The maxilla contacts the **zygoma** or **malar bone** lateral to it; both contribute to the inferior orbital rim and constitute the major component of the lateral portion of the orbital rim. The zygoma also projects upward to contact a short extension of the frontal (the *zygomatic process of the frontal*). Together, these two extensions form a barlike structure that lies lateral and slightly posterior to the orbit. More or less at the same level as the inferior rim of the orbit, the short *zygomatic process of the zygomatic bone* projects posteriorly. This process contacts the anteriorly directed *zygomatic process of the temporal bone.*

The **palatine bones** lie posterior to the maxilla on its oral cavity side and constitute the posteriormost extension of the hard palate. A palatine bone is much wider (mediolaterally) than it is long (anteroposteriorly). At the midline of their posterior border, the abutting palatine bones are distended in concert into a single but variably roundedly blunt to peaked to squared *posterior nasal spine.* On the lateral side, each palatine bone contributes to the formation of a large, somewhat ovoid or elliptical *greater palatine foramen.* The maxilla may contribute to the lateral wall of this foramen. Each palatine bone is bounded posterolaterally by the inferiormost portion of the region at which the medial and lateral pterygoid plates join each other. A small *lesser palatine foramen* and/or an *accessory lesser palatine foramen* may lie just in front of the wall created by the medial and lateral pterygoid plates. The "space" posterior to the palatine bones and pterygoid plates (i.e., the border created by the palatine bones and pterygoid plates on one side and the basiocciput on the other) is the *pterygopalatine fossa,* through which various soft tissue structures pass (e.g., branches of the maxillary, mandibular, nasopalatine, and greater palatine nerves and their attendant arteries). The pterygopalatine fossa is also confluent with the inferior orbital fissure (however, the confluence between the pterygopalatine and infratemporal fossae is greater).

The remaining bones of the face—the **vomer** and **inferior nasal conchae**—are visible only when seen directly through the nasal aperture into the *nasal cavity.* The **vomer** is a thin, vertical sheet of bone in the midline of the nasal cavity that rises from the floor of the nasal cavity to contact an equally thin, sheetlike extension of the ethmoid bone, the *perpendicular plate of the ethmoid.* Together, these contiguous, vertical plates of bone bisect the nasal cavity. The vomer spans the distance between the anterior and posterior nasal spines and thus overlaps both the maxilla and palatine bones. The **inferior nasal conchae** extend down from the lateral walls of the nasal cavity and protrude in toward, but do not touch, the vomer. Each concha takes origin as a relatively thin, smooth-boned, and winglike structure that expands medially and inferiorly into a more rugose ledge and eventually fuses to the wall of the nasal cavity.

The **mandible** consists on each side of a somewhat horizontally oriented *body* or *corpus* and a fairly vertical (*ascending*) *ramus.* The body of the mandible contains teeth (which are anchored in the *alveolar process* or *margin* of the mandible), and its general contour mirrors the curvature of the *dental arcade* of the upper jaw. The juncture of the ramus and corpus is the *gonial region.* The angle that can be measured between the inferior margin of the body and the posterior margin of the ascending ramus is the *gonial* or *mandibular angle*; alternatively, one can measure this angle at the intersection of the imaginary long axes of the body and ascending ramus. More so in males than in females, the external surface of the gonial region is more markedly muscle-scarred and lipped around its external margin. The mandibular angle in young to middle-aged adult males tends to approach 90°; it is more obtuse in females of similar age. In young and very old individuals of both sexes, this angle is more obtuse: in the former case, adult features have not yet been established, while in the latter, the bone has become thinned and remodeled as a result of resorption that normally occurs with age and/or dental/periodontal disease.

Superiorly, the ramus bifurcates broadly and shallowly into a thin anterior projection (which comes to lie within the infratemporal fossa when the upper and lower jaws are in occlusion) and a posterior, wider, barlike end (which articulates directly with the skull). The thin projection is the *coronoid process,* and from it emanates a large muscle of mastication

—the *temporal muscle*—that passes up between the zygomatic arch and the lateral wall of the skull and then fans out over the sidewall of the braincase. The perimeter of this muscle leaves a set of roughly parallel muscle scars, the *superior* and *inferior temporal lines*. These lines (1) course up behind the orbit and along the posterior orbital wall, (2) arc superiorly at their peak just behind the coronal suture and then arc inferiorly toward the lambdoid suture, (3) curve inward beyond asterion as they proceed anteriorly and in parallel with the zygomatic process of the temporal bone, and then (4) fade out. Separate superior and inferior muscle scars are most distinct as they proceed away from the temporal ridge (of the posterior orbital region of the frontal bone) and until the point at which they begin their descent toward asterion.

The part of the ascending ramus that articulates with the base of the skull has been called the *head of the mandible*, the *mandibular condyle*, or the *condyloid* or *condylar process of the mandible*. When the upper and lower jaws are in occlusion, the mandibular condyle sits in a typically shallow but mediolaterally variably ovoid to elliptical depression that lies in front of the external auditory meatus, inferior and medial to the root of the temporal bone's zygomatic process. This depression —the *articular*, *mandibular*, or *glenoid fossa* —is wider (mediolaterally) than it is long (anteroposteriorly).

The frontal and nasal bones articulate along the *frontonasal suture*. The frontal and the frontal process (of the pre-/maxilla) articulate along the *frontomaxillary suture*. The *zygomaticofrontal suture* lies between the frontal and zygoma; the *frontolacrimal suture* separates the frontal and lacrimal bones. The midline suture between the two nasal bones is the *nasal, internasal,* or *nasonasal suture*. The suture between a nasal bone and the maxilla is the *nasomaxillary suture*. The frontal bone's contribution to the sutures with the nasal bones, frontal process of the maxilla, and lacrimal bones is variably arcuate, depending

especially on the degree to which the nasal bones extend superiorly beyond the upper margins of the frontal processes. The contact between the lacrimal and maxilla is the *lacrimomaxillary suture*; that between the lacrimal and ethmoid is the *ethmolacrimal suture*.

The facial skeleton bears a profusion of anthropometric landmarks. Sometimes coincident with the zygomaticofrontal suture is *ectoconchion*, which is defined as the most lateral extent of the orbital wall. The juncture of the internasal and frontonasal sutures is *nasion*. Somewhat lateral to nasion is *maxillofrontale*, or the *anterior lacrimal point*, which is identified as the juncture of the anterior lacrimal crest and the frontomaxillary suture. Just lateral to maxillofrontale is *dacryon*, the point at which the frontomaxillary, frontolacrimal, and maxillolacrimal sutures meet; some authorities (e.g., Vallois, 1965) considered the term "dacryon" to be obsolete and used only "maxillofrontale" when taking measurements that involve the orbital region. The juncture between the posterior lacrimal crest and frontolacrimal suture is *lacrimale*.

Various anthropometric landmarks are located in the median sagittal plane of the facial skeleton. For example, *nasospinale*, which is usually found at the base of the anterior nasal spine, is defined as the midpoint of a line connecting the lowest extents of the inferior margins of the right and left sides of the nasal aperture (technically, the lowest point of the right or left side of the inferior margin of the nasal aperture is called *nariale*). Below nasospinale lies the region of *prosthion*, which used to be subdivided into two landmarks, *hypoprosthion* and *exoprosthion*. Hypoprosthion was defined as the most inferior bony extension (of the alveolar bone or margin) between the upper central incisors. Nowadays, the locus of hypoprosthion is almost exclusively referred to as *alveolare*. *Exoprosthion*, which would be situated slightly superior to hypoprosthion, was defined as the most anterior point or extension of the alveolar margin. The landmarks prosthion and exoprosthion are now taken as the same point. Regardless of the terms used, the distinction between two different landmarks in the medial sagittal plane of the alveolar region is warranted because, in fact, the two may be well separated from one another. "Prosthion" is used for measuring longitudinal diameters and "alveolare" for vertical measurements (see Vallois, 1965).

On the mandible, the anterior- and superiormost extent of the alveolar bone between the lower central incisors is *infradentale*. Farther inferiorly, *pogonion* represents the anteriormost point on the chin, below which is *gnathion*, which is located at the most anterior- and inferiormost point on the mandible. On the palatal side of the maxilla, *orale* is the point where the medial sagittal plane is crossed by a line connecting the most posterior borders of the sockets of the central incisors. Much farther back is *staphylion*, which is the posteriormost extent of the hard palate along the medial sagittal plane; this point usually coincides with the posterior nasal spine. Just in front of, or sometimes coincident with, staphylion is *alveolon*, which is the intersection between the medial sagittal plane and a line drawn between the posteriormost extents of right and left alveolar processes.

Anthropometric landmarks for determining maxillary and mandibular widths are, respectively, *endomolare/ectomolare* and *gonion*. Endomolare is the most medial point of the medial margin of the alveolar bone surrounding the second upper molar. Ectomolare is the most lateral point on the lateral alveolar border of the tooth. Gonion is located externally on the gonial region of the mandible; it is the most lateral point of the gonial or angular region of the mandible.

Well back within the orbit and visible only when peering obliquely onto the medial orbital wall lies the short and vertically oriented *sphenoethmoidal suture*. Superiorly, this suture contributes to a tripartite juncture with the *frontoethmoidal* and *sphenofrontal sutures*. The frontoethmoidal suture courses more or less horizontally posteriorly along the medial orbital wall. The sphenofrontal suture can be located by looking directly into the orbit (when the skull is in norma frontalis). It courses laterally and anteriorly from the tripartite sutural juncture noted above and meanders along the short exposure into the orbit of the lesser wing of the sphenoid, over the superior flexure of the superior orbital fissure, and across the orbital contribution of the greater wing of the sphenoid until it becomes confluent with the zygomaticofrontal suture. The juncture between the sphenofrontal and zygomaticofrontal sutures is delineated by the termination superiorly of the rather vertically oriented, variably undulating *sphenozygomatic suture*, which descends to the slight upward invagination of the anterior end of the inferior orbital fissure.

Variation is noted in the development of an *exsutural anterior ethmoid foramen* (i.e., lying in the medial orbital wall in the frontal alone, rather than in the suture between the frontal and the ethmoid) and a *posterior ethmoid foramen*. The sphenoid and maxilla meet along the *sphenomaxillary suture*, and the sphenoid and the zygoma create the *sphenozygomatic suture*.

Internal Morphology

Skullcap

[Since most study specimens have had the top of the braincase sawn off, it will be discussed as a unit, albeit an incomplete one (Figures 2–6 and 2–7).]

In order to expose the interior of the cranium, one must remove some portion of its surrounding vaultlike structure, the skullcap. Typically, this consists of portions of the frontal, parietal, and occipital bones; the superior extent of the squamous region of the temporal bone is not included in the

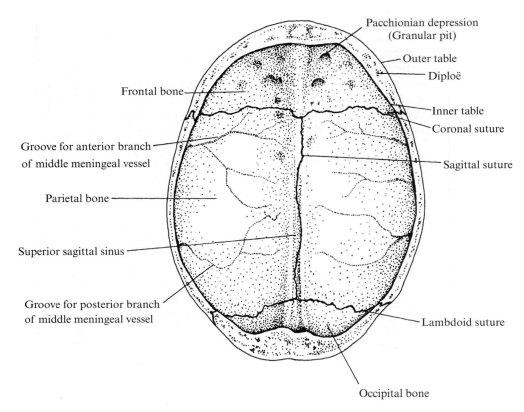

Figure 2–6   Internal morphology of skullcap.

detached skullcap. With the skullcap removed, one can study cranial bones in cross section and appreciate differences in thickness within the same bone as well as between adjacent bones. (For example, the parietal bones are more consistently thinner than the frontal and occipital bones; the occipital bone is the thickest and is expanded farther toward its midline; the frontal bone bears a prominent crest along its midline.)

Most clearly noted in the region of the frontal and occipital bones is the three-layered morphology typical of many cranial bones. There are *outer* and *inner tables* of compact bone and an intervening layer of spongy bone (*diploë*). In regions where the bone is not very thick (e.g., much of the length of the parietal bones, the posterior extents of the frontal bone), the diploic layer may be minimally expressed and the bone appears to be continuously solid.

Among the dominant features of the interior of the skullcap, especially of younger individuals, are the *coronal, sagittal,* and *lambdoid* sutures. Sutural distinctiveness tends to diminish with increasing age as the sutures coalesce and fuse. Sutures generally close from the inside out (*endocranially* to *ectocranially*). The endocranial sequence is usually coronal→sagittal→lambdoid. The lambdoid suture is the most variable in degree of completeness of closure (e.g., it may be minimally fused even in the most senile individuals), whereas the coronal suture is almost always completely closed by 30 years of age and the sagittal suture is completely fused in 67% of

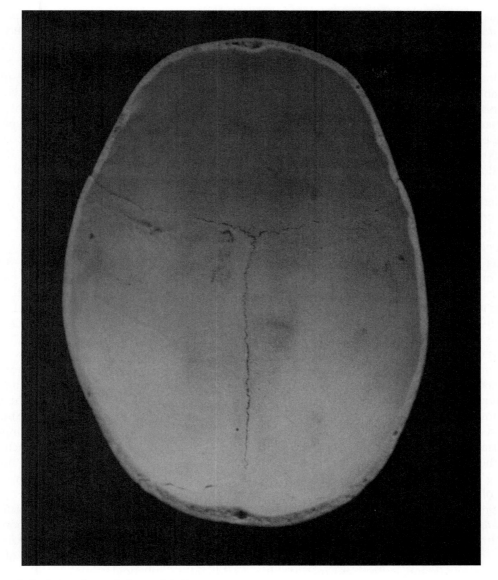

Figure 2–7   Internal morphology of skullcap.

individuals 70 years of age or older. Ectocranially, the sagittal and lambdoid sutures close at about the same time and well before the coronal suture.

Persistent throughout an individual's adult life are the impressions left endocranially by the *meningeal vessels*, which attend the meninges surrounding the brain. The network of grooves these vessels leave is confined primarily to the inner table of the parietal bones. However, a few meningeal branches may cross onto the frontal bone and, less frequently, onto the occipital bone. The *anterior division* or *branch of the middle meningeal vessel* lies anteriorly on the parietal bone, coursing more or less in parallel with the coronal suture. This branch divides into a variable number of smaller branches, the majority of which typically lie posterior to the primary trunk. The *grooves for the posterior divisions* or *branches of the middle meningeal vessels* lie posterior to the anterior branch and its divisions, approximately halfway along the parietal bone. Like the anterior complex, the posterior branches arborize superiorly and posteriorly. The internal surface of the parietal is, therefore, characterized by (1) fewer grooves for the middle meningeal vessels along its inferior portion and (2) an increasing number superiorly that ramify and fan out in a posterosuperior direction (which provides information relevant to siding this bone).

The generally broad and shallow *sagittal sulcus* courses along the internal aspect of the sagittal suture; it is the groove impressed on the inner table of bone by the superior sagittal sinus, which is a conduit of venous drainage associated with the brain and its meninges. This sulcus is most clearly expressed along the posterior portion of the sagittal suture, but traces of it may also occur on the frontal bone. Within the trough and along the perimeter of the sulcus are often pits of small to medium size that are formed by protrusions into the bone of *arachnoid granulations* of *pacchionian bodies* (from the arachnoid meningeal layer). With advancing age, the number, depth, and width of these *granular pits* (granular foveolae) increase. In older individuals, the large pits are identified as *pacchionian depressions*.

The superior portion of the **frontal bone** is essentially featureless, except for the often sharp and prominent *frontal crest*, to which attaches the falx cerebri (which tethers the brain anteriorly to the interior of the skull). The impression of the sulcus of the superior sagittal sinus is usually visible at the "base" (i.e., the superiormost portion) of the frontal crest.

The bit of **occipital bone** that is captured in the skullcap is also typically relatively featureless. However, it tends to bulge along the median sagittal plane (providing a site of attachment for the falx cerebri); the sulcus for the superior sagittal sinus may be visible veering off to one side or the other of this low "crest." In humans, the sulcus for the superior sagittal sinus typically and most visibly courses to the right side of the median "bulge" (*internal occipital protuberance*) of the occipital bone (see Falk and Conroy, 1983; Tobias, 1967; and references therein).

Base of the Skull

The interior of the skull base (Figures 2–8 and 2–9) is subdivided into three major paired "units" or depressions that receive (from front to back) the paired frontal, temporal, and occipital lobes of the cerebrum. These depressions, respectively, are the right and left portions of the **anterior**, **middle**, and **posterior cranial fossae**. The surface of the anterior cranial fossa is the most irregular topographically because of impressions left by the gyri (raised areas of the convolutions) of the brain; the clarity and profusion of these impressions are affected by the extent to which the meninges obscure the surface of the brain. The middle cranial fossa may also bear gyral impressions.

The **anterior cranial fossa** is confined anteriorly and laterally by a platelike portion of the frontal and inferiorly by the orbital portion of the frontal, as well as by parts of the ethmoid and sphenoid behind it. The ethmoid —represented by the upwardly protruding *crista galli* and *cribriform plate* around its base —intervenes midsagittally between the rugose and upwardly bulging frontal contributions to the *orbital plate*. The relatively narrow and anteroposteriorly elongate *crista galli* ("crest of a chicken" or cock's comb, because of its general shape) lies behind and in line with the frontal crest. It is tallest and broadest anteriorly and slopes (sometimes quite

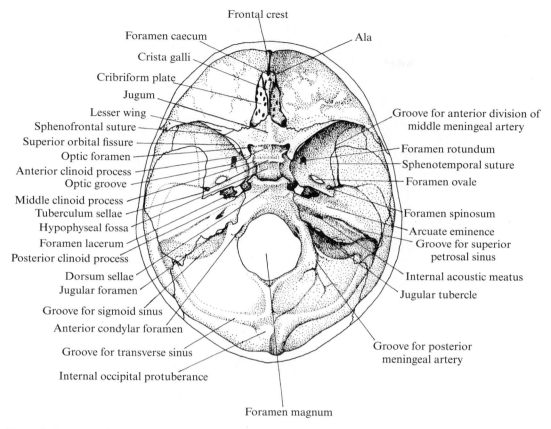

Figure 2–8   Internal morphology of skull base.

markedly) and tapers posteriorly. Anteriorly and inferiorly, it spreads out like a winged mantle, forming a sutural contact with the surrounding frontal bone. The wings or *alae* (also *ethmoidal alae*) of this mantle appear to embrace the frontal crest. The small *foramen caecum* (a caecum is a blind pouch or cul-de-sac) lies between the "mantle" of the crista galli and the base of the frontal crest and typically opens only on the floor of the anterior cranial fossa; sometimes, however, the foramen caecum transmits a vein from the superior sagittal sinus.

On either side of the crista galli and often extending around it posteriorly is a narrow field of thin, perforated bone (the *cribriform plate of the ethmoid*), which lies well below

the level of the frontal contributions to the orbital plate. It forms a shallow trough around the crista galli, isolating it (1) on its sides from the frontal portion of the orbital plate and (2) posteriorly from the body of the sphenoid. The cribriform plate forms the thin roof of the nasal cavity; tiny branches of the olfactory nerve course from the mucous membrane of the nasal cavity, through the perforations or foramina of the cribriform plate, to the olfactory bulbs.

The sphenoid contributes to the posterior margin of the anterior cranial fossa; the laterally tapering lesser wings of the sphenoid embrace from behind the frontal contribution to the orbital plate. Posterior to the cribriform plate, the midsagittal portion between

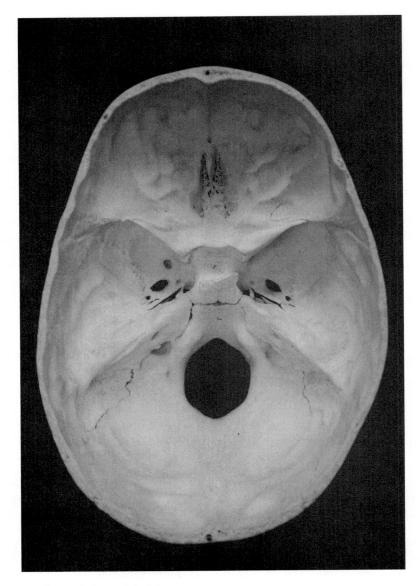

Figure 2–9   Internal morphology of skull base.

the lesser wings (the *jugum* or *jugal region*) is broad, deep, smooth, and relatively flat. In young individuals, the suture between the lesser wings of the sphenoid and orbital plate of the frontal is visible; one can also often delineate the region where the jugum overlaps the ethmoid. Posteriorly, in the macerated skull, the winglike shape of the lesser wings is emphasized further by blunt to pointed projections (*anterior clinoid processes*) that extend into "open space" from either side of the jugum and over the right and left superior orbital fissures. The *optic foramen* lies medially and anteriorly at the base of each anterior clinoid process.

In contrast to the convex surfaces of right and left anterior cranial fossae, the two parts of the **middle cranial fossa** are concave, presenting two deep concavities that expand outward from a narrow, "saddle-shaped" mid-sagittal depression in the sphenoid (the *sella turcica*). The sella is bounded superolaterally by the anterior clinoid processes and superiorly across the midline by the variably incised *optic groove*, which courses between the optic foramina and is bordered anteriorly and posteriorly by variably distinct ridges. The anterior wall or *tuberculum sellae* of the sella descends from the optic groove, terminating in the concave *hypophyseal fossa*, in which the cerebral hypophysis (pituitary gland) lies. Below the level of the optic groove, the superior part of the tuberculum sellae may sometimes bear small, lateral protrusions (*middle* or *medial clinoid processes*). The sella is confined posteriorly by an anteriorly sloping plate of bone (*dorsum sellae*), which is distended upward and often laterally at its superolateral "corners" into *posterior clinoid processes*. Midsagittally, the superior margin of the dorsum sellae is variably excavated. The posterior end of the base of the dorsum sellae contacts the basiocciput. The sides of the base of the sella are indented. These indentations constitute the sphenoidal contributions to right and left *foramina lacera* (s. *foramen lacerum*). The right and left temporal lobes of the cerebrum sit like saddlebags on either side of the sella. Each lobe is cupped in a shallow, depressed *lateral part of the middle cranial fossa*.

The middle cranial fossa is broad anteriorly and, being truncated obliquely along its medial margin, tapers posteriorly. Anteriorly and anteroinferiorly, the fossa is bounded by the interior surface of the greater wing of the sphenoid and the anterior part of the squamous portion of the temporal. The fossa is subtended laterally and, in part, lateroinferiorly by the squamous portion of the temporal. The petrous portion of the temporal bounds the lateral part of the middle cranial fossa medially and medioinferiorly.

Anteriorly, medially, and somewhat superiorly, the arcuate *superior orbital fissure* connects the lateral part of the middle cranial fossa with the orbit. A series of foramina continues the arcuate trajectory of the superior orbital fissure. Situated below and slightly lateral to the inferiormost extent of the superior orbital fissure is the fairly round *foramen rotundum*, whose course through the greater wing of the sphenoid essentially parallels the Frankfort horizontal (it "points" forward). A bristle passed through the foramen rotundum from the middle cranial fossa demonstrates that it opens onto the posterior pole of the maxilla via the pterygopalatine fossa. The maxillary nerve courses from the trigeminal ganglion at the base of the brain through the foramen rotundum to the maxillary region, wherein it sends its sensory branches to teeth and subsequently continues anteriorly as the infraorbital nerve along the floor of the orbit to exit upon the face from the *infraorbital foramen*.

Below and somewhat lateral to the foramen rotundum lies the larger, oval-shaped *foramen ovale*, whose long axis is typically oriented obliquely anteromedial to posterolateral. This foramen is easily located externally, at the base and just lateral to the lateral pterygoid plate. The mandibular nerve, which penetrates the mandible on its medial side (via the mandibular foramen) and then innervates lower dentition, exits the cranial base through the foramen ovale. One sometimes finds accessory foramina ovale or septa that subdivide the primary foramen ovale.

Completing the arcuate arrangement of foramina in the middle cranial fossa is the *foramen spinosum*, which is the smallest of these foramina. It is located behind and lateral to the posterolateral end of the foramen ovale, just medial to the sphenosquamosal suture. Variability in the expression of the foramen spinosum is more common than in the foramen ovale, resulting in, for example, this foramen's incomplete separation from the foramen ovale posteriorly or its continuity with a thin fissure that separates the petrous portion of the temporal anteromedially from the adjacent sphenoid. The middle meningeal artery courses through the foramen spinosum. This artery parallels the middle meningeal veins in arborizing laterally across the squamous portion of the temporal.

Beyond the arcuate arrangement of foramina described above, and medial and sometimes slightly anterior to the foramen ovale, may lie a tiny *emissary sphenoidal* foramen. When it is present, it is usually bilateral. Less frequently, and unilaterally, this foramen may not be closed off completely by bone, thus opening as a notch upon the foramen lacerum. This foramen may also be associated with a variably shallow depression-to-definitive groove that may extend as far as the inferior border of the superior orbital fissure. Occasionally, and unilaterally, a small accessory foramen may lie just below the inferior border of this fissure; if this small accessory foramen is present, the shallow depression/groove will terminate at it.

The remainder of the lateral part of the middle cranial fossa is relatively unremarkable. It is traversed by sutures and grooves for middle meningeal vessels and slightly and variably corrugated by impressions of cerebral gyri; midway along the length of the petrous portion of the temporal lies the typically mound-shaped *arcuate eminence*. At a variable distance laterally away from the foramen spinosum, the groove for the middle meningeal vessel bifurcates into anterior and posterior divisions or branches. The posterior division courses up and usually posteriorly along the squamous portion of the temporal and may cross the squamosal suture before arborizing.

Depending on how far up the squamous portion of the temporal bone the groove for the middle meningeal vessel ascends before bifurcating, there may also be a relatively long anterior division of the middle meningeal vessel coursing upward and anteriorly, roughly paralleling the sphenosquamosal suture. This anterior division may then cross the sphenosquamosal suture to ascend the superiormost portion of the greater wing of the sphenoid, or a short anterior branch of this primary anterior division may course first across the sphenosquamosal suture and then across the superior part of the greater wing of the sphenoid. In either case, the primary anterior division continues up the lateral wall of the middle cranial fossa and eventually along the parietal bone. In addition to these variant patterns, the anterior division may sometimes be enclosed in a bony canal a few centimeters in length that takes origin beneath the lateralmost extent of the lesser wing of the sphenoid—and/or this primary anterior division may bifurcate, with its secondary branch arcing toward the lateral margin of the superior orbital fissure.

The **posterior cranial fossa** dominates the interior of the cranium. From above, its shape is bluntly triangular anteriorly and swollen or bulbous posteriorly. It is narrowest anteriorly, being delimited by the dorsum sellae and basisphenoid. This fossa then broadens laterally and posteriorly in parallel with the crisp superomedial margins of the petrosal portions of the temporals, beneath which it again expands somewhat. The uppermost lateral portion of the posterior cranial fossa is defined by the part of the parietal bone that contributes to the region of the parietal notch and asterion. Much of the base and posterior part of the posterior cranial fossa is "cupped" by the occipital bone.

The apex of the triangular portion of the posterior cranial fossa is formed by the confluence of the basisphenoid and basiocciput, which together create a slightly concave and somewhat vertical wall. Below the contact with the anterior part of the petrosal, this wall broadens inferiorly and thickens bilaterally into the *jugular tubercles* and then continues on to form the anterior border of the *foramen magnum*. The anterior border of the foramen magnum is superoinferiorly tall or thick—because of the verticality of the basioccipital region itself and the near confluence of the jugular tubercles with the anterior border of the foramen—whereas the posterior margin of this foramen is thinner and forms more of an edge. *Occipital condyles* are located bilaterally on the anterolateral margin of the foramen magnum; the inner-most edge of an occipital condyle may extend a bit more medially than the margin of the foramen magnum and may thus be visible from within the cranial base.

In addition to the foramen magnum, the posterior cranial fossa contains other major cranial foramina. Bilaterally and lateral to the jugular tubercle is the moderately large and variably shaped *jugular foramen*, through which the jugular vein exits. Right and left jugular veins are the conduits through which blood drains from the meningeal sinuses. The jugular foramen is bounded above by the inferior margin of the petrosal portion of the temporal and below by the occipital bone. In its most complex configurations, a jugular foramen actually appears as a series of variably deep pockets (Schwartz, unpublished data). Corresponding with the emphasis of the right versus left groove for the transverse sinus, the right jugular foramen is often larger than the left.

The groove for the descending superior sagittal sinus bifurcates at the raised *internal occipital protuberance* into right and left, fairly horizontally oriented *grooves for the transverse sinuses*, which form the border between the fossae for the cerebellar hemispheres (below) and the fossae for the cerebral occipital lobes (above). Beyond asterion, each transverse groove swings under the petrous portion of the temporal and then arcs anteriorly to become confluent with the jugular foramen. The curved part of this sinus drainage system is the *groove for the sigmoid sinus*.

Sometimes, just on the internal side of the occipital margin of the jugular foramen lies a posteroinferiorly coursing smaller foramen that penetrates the thickened rim of the foramen magnum. This foramen is interpreted as a *nonmetric variant* and is usually not present bilaterally. However, a "normally occurring" *anterior condylar foramen* lies in the same region below the thickening of the jugular tubercle, on the superior and inferior surface of the occipital condyle. It perforates the superior portion of the occipital condyle, exiting the posterior cranial fossa above the raised, lateral margin of the condyle. On occasion, there may also be an *accessory anterior condylar foramen*, which is the internal aperture of a *posterior condylar foramen*. The latter typically manifests itself as a variably deep, nonpatent pit behind and lateral to the posterior margin of the occipital condyle. When patent, this foramen opens externally and slightly lateral to the posterior margin of the occipital condyle. In anatomy texts, the anterior condylar foramen may be identified as the *hypoglossal canal* (the "canalis hypoglossi" of the *Nomina Anatomica*) and the posterior condylar foramen as the *condyloid foramen* or *canal* ("canalis condylaris").

Above the jugular foramen the *internal acoustic (auditory) meatus* emerges from the petrous portion of the temporal bone. It is oriented more or less horizontally and opens obliquely anteromedially upon the posterior cranial fossa. The auditory nerve, both roots of the facial nerve, and the auditory vessels course through the internal acoustic meatus.

The last penetration of the posterior cranial fossa is the *mastoid foramen* (mentioned earlier when describing external cranial morphology). Although externally this foramen may be ex- rather than sutural, internally it is typically contained entirely within the squamous portion of the temporal bone. The mastoid foramen is oriented horizontally, faces straight forward, and opens upon the downwardly coursing portion of the sigmoid sinus.

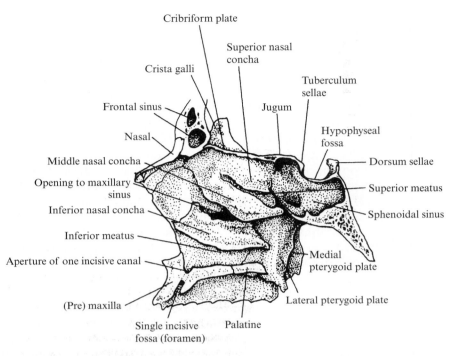

Figure 2–10   Nasal cavity (*right side*).

Nasal Cavity

Additional surface topography in the posterior cranial fossa is seen in the *posterior meningeal grooves* and a variably impressed *groove for the superior petrosal sinus* that courses along the superior and medial edge of the petrous portion of the temporal bone.

Nasal Cavity

Cranial fragments recovered in paleontological, archaeological, and forensic contexts often retain elements that contribute to the nasal cavity (Figures 2–10 and 2–11). Thus, although individual bones are discussed elsewhere, the overall region will be considered first.

The nasal cavity consists of the *vomer*, *ethmoid*, and ethmoidal extensions in the midline; *maxillary* and *palatine* contributions to its "floor"; and *nasal*, *frontal*, *sphenoidal*, and *inferior nasal conchal* contributions to

its "roof" and/or "walls." In addition, some of these bones house *paranasal sinuses*, which are confluent with or open upon the nasal cavity via variably sized orifices.

Within the nasal aperture, the *vomer* (below) and *perpendicular plate of the ethmoid* (above) form a vertical bony plate that bifurcates the nasal cavity anteroposteriorly (in life, a cartilaginous septum that sits in the midline groove of the vomer extends the bifurcation farther anteriorly).

In adults, the *floor of the nasal cavity* is typically smooth and relatively flat. In younger individuals, especially those in whom the successional anterior teeth are forming and still unerupted, the floor of the nasal cavity anteriorly may be swollen and raised. Just behind and on either side of the swollen anterior extremity of the vomer (from which point the anterior nasal spine appears to originate), variably small to moderate foramina [which

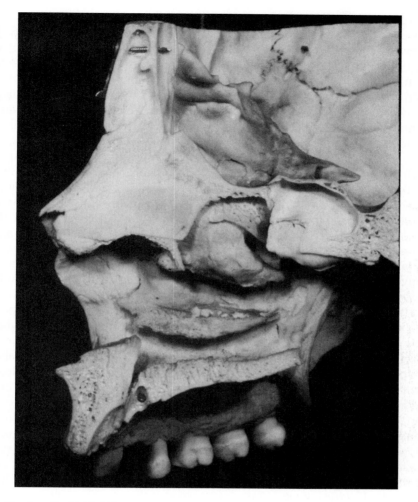

Figure 2–11   Nasal cavity (*right side*).

should be referred to as *incisive fossae* (Schwartz, 2004)] penetrate the floor of the nasal cavity. These foramina lead into separate canals—*incisive canals* or *foramina of Stenson*—that converge toward the midline, where they may coalesce into a single canal. As either one or two closely approximated canals, they flow into the large, single *incisive foramen* that emerges on the oral cavity side of the maxilla, not far behind the upper incisors. At times, the incisive canals are located on the right and left sides of the maxillomaxillary

suture, while at others one canal (corresponding to the left canal) lies anterior to the other. Some branches of the (sensory) nasopalatine (sphenopalatine) nerve and artery, which course anteriorly along the floor of the nasal cavity, continue toward the oral cavity via the incisive canals, through which also pass branches of the greater palatine nerve and artery as they proceed from the oral cavity toward the nasal cavity. Especially in younger individuals, a fine sutural line may extend laterally from the aperture on the floor of the nasal cavity of each incisive canal. When present, this suture represents the contact between the premaxilla and maxilla (Schwartz, 1982, 1983). Sometimes these fine sutures continue for a short distance up the lateral wall of the nasal cavity, delineating internally the boundary of the frontal process of the premaxilla. More consistently observable into adulthood is the palatomaxillary suture.

Although the bony palate is relatively thick and, because of this, is often preserved after interment, many of the thin and friable contributions to the nasal cavity often do not survive. (Note, for example, the sorry state of preservation of the vomer, perpendicular plate, and/or inferior nasal conchae among skulls in any laboratory or museum osteology collection.)

Superiorly and superolaterally, the nasal cavity is bounded by elements of the ethmoid. The perpendicular plate terminates superiorly in the lacelike cribriform plate. On either side of the perpendicular plate lie the variably inflated and thinly waferlike *superior* and *middle conchae*, which emanate from the medial surfaces of the ethmoid. Farther posteriorly and superiorly, where the vomer splays out into the wings or *alae* that embrace the sphenoid from below, the interior of the sphenoid houses two relatively large sinus chambers. The bulk of the left *sphenoidal sinus* lies directly above the alae of the vomer; the right sinus extends under the jugum and hypophyseal fossa. The lateral wall of the nasal cavity is formed primarily by the thin medial wall of the maxilla, which is perforated by a variably moderate to large opening through which the *maxillary sinus* communicates with the nasal cavity. The maxillary sinus occupies much of the interior of the maxilla.

After death, and even without postmortem disturbance, the lower facial skeleton may become dissociated from the rest of the skull. This is understandable. The lacrimals and ethmoid are extremely thin at their contributions to both the nasal cavity and the orbit, and the bone that encases the sphenoidal sinuses is thin. The contacts between the temporal process of the zygoma and zygomatic process of the temporal, the frontal process of the zygoma and zygomatic process of the frontal, and the nasals and frontal process of the premaxilla and the frontal are not robust. In addition, the region of glabella houses the *frontal sinuses*, which develop as a superior expansion of the anteriormost cell of the ethmoidal sinus complex. The specific location within the ethmoidal sinus complex from which right and left frontal sinuses originate is variable. Nevertheless, all sinuses open upon the nasal cavity either directly (the sphenoidal, ethmoidal, or maxillary sinuses) or via other sinuses (frontal via the ethmoid).

Chapter 3

Individual Bones of the Skull

Although it is necessary to begin the study of cranial bones in association, a deeper understanding of their morphological detail comes from studying them separately. Sometimes, especially in the case of young individuals, isolated bones, or even a single bone (or fragment thereof), may be the only remains recovered during archaeological, paleontological, and forensic field-work. Particularly revealing clues to their identity are the orientations of bony features and landmarks (foramina, vascular grooves, sutures). Also important are differences within the same bone in shape, surficial orientation, and cross-sectional thickness. Since a certain amount of detail was presented in the preceding chapter, repetition will be kept to a minimum. The development of each bone is discussed in order to provide an appreciation of morphological similarities between juveniles and adults. This also provides a context for understanding age-related features as well as the etiology of many nonmetric variants and pathologies. Figures 3–1 to 3–6 of fetal and postnatal crania illustrate some of these growth-related changes.

Mandible

Morphology

Additional features on the external surface of the mandible (Figure 3–7) include the *oblique line* and *incisive fossa*. The former is a variably developed ridge that courses somewhat arcuately between the mental tubercle and the inferior portion of the anterior border of the ascending ramus; depressor muscles of the lower lip and the mouth attach anteriorly to the oblique line. Incisive fossae occur bilaterally to the side of the mental protuberance; muscles of the lower lip and mouth take origin there.

Each of a pair of variably distended and pointed *mental spines* (also *genial tubercles*) lies on either side of the midline of the mandible, typically closer to the rounded *inferior border* than to the *alveolar margin*. Muscles of the tongue attach to these spines, which are usually larger in males than females. A muscle coursing between the hyoid bone and mandible attaches on the internal surface beneath

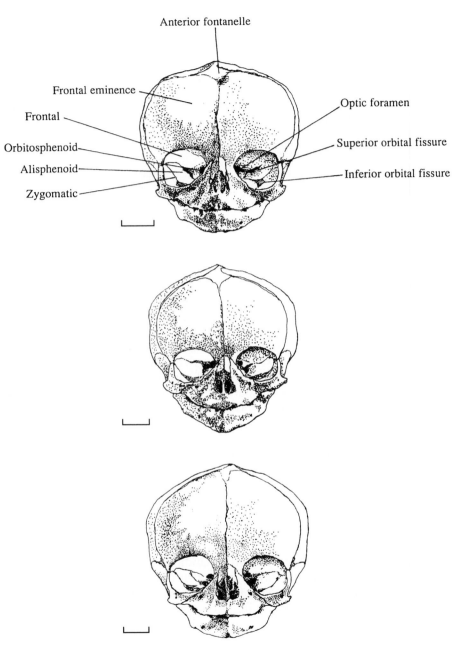

Figure 3–1   Anterior view of fetal skulls: (*from top to bottom*) 6, 7, and 8 months (scale = 1 cm).

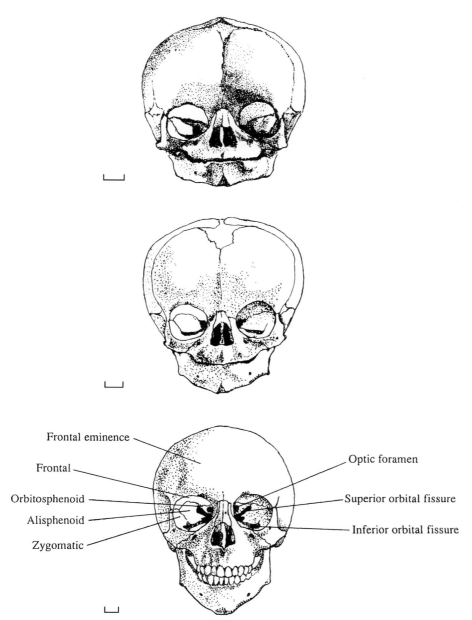

Figure 3–2   Anterior view of postnatal skulls: (*from top to bottom*) neonate, 3 months, and 5 years (scale = 1 cm).

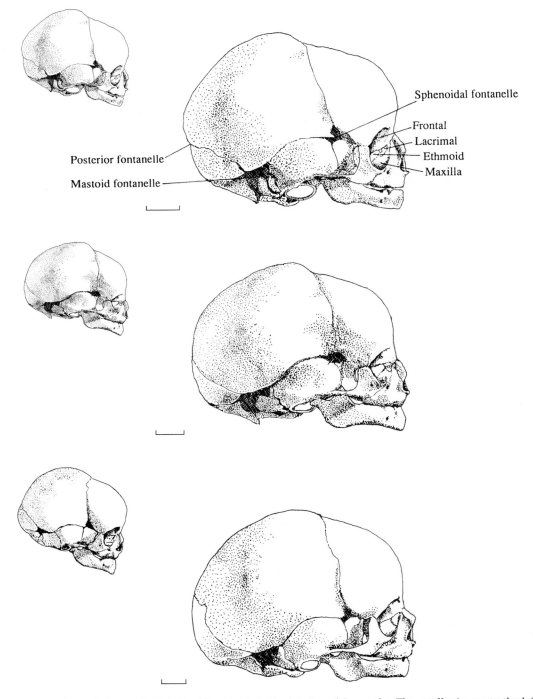

Figure 3–3   Lateral view of fetal skulls: (*from top to bottom*) 6, 7, and 8 months. The smaller image to the left of each skull is that skull oriented in the Frankfort horizontal (scale = 1 cm).

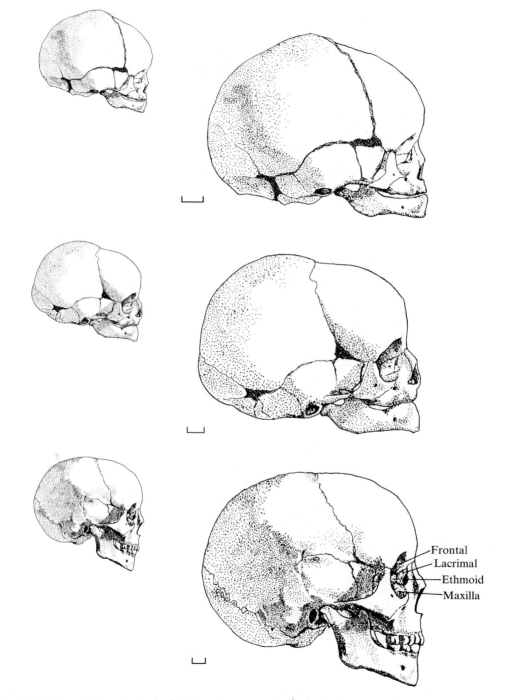

Figure 3–4   Lateral view of postnatal skulls: (*from top to bottom*) neonate, 3 months, and 5 years. The smaller image to the left of each skull is that skull oriented in the Frankfort horizontal (scale = 1 cm).

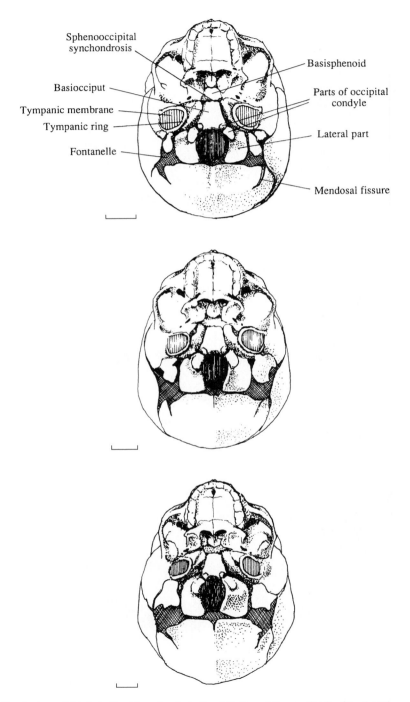

Figure 3–5 Basal view of fetal skulls: (*from top to bottom*) 6, 7, and 8 months (scale = 1 cm).

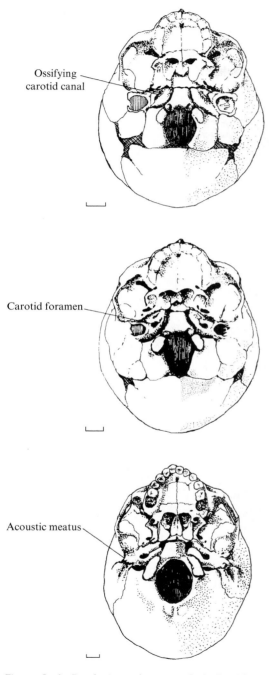

Ossifying carotid canal

Carotid foramen

Acoustic meatus

Figure 3–6 Basal view of postnatal skulls: (*from top to bottom*) neonate, 3 months, and 5 years (scale = 1 cm).

these spines; the (unnamed) scar this muscle leaves may take the form of elevations below each mental spine or a ridge in the midline of the mandible, below and between the spines. A shallow, wider than tall, ovoid *fossa for the sublingual gland* lies lateral to each mental spine. Below each fossa, almost on the mandible's inferior border, is a roughened, variably ovoid *digastric fossa*, to which the anterior belly of the digastric muscle attaches (the posterior belly of this muscle courses through the digastric or mastoid notch medial to the mastoid process, attaching behind the process). A variably rugose line or thin ridge (*mylohyoid line*) usually takes origin below and slightly behind the third molar and then proceeds toward the region of the sublingual fossa and the mental spine. This line may, however, diminish in intensity anteriorly, retain its character throughout its entire length, course above the sublingual fossa and terminate at the mental spine, or terminate at the lateral border of the fossa. An elongate but shallow *submandibular fossa* lies beneath this line and may intervene slightly between the sublingual and digastric fossae.

The body of the mandible is variably distended anteriorly into an often subtriangularly shape *mental protuberance* (*mental trigon* or *triangle*, chin), which is anteroposteriorly thickest inferiorly and may be accentuated laterally along the inferior margin on each side by a *mental tubercle* (typically more prominent in males than females).

Internally on the ascending ramus of the mandible, the *mylohyoid groove* courses inferiorly and somewhat anteriorly from the inferior margin of the mandibular foramen. Anterior to the mandibular foramen is a tonguelike, variably superiorly projecting slip of bone, the *lingula*, to which attaches a ligament that assists in tethering the mandible to the base of the skull.

Overall, a male's mandible is usually larger, bulkier, and more rugose than a female's. Also often more pronounced in males are muscle

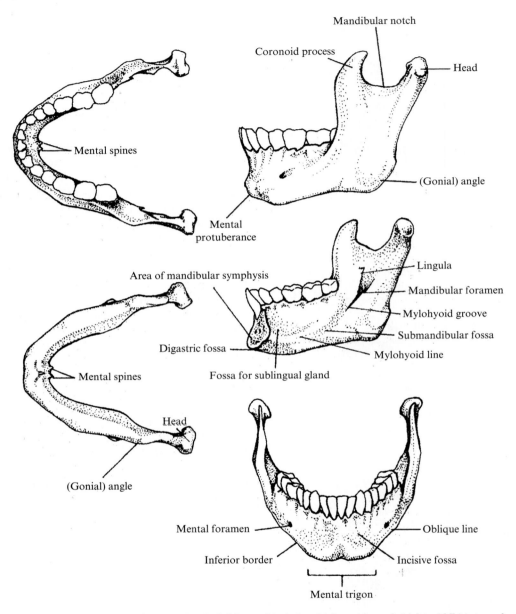

Figure 3–7 The mandible: (*left top*) occlusal, (*left bottom*) inferior, (*right top*) lateral, (*right middle*) internal, and (*right bottom*) anterior views.

scarring and/or distension of the coronoid process, the lingula, the mylohyoid line, the digastric fossae, the mental spines, and, in particular, the gonial region (i.e., the inferior margin of the mandibular angle), the mental region (especially the tubercles), the distance from infradentale to pogonion, and the verticality of the ramus.

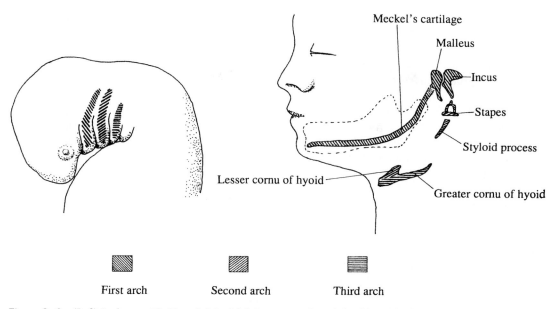

Figure 3–8   (*Left*) Embryo with (*from left to right*) first, second, and third branchial arches emphasized; (*right*) child with branchial arch derivatives delineated (first arch = Meckel's cartilage, incus, and malleus; second = stapes and parts of styloid process and hyoid; third = remainder of hyoid bone) [after Moore, 1974].

Development and Ossification

The mandible is derived from the cartilaginous branchial arch skeleton (Figure 3–8). Ossification begins during the sixth to seventh fetal weeks as two laminae on either side (internally and externally) of Meckel's cartilage (the first arch), which eventually disintegrates. These laminae eventually fuse inferiorly, forming a troughlike structure in which the lower primary teeth and their accompanying alveolar bone develop. The mandible grows rapidly, attaining its recognizable shape by the middle of the third gestational month.

As the mandible ossifies, it encases a large portion of a branch of the mandibular nerve [the inferior alveolar (dental) nerve], which proceeds internally via what will become the *mandibular foramen*. The inferior alveolar nerve extends to the front of the jaw through the *mandibular canal*. Anteriorly, the mental nerve (a branch of the inferior alveolar nerve) exits the mandibular body through what

will become the *mental foramen*; an accessory mental foramen may be present as a uni- and/or bilateral *variant*.

Sometime after the tenth fetal week, accessory cartilaginous nuclei form in the regions of the condylar and coronoid processes as well as in the angle of the mandible and are subsequently replaced by surrounding membranous bone. Cartilaginous nuclei also appear in the presumptive alveolar region (for discussion of tooth development, see Chapter 7).

The telltale morphologies of the "mental trigon" are established in bone before the fifth fetal month (Schwartz and Tattersall, 2000). On each mandibular half there is a raised or everted margin that courses down the anteriormost edge and then around and back along the inferior margin, delineating a *mandibular fossa* lateral to it. At or about this time, right and left mandibular halves begin to come closer together, starting at the presumptive alveolar margin and proceeding inferiorly. As this approximation progresses, the right and left raised/everted midline edges come together as a single elevation or central keel along the now "lost" mandibular symphysis. The laterally directed and raised inferior margins eventually also come together and form the base of an inverted "T" (Schwartz and Tattersall, 2000). Postnatal growth may elaborate or, more often, soften or obscure the crispness of this clearly "inverted T-shaped" developmental precursor of the adult mental trigon.

By birth, growth and ossification have proceeded along the pattern laid down during the first trimester and the primary antemolar teeth have calcified to varying degrees of completeness. Of particular interest in following the course of development of the mandible is that, at birth, the *ascending ramus* is shorter relative to mandibular body size than in the adult and is oriented at approximately a 45° angle, thereby creating an exceedingly obtuse mandibular angle. The *head of the mandible* (*condylar process* or *condyle*) does not typically rise above the highest portion of the mandibular corpus, whereas the large *coronoid process* does. Right and left mandibular halves are not fused across the midline (*mandibular* or *mental symphysis*) but are held together by fibrous mesenchymal tissue. Mineralization in the mandibular symphysis begins with the appearance of one to four centers of ossification, which fuse together by the fifth to sixth month after birth; during the first to second year, these islands of ossification fuse with the mandible.

Other changes in the mandible after birth include (1) lengthening of the mandibular body, especially posteriorly, to accommodate the proliferating molar-class teeth; (2) deepening of the body in conjunction with the development of the primary and then (especially) successional teeth, their roots, and attendant alveolar bone; (3) an increase in length as well as in elevation of the ramus with a concomitant decrease in the obtuseness of the mandibular angle so that, in the adult, the ramus approaches the vertical; and

(4) deepening of the *mandibular* or *sigmoid notch* between the coronoid and condylar processes, in part via their elongation.

With increasing age, resorptive processes can reverse the growth profile of the mandible. Most striking is the thinning and overall remodeling of mandibular bone, in which the body and ramus become more gracile and the gonial angle becomes increasingly obtuse. Remodeling also narrows the coronoid process and condylar neck anteroposteriorly and deepens the sigmoid notch. If tooth loss occurs with aging, alveolar bone will resorb and the affected portion of the body will become extremely thin.

Maxilla

Morphology

The adult "maxilla" (Figure 3–9) contributes to the wall and floor of the nasal cavity, the floor of the orbit, the roof of the oral cavity, and the dental arcade; it also delineates the nasal aperture. It is thus notable because it is not uniform in thickness or overall appearance; for example, the bone associated with the margin of the nasal aperture, floor of the orbit, wall of the nasal cavity, and facial region below the infraorbital canal is quite thin (the latter three surfaces form the walls of the maxillary sinus). The *frontal process* and its continuation (in part via the *anterior lacrimal crest*) into the inferior orbital margin become thicker, and the palate and alveolar process are quite thick.

The region of the short *frontomaxillary suture* (which is typically deeply invaginated and densely packed with tall, fingerlike projections that interdigitate with their counterparts on the frontal bone) is slightly thicker than the frontal process below it. Externally, and to approximately the same degree in which the margin of the orbit (the lateral margin of the frontal process) is concavely arcuate upward, the edge of the *zygomaticomaxillary suture* courses convexly and arcuately

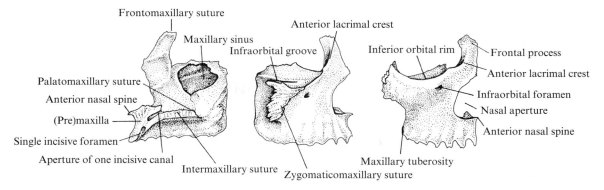

Figure 3–9   Right maxilla: (*left*) internal, (*middle*) lateral, and (*right*) anterior views.

downward from the inferior margin of the orbit. The internal aspect of the zygomatico-maxillary suture, however, follows the plane of the orbital floor. This disparity in height and orientation between the inner and outer aspects of the zygomaticomaxillary suture creates (1) a denticulate and craggy, anteriorly facing articular surface (descending from the internal aspect) that leads (2) to a denticulated furrow or receptacle (bound by the external edge of the zygomaticomaxillary suture) in which the zygomatic bone sits.

The *intermaxillary suture* is seen as two components of somewhat differing morphology separated by the *incisive canals* and *foramen*. In cross section, the maxillary palate is relatively thick posteriorly (where it abuts the palatine bone) and becomes increasingly thick anteriorly toward the incisive canals and foramen. Anterior to the region of the incisive canals and foramen, the palate is distended markedly and steeply downward. The walls of the incisive canals are smooth. Behind the canals, the surface of the intermaxillary suture is pervaded by a network of thin, more or less vertical, and relatively closely spaced "plates" of bone separated by lacunae or spaces; right and left intermaxillary sutural ridges interdigitate. This ridge pattern extends (inferiorly) from the floor of the nasal cavity and terminates at the solid, unridged layer of

bone that forms the roof of the oral cavity. This layer of bone is thickest in the mid-region of this portion of the maxillary palate; it becomes thinner posteriorly (toward the palatine bone) as well as anteriorly (toward the incisive canals). The posterior part of the intermaxillary suture is solid, compact bone. The part of the intermaxillary suture that lies anterior to the incisive canal also is adorned with ridges, but the plates of each ridge are longer, individually more distinct, and separated by longer, deeper grooves. Only toward the neck of the central incisor might the bone become solid and compact.

The maxilla's contribution to the *palato-maxillary suture* is varied. Along the hard palate, the sutural surface is basically a variably thickened and smooth to roughened wall of compact bone against which the thinner palatine bone abuts. Contact between the maxilla and palatine bone becomes less intimate as the palatomaxillary suture "rounds the corner" at the base of the alveolar process and courses toward and eventually beyond the *greater palatine foramen*. In the region of the *maxillary tuberosity*, the portion of the palatine bone that overlaps and caps inferiorly the lateral pterygoid plate of the sphenoid is variably appressed to the smooth surface of the maxilla.

For clues to siding the maxilla, see Table 3–1.

Table 3–1   Clues to Siding the Maxilla

Feature	Position
Alveolar process	Inferior
Anterior lacrimal crest	Laterosuperior
Anterior nasal spine	Anteromedial
Canine fossa	Lateroanterior
Conchal crest	Medioanterior
Ethmoidal crest	Mediosuperior
Frontal process	Superior
Greater palatine canal	Posteromedial
Incisive canal	Medioanterior
Infraorbital foramen	Anterolateral
Lacrimal groove	Laterosuperior
Maxillary sinus	Medioposterior
Maxillary tuberosity	Posteroinferior
Nasal crest	Medial
Nasal notch	Medial
Orbital surface	Superolateral
Palatine process	Inferomedial
Zygomatic process	Lateral

Development and Ossification

The *maxilla* develops from the upper portion of the first branchial arch, which is created when the arch folds in half. It ossifies as membranous bone derived from the first pharyngeal arch. By the sixth to seventh fetal week, two to three ossification centers appear (one or two in the *maxillary prominence* and one in the *median nasal prominence*). The posterior center(s) of ossification coalesces into the presumptive maxilla while the anteriormost center represents the presumptive *premaxilla*. By the fourth fetal month, the presumptive maxilla and premaxilla are partially united. The details of the union of maxilla and premaxilla are still debated.

The traditional interpretation has been that two separate bones fuse together along a plane of contact. However, embryological and histochemical studies (Andersen and Matthiessen, 1967) reveal that the presumptive human maxilla and premaxilla become united via ossification of a bridge—*maxillary isthmus*—that maintains mesenchymal continuity between the maxillary and median nasal prominences from the onset of facial development. The floor of the infraorbital canal and foramen represents the region of the embryonic maxillary isthmus. A *premaxillary–maxillary suture*, formed between contributions of equal size from the premaxilla and maxilla (persisting even into the young adult), may bisect the superior margin of the infraorbital foramen vertically. *Variation* does exist, however, in the extent to which the maxillary contribution to the superior margin of the infraorbital foramen (1) extends to the midline of the foramen, (2) forms the entire superior margin of the foramen, or (3) expands beyond the superior margin of the foramen to contact the frontal process of the premaxilla (Schwartz, 1982). Evidence of a premaxillary–maxillary suture may persist along the oral cavity side of the palate, with right and left sutural components emanating laterally from the incisive foramen and coursing to the general regions of the canines. As early ossification of the premaxilla creates a sheet of woven bone that extends down from the nasal bone to overlie the developing tooth germs (Schwartz and Langdon, unpublished data) and ossification in its palatal region is confined to thin septa growing between and lingually around tooth germs, it appears that the bone of the premaxilla palatally is alveolar bone in origin (also see argument in Schwartz, 1982). As such, it is obvious why the premaxillary–maxillary suture in humans is not expressed as it is in other mammals.

The *hard palate* of the maxilla begins embryonically with the emergence of ossification centers in the palatine processes, which are initially vertically oriented downward. With hydration and under proper developmental conditions, the palatine processes become elevated and meet in the midline of the presumptive palate. The embryonically distinct dental ridge is "sandwiched" between the median nasal and maxillary prominences and the palatine process (e.g., Andersen and

Matthiessen, 1967; Nery et al., 1970). By the end of the third trimester, the "maxilla" plus the premaxilla, dental arcade, and palate are sufficiently joined (gomphosed) so that they appear as a single bony unit. The *inferior orbital rim*, which lies close to the presumptive alveolar margin, dominates this maxilla. During the third month after birth, the *maxillary sinus*, which had been a groove along the wall of the nasal cavity, enlarges. At about the fourth month after birth, the alveolar process begins to enlarge. Both events are apparently correlated with tooth development. With increasing age, the maxilla deepens and lengthens, especially posteriorly, in concert with growth and eruption of primary and successional teeth. With senescence, bone resorption reduces the bulk of the maxilla. If tooth loss also occurs with increasing age, concomitant resorption of alveolar bone will severely reduce maxillary depth.

The Palatine Bone

The palatine bone comprises *horizontal* and *perpendicular plates*; the former contributes to the hard palate, and the latter projects into the nasal cavity.

Morphology

The *horizontal plate* (Figure 3–10) is a relatively thin, medially projecting sheet of bone. The paired palatine bones thicken markedly along their shared *median palatine suture*, which is elevated in correspondence with the arcuate elevation of the floor of the nasal cavity. The median palatine sutural surface itself is an irregular swirl and/or layering of thin, sheer ridges separated by variably deep spaces. The edge of the *palatomaxillary suture* (the anterior and lateral margins of the bone) may, however, have a more regularly layered appearance. The points of contact (usually one large area as well as one or more accessory areas) that the *pyramidal process* maintains with the maxillary tuberosity are variably thickened and rugose.

The morphologically more complicated *perpendicular plate* extends anteriorly and superiorly. It forms an arcuate "corner" laterally with the horizontal plate. Anteriorly and slightly above the floor of the nasal cavity, the maxilla overlaps the *maxillary process*; together, these bones form the broadly "V"-shaped, inferiorly pointed margin of the opening of the maxillary sinus. Approximately at the

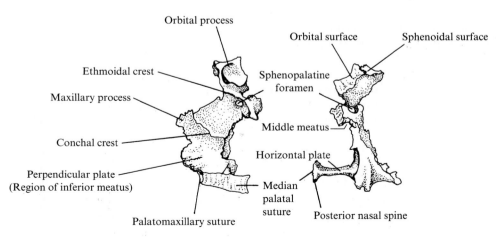

Figure 3–10   Right palatine bone: (*left*) medial and (*right*) posterior views.

Table 3–2   Clues to Siding the Palatine Bone

Feature	Position
Conchal crest	Medial
Ethmoidal crest	Mediosuperior
Greater palatine foramen	Inferolateral
Greater palatine groove	Lateroposterior
Horizontal plate	Inferior
Lesser palatine foramen	Inferolateral
Maxillary process	Anterior
Nasal crest	Inferomedial
Orbital process	Superior
Perpendicular plate	Laterosuperior
Posterior nasal spine	Medioposterior
Pyramidal process	Lateral
Sphenoidal process	Posterosuperior
Sphenopalatine notch	Posterosuperior

level at which the maxilla overlaps the maxillary process of the palatine bone, the band- or ridgelike *conchal crest*, with which the *inferior nasal concha* (*maxillary turbinate* or *concha*) articulates, courses horizontally across the medial surface of the palatine bone. The conchal crest and inferior nasal concha delineate below them the *inferior meatus*. The region of the *middle meatus* is an exceedingly thin, variably concave sheet of bone that lies between the conchal and *ethmoidal crests*. The *ethmoidal crest* is identified where this sheet of bone thickens and flares somewhat medially; the *middle nasal concha* articulates with it. Together, the middle nasal concha and the superior portion of the perpendicular plate of the palatine bone form the *superior meatus*. The *sphenopalatine foramen* lies below and sometimes even behind the posterior end of the ethmoidal crest.

The superiormost part of the perpendicular plate, *orbital process*, articulates anteriorly with the maxilla, posteriorly with the sphenoid, and medially with the ethmoidal labyrinth. It also presents a *lateral surface*, which lies obliquely under the border of the inferior orbital fissure but is actually the region of

flexure that delineates the *orbital surface* (the posteriormost contribution to the orbital floor).

For clues to siding the palatine, see Table 3–2.

Development and Ossification

During (approximately) the eighth fetal week, a single center of ossification appears intramembranously in the angle of what will develop into the horizontal and perpendicular plates. First, ossification spreads medially into the horizontal plate as well as laterally and inferiorly into the presumptive *pyramidal process*. The pyramidal process eventually wraps partially around the maxillary tuberosity posteriorly, embracing the inferior extremity of the lateral pterygoid plate. The *lesser palatine foramen* (or foramina) forms in the region of the "neck" of the pyramidal process. A thinly triangular, concave or furrowed slip of bone extends superiorly from the pyramidal process. It separates the *medial* and *lateral pterygoid plates* inferiorly and contributes to the *pterygoid fossa* that intervenes between the two plates. Within 2 weeks of its onset, ossification has proceeded vertically into the presumptive perpendicular plate and its *orbital* and *sphenoidal processes*. In the adult, the height of the perpendicular plate is approximately twice the mediolateral width of the horizontal plate. At birth, however, the two plates are subequal in these dimensions.

Nasal Bone

Morphology

In cross section, a nasal bone (Figure 3–11) is thickest superiorly (at the frontonasal suture); it tapers and thins quite drastically toward its inferior edge. At the midline on the internal surface lies the *nasal crest*, lateral to which the bone is gently concave; the external surface of the bone is concomitantly convex. Again internally, but on or about the midline of each nasal bone, the groove for the external nasal branch of the anterior ethmoidal nerve,

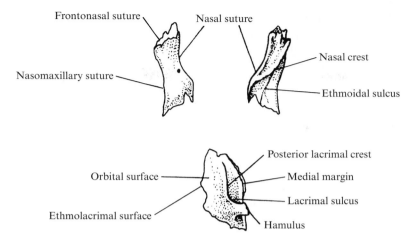

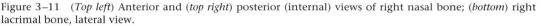

Figure 3–11   (*Top left*) Anterior and (*top right*) posterior (internal) views of right nasal bone; (*bottom*) right lacrimal bone, lateral view.

ethmoidal sulcus, courses prominently for most of the length of the bone. Within the nasal cavity, the ethmoid overlaps slightly the superior portion of the nasal bone.

The essentially straight *internasal* or *nasonasal suture* lies in the median sagittal plane. Its surface is variably roughened along its length but is most corrugated superiorly, in the region of the frontonasal suture. The latter is incised by variably (but often very) deep, thin fingerlike projections on the nasal and frontal bone's contributions to it. The surface of the *nasomaxillary suture* bears less severe sutural interdigitation.

For clues to siding a nasal bone, see Table 3–3.

Development and Ossification

During the eighth to ninth fetal week, the single ossification center that gives rise to each nasal bone appears intramembranously. The paired nasal bones together are broader inferiorly than superiorly. In the adult, they are somewhat "pinched" or "waisted" laterally near the midpoint of their length (forming an odd hourglass shape). But in the newborn, these sides are relatively straight and thus

Table 3–3   Clues to Siding the Nasal Bone

Feature	Position
External surface (convex)	Lateroanterior
Groove for anterior ethmoidal nerve	Posterior (internal)
Inferior border (thin, sharp)	Inferior
Internal surface (concave)	Medioposterior
Internasal suture	Medial
Nasal crest	Posteromedial (internal)
Nasofrontal suture (deeply serrated)	Superior
Nasomaxillary suture (longest border)	Lateral
Vascular foramen	Anterior (external)

give the paired bones a more trapezoidal configuration. The nasal bones also appear squatter and straighter across the superior border of the nasal aperture in the newborn but become more elongated and variably distended inferolaterally with increasing age.

Lacrimal Bone

The ossification center for the lacrimal bone (Figure 3–11), which contributes to the *medial orbital wall* and is the smallest bone of the facial skeleton, appears during the twelfth fetal week in the membrane surrounding the cartilaginous nasal capsule. It is characterized by a relatively shallow and (for the bone) relatively broad groove, *lacrimal sulcus*, that extends along the entire superoinferior length of the bone. This groove is oriented slightly obliquely (facing somewhat anteriorly and laterally). Thus, the thin medial margin is situated just anterior to the more robust *posterior lacrimal crest*, which forms the lateral border of the sulcus (recall, the *anterior lacrimal crest* lies on the *frontal process of the maxilla*). The thin medial margin of the lacrimal sulcus overlaps the edge of the frontal process of the maxilla along the *lacrimomaxillary suture*; on the frontal process, an indentation marks the region in which the thin medial margin of the lacrimal sulcus rests. The posterior lacrimal crest (which may be a low blunt project or a medially projecting sheet of bone) terminates in a variably small, hooklike, anteriorly facing projection [*lacrimal hamulus* ("hamulus" meaning a "hooklike process")] that nestles into the maxilla along the floor of the orbit that to some extent circumscribes further the inferior border of the lacrimal sulcus. The lacrimal hamulus sometimes arises from its own center of ossification and then fuses to the lacrimal bone.

The region of the lacrimal groove is the thickest portion of the lacrimal bone. Beyond the groove, the lateral surface of the lacrimal bone (which faces upon the medial orbital wall) and its medial surface (which faces upon the middle meatus of the nasal cavity and complements the ethmoidal air cells) taper as they converge posteriorly; consequently, the *ethmolacrimal suture* between the lacrimal and ethmoid behind is relatively thin. The orbital surface of the lacrimal bone is slightly concave and smooth, whereas the medial surface is topographically diverse (it may even bear a furrow that mirrors the lacrimal crest). By identifying the sulcus and the hamulus and orienting the thinnest part of the bone poste-

Table 3–4   Clues to Siding the Lacrimal Bone

Feature	Position
Lacrimal groove	Lateroanterior
Lacrimal hamulus	Inferoanterior
Nasal surface	Medial (internal)
Orbital surface	Lateroposterior (external)
Posterior lacrimal crest	Lateral

riorly, one should be able to distinguish right from left lacrimal bones.

For clues to siding lacrimal bones, see Table 3–4.

Frontal Bone

Morphology

Following convention, the *orbital portion* of the frontal bone (Figures 3–12 and 3–13) encompasses the roofs of the orbits and the nasal cavity. The *squama* constitutes everything else, beginning at the superior orbital or *supraorbital margins* and proceeding upward. The contour of the supraorbital margin is typically more arcuate and its edge crisper and sharper to the touch in females than males; in the latter, the edge of the margin is usually thicker and blunter and its midsection may be straighter.

Invariably, just medial to the orbital midline, each supraorbital margin bears either a *supraorbital notch* or *supraorbital foramen*, in which the supraorbital vessels and nerve lie; the foramen results from ossification of soft tissue subtending the notch. Although the "notch" has been considered the variant condition (Berry, 1968), a survey of over 500 globally representative recent human skulls, 300 nonhuman anthropoid skulls, and virtually all Miocene hominoids and fossil hominids demonstrates that, if there is an identifiable feature on the inferior margin of the supraorbital region, it is a notch, not a foramen (Schwartz, unpublished data).

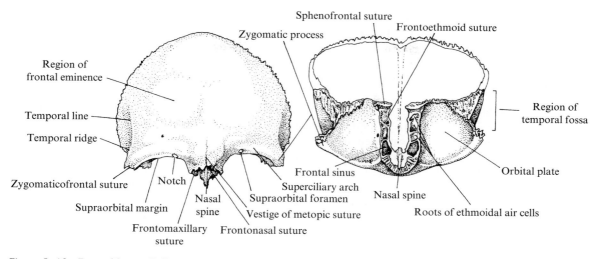

Figure 3–12   Frontal bone: (*left*) anterior (external) and (*right*) inferior views.

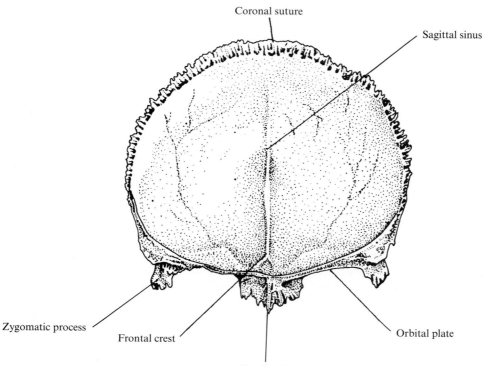

Figure 3–13   Internal morphology of frontal bone.

Therefore, the latter is the derived condition. *Variation* is noted in (1) the notch being bound by a spit or converging spits of bone, (2) the condition being bilateral, and (3) the presence of accessory or additional notches or foramina.

A *frontal foramen* is often confused with a supraorbital foramen; it may be variably small to moderate in size and situated well above the supraorbital margin (e.g., 0.25–0.50 cm above the margin). On occasion, a frontal foramen takes the form of a deeply invaginated notch. Frontal foramina may be single or multiple and/or uni- or bilaterally expressed and may occur in 45% of the population under study.

Slightly above the supraorbital margin and arcing upward and outward on each side from the region of glabella is a variably prominent swelling, *superciliary arch*, which may become even more swollen as it proceeds laterally away from the region of glabella and peaks in both height and distension over the area of the supraorbital notch/foramen. Lateral to the notch/foramen, the bone above the margin is flatter, more platelike, and slanted somewhat posteriorly. This particular configuration of the human supraorbital region can be described as "bipartite" (e.g., Stringer et al., 1984). The degrees to which glabella and the superciliary arch are "male" (more swollen or distended) or "female" (flatter) do not appear to be correlated.

Lateral to the midline of the frontal bone on each side, there is a *frontal eminence* that lies approximately 3 cm above the supraorbital margin and above and slightly lateral to the peaked elevation of each superciliary arch. Frontal eminences are typically more pronounced in females than in males. In females, they may be so prominent that they appear to create "corners" on the bone. In males, the often more backwardly sloping frontal bone bears only faintly developed frontal eminences.

An isolated frontal bone is distinguished (1) by the semicupped and winglike *orbital plates* that bound inferiorly on either side the long, narrow "space" in which the superior portion of the ethmoid nestles, (2) by the spikiness of much of its parietal border (that contributes to the *coronal suture*), and (3) by the sweep of the *temporal ridge* as it rises from the strutlike zygomatic process and arcs back over the postorbital constriction that contributes to the formation of the temporal fossa.

The coronal suture is minimally undulating within a few centimeters of bregma and is most jagged in the region that is crossed by the temporal line. Throughout most of the frontal bone's parietal border, the "spikes" or digitations that contribute to the coronal suture are longer along the inner than the outer table of bone. This gives an oblique configuration to the frontal's contribution to the coronal suture because the bone's inner table is overlain by the long digitations of the parietal's outer table. Bilaterally, a few centimeters lateral to bregma, the sutural digitations of the inner and outer tables become more equal in length and tend to be much shorter than elsewhere along the suture. Bilaterally, within 1 cm of bregma, the sutural digitations may be longer along the outer than along the inner table, although on both sides they are quite truncated.

Although small, the region of the frontal bone that is uniquely delineated superiorly by the temporal ridge/line, inferiorly by the zygomatic process, and posteriorly by the coronal suture is distinguished further by its contribution to the sphenofrontal suture. The contact between the frontal and greater wing of the sphenoid below is essentially triangular and its surface riddled by a profusion of spikelike projections (reminiscent of the roof of a cave adorned with short, pointed, and closely packed stalactites). From the anterior corner of this triangular surface, a thin edge of bone extends forward and laterally and then swells into the area of sutural contact between the zygomatic process of the frontal bone and the zygoma itself. This stubby contribution to the *frontozygomatic suture* is irregularly distended inferiorly and adorned with platelike sutural digitations. Extending inward from the apex of its triangular contact with the greater wing of the sphenoid, the surface of the frontal bone that contributes to the *sphenofrontal suture* is relatively thin and its edge finely jagged. This region represents the contact between the posterior border of the orbital plate and the anterior margin of the lesser wing of the sphenoid. Where the jugum of the sphenoid protrudes anteriorly into the cribriform plate, the orbital plate is distended into a thin, platelike, severely angled "corner" that overlaps the sphenoid. The surface of this portion of the frontal bone's contribution to the sphenofrontal suture is rough and irregular.

The *frontoethmoid suture* courses anteriorly beyond the severely angled medial corner of the orbital plate. It is distinguished by thin medial and lateral edges, which (1) are distended somewhat inferiorly and (2) form walls that subtend a longitudinal furrow that is subdivided unevenly by thin bony septa (the roofs of primarily ethmoidal air cells). The *frontal sinus* is the largest and anteriormost of these air cells; it invades the region of glabella and extends above the variably semilunar to semicircular contact between the frontal bone and the frontal process of the maxilla and nasal bones. The contact zone of the *frontolacrimal suture* is thin. The frontal bone's contributions to the *frontomaxillary* and *frontonasal sutures* are much broader and variably deep but finely pitted with fine digitations that mesh with both the maxilla's frontal process and the nasal bones.

Development and Ossification

Ossification of the frontal bone begins by the end of the second fetal month and radiates out from the intramembranous center that appears at or near the midline of the (presumptive) *superior orbital margin*. Although ossification proceeds throughout the platelike portion of the bone, it is concentrated around the orbital margin, particularly its lateral extremity. A week or so later, the ossification center that yields the frontal eminence appears. By the end of the third fetal month, the orbital region is relatively dense but the surrounding bone is still very thin and porously woven. At this time, the frontal bone is wider than it is tall; but these proportions change by the middle of the fourth month, at which time the presumptive *zygomatic process* can be identified. Within a few weeks thereafter, the frontal eminences are established and the surrounding bone is increasingly more arcuate posteriorly and laterally. Toward the end of the second trimester, right and left frontal bones are closely approximated inferiorly but separated from one another superiorly by the narrowly triangular *anterior fontanelle* (Figure 3–14). The medial and superior portion of the presumptive frontal is bluntly pointed and remains so through term. Late in the third trimester, the frontal's zygomatic process makes contact with the zygomatic bone. At birth, right and left frontal bones are (normally) still separated by the *frontal* or *metopic suture*; they begin to coalesce by the eighth year along all but the inferiormost part of the suture and are completely fused between the ages of 10 and 12 years. More or less coincident with the closure of the *metopic (frontal) suture*, secondary centers of ossification, which eventually give rise to the *frontal spine*, arise at the suture's inferior end. In approximately 10% of humans, however, the metopic suture remains unfused (see Figure 9–6).

In one study (Brown et al., 1984), the median age at first appearance of the frontal sinus was 3.25 years for boys and 4.58 years for girls. Frontal sinus size was achieved at a median age of 15.68 years for males and 13.72 years for females. Thus, frontal sinus growth begins earlier in males than in females and the period of frontal sinus growth is longer on average in males than in females, with the result that these sinuses are larger in adult males than adult females. Brown et al. (1984) also suggested that frontal sinus enlargement parallels the pattern of annual growth increments (e.g., bone length) in boys and girls.

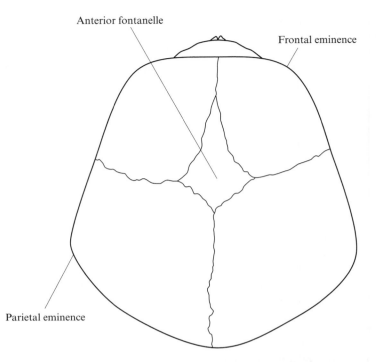

Anterior fontanelle

Frontal eminence

Parietal eminence

Figure 3–14 Skull of neonate (superior view) illustrating anterior fontanelle intervening partway between separate right and left frontal bones (the posterior fontanelle, between the parietal and occipital bones, is not visible in this view).

Parietal Bone

Morphology

Externally, the parietal bone (Figure 3–15) is distinguished by (1) the temporal lines; (2) the parietal eminence (typically more angular and pronounced in females than males); (3) great stretches of relatively smooth, morphologically unadorned bone; and (4) most telling, its sutural configurations.

The temporal lines are most noticeable in adults, especially in males (although diet and tooth use—e.g., in hide or tool preparation—can reverse this generalization). When discernible, the temporal lines are typically confined to the lateral wall of the parietal, but they may on occasion be located more superiorly. Although *superior* and *inferior temporal lines* are identified in textbooks, these "lines"

are visually the upper and lower borders of a smooth, arcing band on the side of the parietal bone. At its widest, this band is approximately 1 cm or so thick; its arc typically parallels the curvature of the squamosal border of the parietal bone.

The inferior or *squamosal border* of the parietal makes contact with the greater wing of the sphenoid for a short distance anteriorly, but it mostly articulates with the temporal. For the length of the short *sphenoparietal suture* and the anterior half or so of the *squamosal suture*, the squamosal border of the parietal is wedge-shaped, smooth or very finely ridged, and quite sharp-edged, especially anteriorly. Proceeding posteriorly from the region of the sphenoparietal suture (i.e., the *sphenoidal angle*), the surface of the squamosal border thickens and may become denticulate. These features—especially thickness—are more

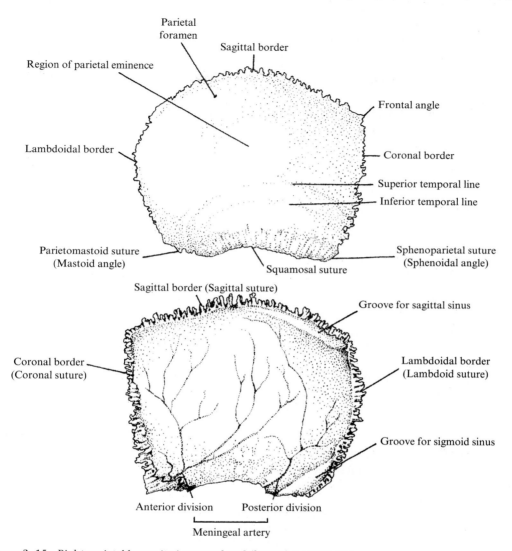

Figure 3–15   Right parietal bone: (*top*) external and (*bottom*) internal views.

marked in the posterior half of the squamosal suture and remain so down to the *parietal notch*. The external surface of the thickened, wedge-shaped posterior portion of the squamosal sutural border is adorned with ridges and grooves that cut across its surface. The wedged shape of the sphenoparietal and squamosal sutural portion of the squamosal border complements the squamosal border

of the temporal, which overrides (and thus overlies) the parietal bone.

The posterior thickening of the squamosal border precedes the change of orientation at the region of the parietal notch and persists for the length of the possibly slightly arcuate *parietomastoid suture* posterior to the notch. Along and somewhat internal to its inferior margin, the parietomastoid suture bears obliquely oriented ridges and grooves that abut and interlock with the sutural border of the mastoid process below. The internally jutting and distended corner of the parietal—at the juncture of the squamosal and parietomastoid sutures—inserts into the parietal notch. The posterior portion of the parietomastoid suture is called the *mastoid angle*.

Although the *frontal, sagittal,* and *occipital borders* of the parietal bone are all denticulate, each has a characteristic configuration. The frontal border's denticulations are typically the finest, while those of the sagittal suture are the thickest, longest, and most deeply invaginated on their sides. The denticulations along the occipital border are as long as in the frontal border but more robust. Since the frontal border of the parietal bone mirrors the opposing border of the frontal bone, the denticulations of the inner table are often more emphasized toward the midline; laterally, the denticulations of the outer table are dominant and override the border of the frontal bone. The denticulations within 2–3 cm of the *sphenoidal angle* are less numerous and projecting.

For the first centimeter or so posterior to the *frontal angle* (the contribution to bregma), the sagittal border bears a series of short, closely packed denticulations that are finer in structure than along the frontal border. Away from this angle, the denticulations become increasingly longer, thicker, and separated. Approximately two-thirds along the length of the sagittal border is a 2 cm or so stretch along which the thick denticulations become shorter and set farther apart. On either side of this portion of the sagittal border, the bone is somewhat flattened and may be perforated by a *parietal foramen* (which, either uni- or bilaterally, is considered to represent a nonmetric *variant*). From this short segment to the *occipital angle* (which contributes to the anthropometric landmark *lambda*), the denticulations assume their "typical" length and spacing. From the occipital angle to the region of asterion (near the mastoid angle), the occipital border is characterized by the relative uniformity of

its denticulations and its "edge" tends to be slightly wavy.

The internal features of the parietal bone are described in Chapter 2. Of special note, though, in identifying parietal bone are the *grooves for the divisions of the middle meningeal artery* and the *sulcus for the sagittal sinus*, with its attendant *granular foveolae* (pits for arachnoid granulations) and, in older individuals, *pacchionian depressions*. The superiorly situated sagittal sinus courses anteroposteriorly along the sagittal border. The grooves for the divisions of the middle meningeal artery emanate from the inferior border and branch posteriorly and superiorly. Parietal bone fragments usually preserve at least one of the sutural borders and enough of the meningeal branching pattern to make identification possible.

[A simple technique for determining right from left parietal bones is to imagine that your wrist corresponds to the middle meningeal artery as it arises from the squamosal border and that your fingers represent its subsequent branching. By placing your hand on the side of your head, you will emulate the pattern of branching. Orient the parietal bone (or fragment thereof) to match your right or left hand.]

For additional clues to siding parietal bones, see Table 3–5.

Table 3–5   Clues to Siding the Parietal Bone

Feature	Position
Frontal angle	Anterosuperior
Frontal border	Anterior
Grooves for middle meningeal vessels	Medial (internal, arborize superoposteriorly)
Inferior temporal line	Lateroinferior (external)
Mastoid angle	Posteroinferior
Occipital angle	Posterosuperior
Occipital border	Posterior
Parietal eminence	Lateral
Sagittal border	Superior
Sigmoid sinus	Posteroinferior (internal)
Sphenoidal angle	Anteroinferior
Superior temporal line	Lateroinferior (external)
Temporal border	Inferior

Development and Ossification

By the end of the second fetal month, each parietal bone has begun to mineralize intramembranously via two ossification centers that eventually coalesce during the fourth month into the presumptive *parietal eminence*. Before these two centers fuse, however, bony trabeculae fan out radially from the region of the presumptive parietal eminence and create a broad-based, cone-shaped structure. This radiating pattern continues until the sixth month, during which time the *coronal*, *sagittal*, and *lambdoidal borders* become recognizable and the "corners" of the coronal and sagittal, as well as of the sagittal and lambdoid, sutures become angular. Retardation in ossification of the superior angular regions (particularly the anterior one) contributes to the formation of the *fontanelles* (or *fonticuli*), which persist postnatally. The *anterior fontanelle (fonticulus)* is larger at birth than the *posterior fontanelle*. By the middle of the third trimester the anteroinferior portion of the enlarging parietal bone has become not only more angular but also reminiscent of its final shape (and thus of the configuration of its contribution to the *pterion*). But the posterior portion remains rather rounded. At birth, the inferior fontanelles are the *sphenoid* and *mastoid fontanelles*. Because the parietal bones are separated at birth from one another and from other bones, sutural features (including the overlapping nature of the squamosal suture) have not yet developed.

Zygoma

Morphology

[An isolated zygomatic bone (Figure 3–16) is similar in general outline to the inverted haft of a sword. The thick vertical strut that encloses the orbit laterally and proceeds superiorly as the *frontal process* corresponds to the sword's handle. The blunt *temporal process* that extends posteriorly, together with the more pointed, medially arcing *maxillary process*, represents the sword's crosspiece.]

Variation is noted, for example, in a temporal process that (1) may be much shorter than the maxillary process and/or (2) is distended inferiorly into a pointed slip of bone (instead of being blunt) that articulates with, and embraces from below, the temporal's zygomatic process. Variation may be asymmetrically expressed in the same individual.

The articular surface of the zygoma's *maxillary process* mirrors its counterpart on the maxilla. On the zygoma, the edge contributing to the facial expression of the *zygomaticomaxillary suture* is variably jagged and obliquely oriented and courses inferiorly as well as laterally away from the inferior orbital margin. The inner surface of the maxillary process is rugose and corrugated and meshes

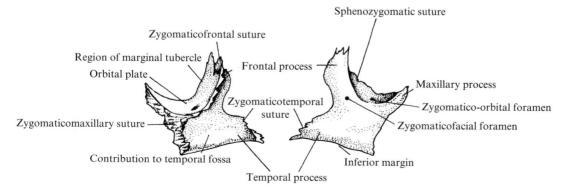

Figure 3–16   Right zygomatic bone: (*left*) internal and (*right*) facial views.

with the outwardly facing articular surface of the maxilla. The *temporal process'* contribution to the *zygomaticotemporal suture* is variably denticulate; the denticulations tend to be fine, short, and closely approximated. The temporal process is compressed laterally (it is much deeper superoinferiorly than thick mediolaterally); its inner and outer surfaces are fairly smooth. The *inferior margin* of the temporal process bears the masseter muscle's attachment. It is variably thin to rugose and stouter, thicker, and more muscle-marked in males than females in populations in which females do not use their jaws and teeth as tools or vises. The zygoma is deepest inferiorly at the juncture of the temporal and maxillary processes, where it is often peaked or pointed. A small *zygomaticofacial foramen*, or, more rarely, multiple zygomaticofacial foramina, may lie more or less directly above this point, level with or slightly above the superior margin of the temporal process, and often above the inferior margin of the orbit. These foramina may occur asymmetrically and communicate with the *zygomatico-orbital foramina* that perforate the floor of the orbit.

The zygoma may also be identified on the basis of the morphology of the frontal process and its relation medially to the orbit and posteriorly to the temporal fossa. The orbital margin is rounded, blunt, and arcuately "L"-shaped (somewhat like a hockey stick); its smooth inner angle faces medially. The contribution of the temporal process to the lateral orbital wall is somewhat thicker and definitely deeper than the reflection of the maxillary process upon the orbital floor. Just inside the rim of the orbit and just below the *frontozygomatic suture*, the orbital border is variably thickened or swollen into the *marginal tubercle (tuberculum marginale)*. The exact shape of the orbital portion of the temporal process may differ asymmetrically within the same individual, due in part to the extent to which the bone overlaps the orbital contribution of the greater wing of the sphenoid. As a result, the edge of the

sphenozygomatic suture is not morphologically homogeneous. The length of the thin, tapering portion of the frontozygomatic suture may also vary. The surface of the more consistently triangular portion of the frontozygomatic suture is cross-hatched with a series of parallel to subparallel ridges and grooves that interdigitate with their counterparts on the zygomatic process of the frontal bone. There is either a small projection of bone, an indentation, or both (the small spit of bone lying over the indentation) at the intersection of the superior and inferior orbital flanges of the zygoma. The latter juncture marks the anteriormost extent of the inferior orbital fissure. This small stretch of bone does not bear any sutural morphology and represents a break between the orbital aspects of the sphenozygomatic and zygomaticomaxillary sutures.

The relatively thin posterior extension of the zygoma's frontal process is oriented at approximately 90° to the orbital projection of the process itself. These two flanges of bone form the anterior wall of the *temporal fossa*. The posterior margin of the frontal process may either transcribe a relatively smooth arc down toward the temporal process or, c. 1 cm below the frontozygomatic suture, bulge out and then curve inward toward its juncture with the temporal process. (This bulge is often misidentified as the marginal tubercle.)

For clues to siding zygomatic bones, see Table 3–6.

Table 3–6   Clues to Siding the Zygomatic Bone

Feature	Position
Frontal process	Superior
Malar tubercle	Inferior
Maxillary border (craggy surface)	Medioinferior
Maxillary process	Anteromedial
Orbital surface	Mediosuperior
Temporal process	Posterolateral
Zygomaticofacial foramen	Anterior
Zygomatico-orbital foramen	Medial
Zygomaticotemporal foramen	Posterior

Development and Ossification

During the latter part of the second fetal month, the zygoma arises intramembranously beneath and lateral to the orbital region. Mineralization begins often only in one center but sometimes in as many as three centers of ossification that typically fuse within a few weeks into a single mass from which further ossification spreads outward. Incomplete coalescence of multiple centers produces a horizontal suture that divides the bone into a large superior and a smaller inferior moiety (*os japonicum*).

Vomer

The vomer (Figure 3–17) develops during the eighth fetal week as right and left plates or lamellae that arise from ossification centers appearing in the posterior and inferior portions of the membrane covering the cartilaginous septum that bisects the nasal cavity. These lamellae begin to coalesce by the third fetal month but do not completely fuse until the onset of puberty. As they unite, a groove (obvious in adults) develops between them in which the cartilaginous nasal septum rests. One can also observe in adults that (1) the platelike "walls" of the groove are not symmetrically tall (e.g., the right lamella may be taller than the left) and (2) the groove is usually broadest at its anterior and inferiormost extremity, where it flares out beyond the thin median sheet of bone that comprises the corpus of the vomer.

At birth, the vomer is relatively long anteroposteriorly but not very tall superoinferiorly; as the facial skeletal deepens and lengthens, it becomes taller and longer. At its juncture with the sphenoid, the vomer splays laterally, embracing the midregion of the basisphenoid with its *alae* (vestiges of its bilamellar origin). *Variation* is noted in the degree to which the alae are confined to, or extend laterally beyond, the side of the basisphenoid. On an isolated vomer, one can see that the two alae flare laterally and are separated by a deep cleft. (A damaged vomer reveals that the "single" bone is composed of two lamellae with narrow air spaces between them.) The contacts between the vomer and the perpendicular plate of the ethmoid above it, as well as between it and the maxilla and palatine bones below, are relatively thin.

Inferior Nasal Concha

The inferior nasal concha (Figure 3–18) begins to develop during the fifth fetal month from an ossification center that arises endochondrally in the lateral wall of the nasal capsule. The fully formed inferior nasal concha protrudes medially and inferiorly into the nasal cavity. It articulates along its *superior border*, respectively and proceeding anteroposteriorly, with the maxilla, lacrimal, ethmoid, and palatine bones. It articulates thinly with the lacrimal and ethmoid via its *lacrimal* and *ethmoidal processes*. The typically horizontally oriented ridges of contact (*conchal crests*) with the maxilla and the palatine are longer and more substantial; the inferior nasal concha may fuse with these bones, making it seem as if the former is an outgrowth of the latter. The rugose and variegated medial surface of the

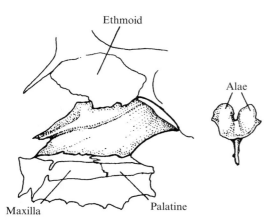

Figure 3–17 Vomer (*stippled*): (*left*) left side and (*right*) posterior aspect.

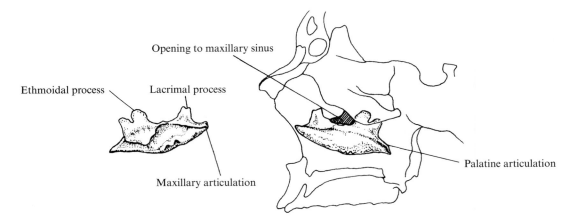

Figure 3–18   Right inferior nasal concha: (*left*) medial and (*right*) lateral views.

Table 3–7   Clues to Siding the Inferior Nasal Concha

Feature	Position
Anterior end (stubby)	Anterior
Ethmoidal process	Superoanterior
Lacrimal process	Superoposterior
Maxillary process	Lateral
Posterior end (elongate)	Posterior

inferior nasal concha is convex. Its relatively smooth lateral surface is concave. The most medial portion of the inferior nasal concha is expanded and spongy in texture. *Variation* is noted in the degree to which the inferior nasal concha protrudes into the nasal cavity and its medial extremity is expanded and pneumaticized (Schwartz et al., 1999).

For clues to siding inferior nasal conchae, see Table 3–7.

Ethmoid

Morphology

[Since completely intact, isolated human ethmoids are not frequently encountered in archaeological, forensic, or paleontological collections, only the bone's major features are reviewed here.]

Because the ethmoid (Figure 3–19) is largely composed of exceedingly thin bony walls surrounding air cells and a thin perpendicular plate, it typically fragments into small, thin flakes. Identifiable ethmoidal fragments usually form parts of a larger fragment (e.g., the medial orbital wall).

Two air cell-riddled labyrinths hang down like saddlebags on either side of an anteroposteriorly long, thin perpendicular plate. Their lateralmost surfaces are irregularly shaped rectangles of smooth bone whose long superior and inferior edges and much shorter anterior and posterior edges form thin, finely denticulate sutural contacts with neighboring bones. The smooth-boned portion contributes to the medial orbital wall as well as to the following sutures: *frontoethmoidal* (superiorly), *ethmoidomaxillary* (inferiorly), *ethmoidolacrimal* (anteriorly), and *sphenoethmoidal* (posteriorly). The somewhat inflated *superior nasal concha* lies medially below the labyrinthine bundle of air cells; it is separated from the scrolled *middle nasal concha* by the channel-like *superior meatus*. The seemingly freely hanging, medially situated medial nasal concha is separated from the perpendicular plate. The thin, variably folded, swollen, or pointed downwardly projecting *uncinate process* lies laterally opposite the middle nasal concha,

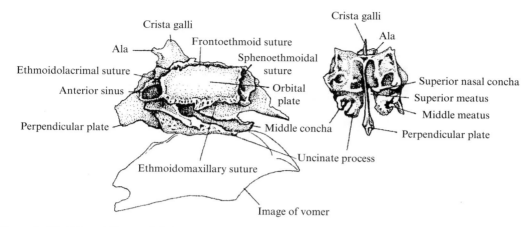

Figure 3–19   Ethmoid bone: (*left*) lateral and (*right*) posterior views.

from which it is separated by the relatively deep *middle meatus.*

The thin *perpendicular plate* descends (approximately) in the midline of the nasal cavity, and the crista galli arises more or less from the midsection of the porous *cribriform plate.* The upper portion of the anterior edge of the perpendicular plate articulates with the *spine of the frontal bone* and the lower portion with the *crest of the nasal bones.* From the nasal crest, the perpendicular plate's long inferior edge courses down and back, forming a deep, anteriorly opening "V" with the vomer, which it contacts farther back in the nasal cavity. The nasal septal cartilage lies in this "V." The perpendicular plate's articulation with the vomer angles upward as it courses posteriorly. The posterior border of the perpendicular plate articulates with the *sphenoidal crest.*

Development and Ossification

The ethmoid arises endochondrally during the sixteenth to eighteenth fetal weeks, with its two labyrinths appearing as bony granules in the orbital laminae. Osteoclastic activity in the developing labyrinths results in the formation of the ethmoidal air cells, which are identifiable before birth and continue to

expand as the ethmoid enlarges. Most of the ethmoid is still entirely cartilaginous at birth, but the walls of the small labyrinths are partially ossified. The ossification center that gives rise to both the crista galli and the perpendicular plate does not appear until the first year. The crista galli and perpendicular plate coalesce with the labyrinths in the second year.

Temporal Bone

Morphology

Because it is composed of extremely dense bone, the petromastoid portion of the temporal bone (Figure 3–20) is preserved archaeologically and paleontologically more frequently than many other parts of the cranium. Its two major elements are the *mastoid process* and the *petrous portion (petrosal)*, from which the *tubular ectotympanic (tympanic tube)* extends, terminating in the *external acoustic meatus (external auditory meatus).* Humans are the only extant primates in whom the mastoid process typically protrudes inferiorly below the skull base, sometimes as far as or even below the level of the occipital condyles. The mastoid process is usually long anteroposteriorly and deep mediolaterally at its base, and

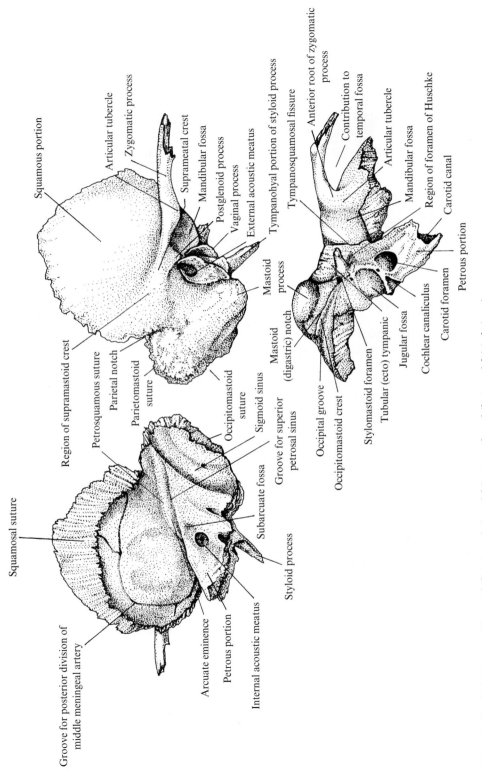

Figure 3–20 Right temporal bone: (*left*) internal, (*top right*) external, and (*bottom right*) inferior views.

its lateral surface is roughened. Its tip tends to be narrower than its base. The male mastoid process is characteristically stubbier and blunter than the female's, in whom it is often thinner or narrower. Sexual differences in mastoid process length are less consistent.

The mastoid process' medial surface is usually smoother than its lateral surface. The former forms a variably steep to sloping and tall to short wall that faces upon the obliquely oriented, typically shallow *mastoid (digastric) notch* or *groove*. Viewed from below, this notch's orientation gives the mastoid process a somewhat triangular outline (i.e., broad at its abutment with the external acoustic meatus with a posterior taper). The *occipital groove* lies just medial to this notch and is sometimes separated for some part of its length from it by the variably thick and ridgelike *occipito-mastoid crest*. The occipital groove tends to be confluent with the mastoid notch anteriorly, but the two features may be inseparable for their entire lengths. When it is not impressed upon the occipitomastoid suture itself, the occipital groove lies just on the edge of the mastoid region.

Starting at the mastoid process and proceeding anteromedially along the petrosal, one encounters, respectively, (1) the mastoid (digastric) notch; (2) the *stylomastoid foramen*; (3) the styloid process (or at least its base, as this slender structure is easily broken); (4) the vacuous, posteromedially directed contribution to the *jugular fossa*; (5) the somewhat smaller but still large, teardrop-shaped *carotid foramen*; and (6) the craggy apex of the petrosal, from which the carotid canal emanates. Appressed to the anterior surface of the mastoid process is the posterior wall of the tubular ectotympanic. From this juncture and proceeding along the petrosal to or just beyond the anterior side of the carotid foramen, the tubular ectotympanic appears creased and "thrown up" as a wall against the mastoid notch, stylomastoid foramen, and styloid process. Wrapping around the posterior side of the styloid process is the crestlike

vaginal process, which appears to diverge from the crestlike edge of the external acoustic meatus [a configuration characteristic of *Homo sapiens* (Schwartz and Tattersall, 2003)].

The rather steeply inclined, anterolaterally facing anterior surface of the tubular ectotympanic is relatively smooth. It forms the posterior wall of the basinlike *mandibular fossa*. The *tympanosquamosal fissure* lies at the juncture of the tubular ectotympanic and the temporal, posterior to the deepest part of the mandibular fossa. Although characteristically wide from side to side, the mandibular fossa is variably long anteroposteriorly. Thus, the fossa is variably deep to shallow from one individual to another.

The mandibular fossa is bounded laterally and posteriorly by the short, stout, and downwardly projecting *postglenoid process (plate)*, which is appressed to the edge of the tubular ectotympanic. It is constrained anteriorly by the superoinferiorly taller *articular tubercle (eminence)*, which extends fully mediolaterally in front of the fossa. Laterally, this tubercle thickens somewhat and projects away from the skull's side as the *anterior root of the zygomatic process*. The posterior portion of the mandibular fossa also projects from the side of the cranium; its roof is formed by a thinner strut or bridge between the postglenoid process and the articular tubercle. There appears to be a correlation between the degree of lateral projection of the fossa and tubercle, and to some extent the process, and the degree of development of the *supramastoid* and *suprameatal crests*. The supramastoid crest, which is a continuation of the posterior arc of the temporal lines, courses down toward the mastoid region from just above the parietal notch and curves toward the external acoustic meatus, over which, as the suprameatal crest, it courses to the postglenoid process; there, it expands to become the ledge into which the mandibular fossa is impressed. The *posterior root of the zygomatic process* is the crest or thickening that proceeds posteriorly from the postglenoid process and passes above the external acoustic meatus. The degree of robusticity of the supramastoid/suprameatal crest is a function of the demands of the temporal muscles and may also reflect differences between males (thicker, more marked crests) and females. The small, semitriangular, slightly raised region *suprameatal triangle* is bordered superiorly by the supramastoid/suprameatal crest, which forms the "base" of the triangle.

Damage to the mastoid region exposes its internal pneumatization. *Mastoid air cells* are largest and most vacuous toward the center of the process and generally smaller and more numerous inferiorly and around the perimeter. (The air cells, parietal notch, and thick and craggy occipitomastoid suture are particularly useful features to look for when identifying mastoid fragments.)

Although the relatively slender *zygomatic process*, which projects forward as well as laterally away from the side of the cranium, is often damaged or broken off near its anterior root; vestiges of this root, in conjunction with features in and around the mandibular fossa, provide clues to identifying fragments and determining side. When a piece of zygomatic process is found, the clear distinctions between the anterior root and the backwardly angled zygomaticotemporal suture—as well as between the slightly thicker, muscle-scarred inferior surface and the finer, more edgelike superior margin—make identification possible.

The relatively flat external surface of the *squamosal* or *squamous portion* of the temporal is further distinguished by its fanlike appearance (which appears to spread out and rotate around the posterior root of the zygomatic process). Fragments of the thin squamous portion can be identified by internal features as well as by the *squamosal suture*, which (1) thins as it overlaps and (2) bears fine, radiating grooves and ridges that interdigitate with their counterparts on the parietal bone.

Internally, the anteromedially oriented, somewhat tapering petrosal is creased superoposteriorly along most of its length into an edgelike border that may bear a shallow *groove for the superior petrosal sinus*. When present, this groove is usually most pronounced in the petrosal's midregion, lateral to which the *arcuate eminence* arises from the bone's superior surface. Farther along the superior surface of the petrosal and still toward its edge lies the variably depressed *subarcuate fossa*, which is situated above the posteroanteriorly obliquely oriented *internal acoustic meatus*. Arcing into the middle cranial fossa and laterally around the petrosal is the *petrosquamous suture* (or at least traces of it), which is typically obliterated externally. The squamous portion of the middle cranial fossa bears a characteristic pattern of shallow, broad grooves or depressions and low ridges that reflects the gyral and convolutional pattern of the brain; regardless of variation anteriorly, the *groove for the posterior division of the middle meningeal vessel* courses posteriorly and somewhat superiorly (up and back). The *sigmoid (sigmoidal) sulcus*, which lies below the root of the petrosal, provides another clue to identifying temporal bone fragments.

For clues to siding temporal bones, see Table 3–8.

Nonmetric traits commonly recorded for the temporal bone include (1) exsutural mastoid foramen (it does not lie in the occipitomastoid suture but penetrates the mastoid region instead); (2) absence of mastoid foramen; (3) persistence of a foramen of Huschke; (4) development of *exostoses* in the tubular ectotympanic [*auditory* or *acoustic tori (tori auditivi)*], usually along its posterior wall or floor; or (5) congenital absence of the entire tube. Acoustic exostoses may be minor bony thickenings or so large that they virtually occlude the meatal openings. Development of an auditory torus may be genetically based but may also be a response to cold and pressure of habitual diving in deep coastal waters.

Development and Ossification

The temporal bone consists initially of three entities: the *squamous, tympanic,* and *petrous*

Table 3–8   Clues to Siding the Temporal Bone

Feature	Position
Articular tubercle	Inferoanterior
Carotid canal	Inferomedial
External acoustic meatus	Lateroinferior
Groove for middle meningeal vessel	Mediosuperior (internal)
Groove for middle temporal artery	Lateral
Internal acoustic meatus	Medioinferior (internal)
Mandibular fossa	Inferolateral
Mastoid notch	Inferoposterior
Mastoid process	Lateroposterior
Occipital groove	Posteroinferior
Parietal border (beveled suture)	Superior
Parietal notch	Superoposterior
Postglenoid tubercle	Inferolateral
Sigmoid sinus	Medioposterior (internal)
Styloid process	Inferior
Stylomastoid foramen	Inferior
Suprameatal triangle	Lateral
Zygomatic process	Lateroanterior

portions. The former two arise intramembranously, whereas the petromastoid ossifies endochondrally.

Between the end of the second fetal month and the middle of the third, ossification centers appear intramembranously: first, for the *zygomatic process* (at its presumptive root), and then, respectively, for the anterior and posterior "halves" of the *squamous portion*, which continues to ossify in a radial manner. These three centers quickly coalesce, but a fissure delineating the anterior from the posterior squamosal moiety may persist through term. By the end of the eighth fetal month but perhaps at the beginning of the third trimester in (some) males, the squama fuses with the tympanic part.

The *tympanic ring* begins to ossify intramembranously during the ninth to tenth fetal weeks at its superior ends as well as inferiorly. Within a few weeks, these three centers

of ossification coalesce. Toward the end of the third trimester, the tympanic ring begins to fuse to the squamosal and petrosal. At birth, the still incomplete tympanic ring bears a distinct groove (*tympanic sulcus*) in which the tympanic membrane is anchored.

Ossification of the petrosal is complex, with multiple centers arising endochondrally prior to or during the fifth fetal month. The first centers appear medially in the *tympanic cavity (otic capsule)*, which lies in the region of the presumptive *arcuate eminence*. These centers contribute to the medial and posterior walls of the tympanic cavity as well as to the *cochlea*, *vestibule*, and *internal acoustic meatus*. The second set of centers arises inferiorly, at the *promontory* on the medial wall of the tympanic cavity, and contributes to the lower portions of the tympanic cavity, vestibule, internal acoustic meatus, *cochlear window*, and *carotid canal*. The third set of ossification centers emerges superiorly, over the *lateral semicircular canal*; as they spread out, they cap the tympanic cavity and *antrum* (via the tegmen tympani).

During the sixth fetal month, (1) the *superior semicircular canal* ossifies, (2) a fissure marking the development of the carotid canal appears superiorly in the petrosal, and (3) the *jugular notch* becomes distinct. By the middle of the third trimester, the "carotid fissure" has enlarged and ossified inferiorly into the carotid canal. The *mastoid region* arises from a center of ossification that emerges near the posterior semicircular canal; the primary growth of the mastoid process occurs postnatally. The *tympanohyal portion of the styloid process* arises prenatally; during the first postnatal year, it fuses with the temporal posterior to the tympanic ring. The petrosal and squamosal also coalesce during the first postnatal year, often by the end of the first postnatal month; the persistence of the *petrosquamous suture* reflects the ontogenetic distinctiveness of these bones.

Postnatally, as skull breadth increases, (1) the *mandibular fossa* and *zygomatic process* become more inferiorly oriented, (2) the mastoid region enlarges and becomes pneumaticized, (3) the natally large and patent subarcuate fossa closes over, (4) the stylohyal portion of the styloid process begins to ossify (year 2) but may not fuse with the temporal until after puberty (if at all), and (5) the tympanic ring begins to grow outward to eventually become the tubular ectotympanic. Uneven ossification of the expanding tubular ectotympanic leaves a *foramen of Huschke* in its floor that usually fills in by the fifth year but may persist uni- or bilaterally and to varying degrees of patency into adulthood (and recorded as a *nonmetric variant*).

Occipital Bone

Morphology

Viewed from behind, the isolated occipital bone (Figure 3–21) is reminiscent of a broad leaf with jagged edges. Superior to the horizontal plane that passes through the *external occipital protuberance* (which coincides with the plane passing through both asterions) lies the *planum occipitale* (occipital plane), which is somewhat convexly swollen outward and relatively smooth-surfaced. Technically, the planum occipitale begins at the level of the *highest nuchal lines*, but these lines are not always discernible. A fully expressed highest nuchal line arises laterally and slightly above the level of asterion, arcs upward, and then curves gently down toward the external occipital protuberance, with which it may be confluent. (Because development of the highest nuchal line is quite variable, the most distinct feature in the general region is often the external occipital protuberance.)

The densely and strongly denticulate *lambdoid border* begins at asterion and, in two stages, courses upward and toward the midsagittal plane. The lower half of the lambdoid border is more vertical than the upper, which angles inward more severely toward the peak (or at least narrower and superiormost portion) of the occipital bone (*superior angle*), which nestles in between the parietals at *lambda*. The *lateral angle* is the most inferior point of the lambdoid border (which contributes to asterion).

The lambdoid border bears two parallel rows or layers of articular denticulations. One protrudes from the outer and the other protrudes from the inner table of the bone. Within 2–3 cm of asterion, the denticulations of these two rows are similar in length. Toward asterion, the denticulations of the inner table may become the more predominant.

The *planum nuchale* bears more surface topography than the occipital plane. The *superior nuchal lines* originate from the region below asterion and curve up toward the external occipital protuberance, with which they may become confluent. The superior nuchal lines increase in thickness and distinctiveness toward the midsagittal region of the bone. At times, the external occipital protuberance appears to be a natural extension of the increasingly thickening lines. If vestiges of the (fetal) lateral occipital fissures persist into the adult, they lie slightly below the lateral angle and delineate further the lateral portion of the superior margin of the superior nuchal lines.

The variably distinct and ridgelike *external occipital crest (median occipital crest)* emanates from the external occipital protuberance and courses anteriorly along the midline of the nuchal plane to the posterior margin of the foramen magnum. Especially in young individuals, this crest may be more pronounced than the external occipital protuberance; in young individuals, the latter may only be a smooth area that is "etched" on its sides by the superior nuchal lines. As this crest proceeds toward the foramen magnum, it courses between two pairs of variably ovoid, roughened patches of bone (one above the other).

The superior pair of roughened areas is bordered above by the superior nuchal lines and below by the *inferior nuchal lines*. In some individuals, the inferior nuchal lines may be restricted to the inferior border of the

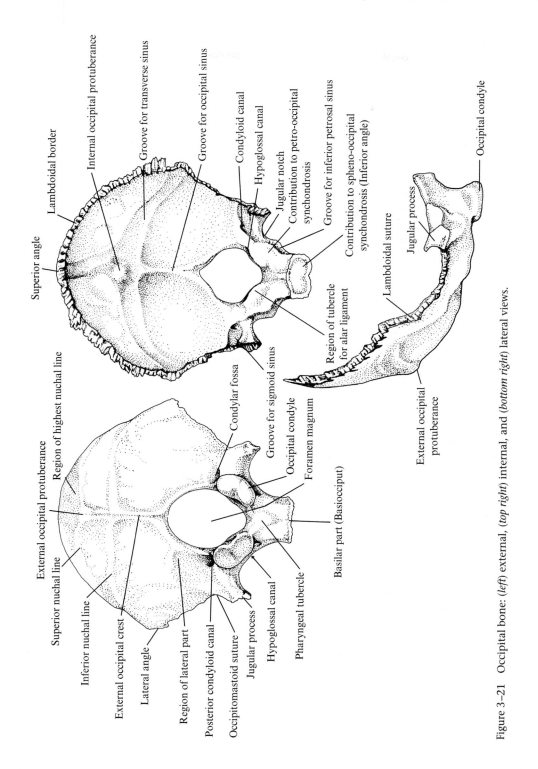

Figure 3–21   Occipital bone: (*left*) external, (*top right*) internal, and (*bottom right*) lateral views.

roughened areas alone. In others, they may arc craggily or present themselves as a series of scallop shapes around the foramen magnum, eventually terminating medial to the *occipitomastoid crest* and *suture*. In yet other individuals, the most prominent features are the portion of the inferior nuchal line that lies just below the semi-ovoid roughened area and the stretch of bone that extends for a few centimeters just in front of the occipitomastoid crest/suture. Rarely does the inferior nuchal line extend around the foramen magnum as far as the *jugular process*, which projects laterally from the lateral margin of the occipital condyle, reaching its greatest depth before or at the *occipitomastoid suture.*

The inferior pair of semi-ovoid, roughened depressions constitute the thinnest portions of the entire occipital bone and are delineated anteriorly by the rim of the foramen magnum and posteriorly by the inferior nuchal line. They are also bounded laterally by another pair of longer, arcing depressions, which correspond to the developmentally distinct lateral parts of the occipital bone. In some individuals, this general region bears three or more pairs of roughened depressions.

Viewed from the side, the nuchal plane exhibits one of two basic configurations: it may angle inward along a single plane, or it may be angled along the "horizontal axis" of the inferior nuchal line into a second plane. In either case, the inferiorly placed, typically ovoid *foramen magnum* remains essentially horizontally oriented, with its anterior margin lying only slightly below the level of its posterior margin. From the foramen magnum, the basilar part of the occipital bone angles upward sharply as it extends to meet the base of the sphenoid. With the exception of the degree to which the basilar part is flexed upward, the other planes, angulations, and orientations of the different regions of the occipital bone are established by the time of birth.

In adults, the *occipital condyles* (which arise from the fusion of contributions from the lateral and basilar parts) lie along the margins of the anterior portion of the foramen magnum. Details of occipital condylar morphology are quite variable, although an occipital condyle is typically elliptical in outline and longer anteroposteriorly than it is wide mediolaterally. An occipital condyle's long axis is oriented obliquely in two planes. In one, its anterior end lies more medially than its posterior extremity. In the other, and when viewed looking down upon it, the medial edge of the condyle projects well below its lateral margin. Thus, the smooth articular surface, which faces away from the opening of the foramen magnum, is oriented obliquely and laterally. The medial edge of an occipital condyle may also extend slightly over and thus protrude somewhat into the space of the foramen magnum. As an occipital condyle roughly parallels the inwardly curving arc of the rim of the foramen, it is convex in the plane of its long axis, arcing upward and outward from its anterior and posterior ends.

The edges of an occipital condyle's articular surface are usually raised and delineated. The posterior end of the condyle, however, is not always crisply delineated from the *condylar fossa* into which it may project. A *posterior condyloid canal* may lie in the region of the condylar fossa (sometimes close to the posterior extremity of the occipital condyle). The rim of the foramen magnum posterior to the condyle is irregularly and variably thickened and roughened. The posteriormost portion of the foramen's rim, which corresponds to the portion of the nuchal plane that intervenes in the fetus between the two lateral parts, is typically the smoothest and thinnest part of the rim. The "Y" of the basilar part of the occipital bone that separates the two condyles and forms the anterior section of the rim of the foramen magnum is the thickest portion of the rim.

Variation in the region of the foramen magnum is noted in the development of a patent posterior condyloid canal and in the extent to which the occipital condyles are partitioned into two (usually unequal) moieties. If this partitioning is incomplete, it either creates the so-called twinned, demiarticular facets or may produce only a notch in the side of the condyle (usually medial). The range of this variation appears to be related to the degree to which the contributions from the lateral and basilar parts of the occipital bone coalesce to form an occipital condyle.

Inferior to the *lateral angle* (the contribution to asterion), the lateral borders of the occipital bone face inward; thus, the rather straight-sided region of the occipital in which the foramen magnum sits becomes constricted and narrower. The *mastoid border* of the occipital bone bears oblique ridges that contribute to the formation of the *occipitomastoid suture*. The edge of the mastoid border may also bear a portion of the occipital groove, which, in some individuals, may be captured between the occipital bone and the mastoid portion of the temporal. At the *jugular process*, the mastoid border of the occipital is variably distended inferiorly. Thus, the mastoid border can contribute variably to the lateral margin of the *jugular notch*, which forms the inferior border of the jugular foramen (the superior border of the jugular foramen is formed by the petrosal portion of the temporal). The aperture of the jugular foramen is oriented anteroposteriorly. The upwardly constricted strut of bone through which the anteroposteriorly oriented *hypoglossal canal (anterior condylar canal)* courses extends between the jugular notch and the occipital condyle.

An unnamed, small, tuberclelike elevation of bone onto which the rectus capitis anterior muscle inserts lies slightly anterior and medial to this canal (if not at the anteromedial edge of the mouth of the canal). The basilar part of the occipital narrows medial to this "tubercle." The anteroposteriorly oriented, low, thin to moderately swollen *pharyngeal tubercle* lies either midsagittally between the right and left of these two small tubercles or just anterior to a line drawn between them.

On either side of, and slightly anterior to, this tubercle is another (unnamed) low tubercle to which the longus capitis muscle attaches. The anteriormost portion of the basilar part, which contributes to the spheno-occipital synchondrosis, is referred to as the *inferior angle*.

Of internal note on the *squamous portion* of the occipital bone is its subdivision by the various grooves for the meningeal sinuses into four quadrants. These quadrants consist of two, paired depressions in which the occipital lobes of the cerebrum (superiorly) and the cerebellar lobes (inferiorly) sit. The variably raised *internal occipital protuberance* lies in the middle of this cruciate pattern. The *groove for the superior sagittal sinus* descends to the internal occipital protuberance and typically veers more markedly to the right than to the left (thus giving the impression that the latter takes origin directly from the internal occipital protuberance). The groove on either side of the protuberance is identified as the *groove for the transverse sinus*, which begins to arc inferiorly at or about the region of asterion. Occasionally, the transition between the superior sagittal sinus and the left transverse sinus is emphasized. Although the general downward curvature of the groove for the transverse sinus begins on the occipital bone, it is identified as the *groove for the sigmoid (sigmoidal) sinus* only when it crosses onto the temporal bone.

In addition to these grooves, this portion of the occipital bone may bear a variably distinct crest, *groove for the occipital sinus*, that courses between the internal occipital protuberance and the midline of the posterior margin of the foramen magnum. It might broaden slightly toward the rim of the foramen magnum; be a rather broad, shallow, groovelike depression; or assume any intermediate configuration.

Proceeding anteriorly around the foramen magnum (internally), the bone thickens and becomes elevated (due to the upward deflection of the basilar part). Parallel (approximately) to the midpoint of the length of the foramen magnum, the *groove for the sigmoidal sinus* courses across the occipitomastoid suture and onto the occipital bone. It curves upward and anteriorly to terminate at the *jugular notch*, which is typically larger on the right side than on the left. The terminus of this groove is also bordered anteriorly and medially by the fairly large and swollen *jugular tubercle*, which is distended medially, toward the foramen magnum. The somewhat posteriorly and slightly inferiorly oriented *hypoglossal canal* lies below and just behind the most swollen portion of the jugular tubercle. The variably roughened and distended *tubercle for the alar ligament* arises from the latter region and projects upward and slightly medially; it may range in expression from being barely visible to quite prominent. The anterior ends of the jugular tubercles are separated from one another by the rather broad and shallow *groove for the medulla oblongata*. This groove follows the perimeter of the anterior portion of the rim of the foramen magnum (becoming splayed inferior to the jugular tubercles) and traverses the length of the basilar part.

The basilar part narrows toward the inferior angle, whose precoalesced surface (at the end of the basilar part that contributes to the *spheno-occipital synchondrosis*) is rough and pitted and may be surrounded by a raised margin. The sides of the basilar part that contribute to the *petro-occipital synchondrosis* are also roughened and may be traversed by an inconsistently distinct *groove for the inferior petrosal sinus*, which variably reaches the jugular notch and/or groove for the sigmoidal sinus. The *superior petrosal sinus* lies on the medial, superior margin of the petrosal.

Development and Ossification

During the seventh to ninth fetal weeks, the occipital begins to ossify from multiple centers, most of which are endochondral (i.e., in the chondrocranium). By the end of the second or the beginning of the third fetal month, two intramembranous centers appear toward the midline of, but well above, the large chondrocranial portion of the presumptive occipital. The former are followed within a week or so by two other intramembranous centers. Eventually, all intramembranous centers merge almost completely, and the resultant presumptive *occipital plane (planum occipitale)* unites almost completely with the ossifying lower squamous portion of the occipital bone, the presumptive *nuchal plane (planum nuchale)*. In the adult, the nuchal lines represent their boundary. The lateral areas of the occipital remain open due to the persistence of a single midline fissure superiorly and the pervasion of an inferiorly oblique fissure on each side of the squama. The superior fissure starts to close slowly after birth, whereas the inferior *transverse occipital fissures (mendosal fissures* or *sutures)* begin to close only in the third or even fourth year.

With regard to *variation*, incomplete union of the occipital and nuchal planes or the occasional occurrence of a supernumerary ossification center in the midline produces the *Inca (interparietal) bone*. Incomplete coalescence of any of the initial intramembranous centers yields *ossicles along the lambdoid suture (lambdoidal ossicles)*.

Toward the end of the second fetal month or by the early part of the third, two ossification centers (but occasionally one transversely elongate center) appear endochondrally in the midsagittal region of the presumptive nuchal plane. By the middle of the third month, these centers unite and form the core from which trabeculae spread laterally and superiorly. Until the middle of the third trimester, the enlarging occipital squama is wider than it is tall (being typically widest at the lateral plates above the transverse occipital fissures). Subsequent rapid vertical growth of the squama eventually (more or less) equalizes these dimensions.

The portions of the occipital that subtend the foramen magnum anteriorly and laterally (the *basilar* and *lateral parts*, respectively) arise from separate endochondral centers. Mineralization of the basilar part (the adult *basiocciput*) begins during the middle of the third fetal month in one or sometimes two centers. During the fourth month, the initially spindle-shaped presumptive basilar part tapers at its ends and nearly doubles in size. During the fifth fetal month, the posterior end of the basilar part starts to bifurcate and assume its characteristic "Y" shape; the arms of this "Y" subtend the foramen magnum anteriorly. The anterior end also broadens and flattens to form its contribution to the *spheno-occipital synchondrosis*. The lateral parts of the occipital begin to ossify at approximately the same time as the basilar part. Each lateral part ossifies from a single center that arises as a thin lamina lateral to the presumptive foramen magnum. With growth, the anterior end of the presumptive lateral part thickens and then bifurcates into two somewhat closely approximated arms that create a "U" shape, with one arm lying below the other. These arms contribute to the formation of the *hypoglossal canal (anterior condylar canal)*. The basicranially exposed arm of the lateral part contributes to the formation of an occipital condyle. The arm of the basilar part that abuts and eventually fuses with the arm of the lateral part contributes to the anterior portion of an occipital condyle. Posterior to its "U"-shaped anterior end, the lateral part of the occipital broadens, flattens, and assumes a slightly arcuate shape because the lateral side of the bone grows more quickly than its medial side. This differential growth causes the medial border of the lateral part, which subtends the foramen magnum, to become arcuate. Within the posterior cranial fossa is a thickened band that emanates from the arm of the lateral part that subtends the hypoglossal canal internally. The *condyloid canal* arises lateral to this thickened band.

The lateral parts do not fuse with the squama until the fourth postnatal year. The occiput does not become a single bone until (approximately) the sixth year. The *spheno-occipital synchondrosis* usually remains patent until about the seventeenth year, typically closing by the twenty-second year (but it may remain unfused until the twenty-fifth year). Closure of the spheno-occipital synchondrosis is sometimes used to define the beginning of "juvenile age."

Sphenoid

Morphology

Intracranially, the dominant features of the sphenoid (Figure 3–22) are its thick, deeply indented *hypophyseal fossa (sella turcica)* and *greater* and *lesser wings*. Proper orientation of this fossa is dictated by the wall-like, posteriorly situated *dorsum sellae*, whose posterior surface is essentially vertical, while its anterior surface (facing the fossa) is concave; (the *posterior clinoid processes* may be useful in identification because they are easily damaged and project up or outward from the superior corners of the hypophyseal fossa). The dorsum sellae is accentuated further by the *carotid sulcus*, which produces an almost vertical indentation in front of it on each side. This sulcus may be partially or completely enclosed laterally (creating more of a foramen than a sulcus) by a variably developed spit of bone, the *lingula*. The hypophyseal fossa is bulkier and broader anteriorly than posteriorly and bears various landmarks.

Immediately opposite the dorsum sellae, a variably distinct, horizontal ridge (*tuberculum sellae*) courses between the inferior margins of the right and left *optic foramina*. This ridge may be distended slightly into weak lateral "corners," the *middle clinoid processes*. Another variably distinct horizontal ridge courses between the superior margins of the optic foramina and delineates between it and the tuberculum sellae a shallow groove for the optic chiasm, the *chiasmatic groove*. A stout, quickly tapering, pointed extension—*anterior clinoid process*—embraces each optic foramen laterally and projects posteriorly along each

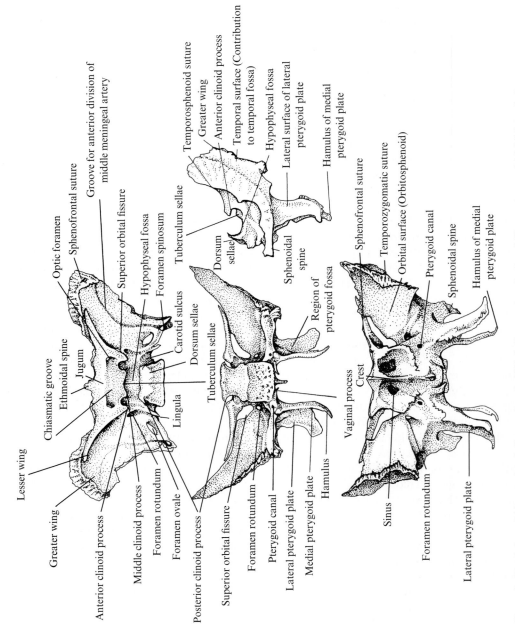

Figure 3–22 Sphenoid bone: (*top left*) superior, (*middle left*) posterior, (*bottom left*) anterior, and (*right*) lateral views.

side of the hypophyseal fossa (an anterior clinoid process is thick enough at its base that it will not be entirely destroyed). The anterior clinoid process is the "back end of the wing" of the lesser wing, which tapers laterally along its rather crisp posterior edge, terminating as a laterally directed point. The *sphenofrontal suture* (between the lesser wing and frontal bone) is mildly undulating; its margin bears fine denticulations that extend from both the inner and outer tables of bone. The variably broad, thin-boned *ethmoid spine* projects anteriorly from the region of the jugum to meet the cribriform plate (this spine may survive even if the ethmoid is damaged).

Lateral and anterior to the body of the sphenoid, the *greater wing of the sphenoid* contributes, respectively, to the base and lateral wall of the cranium and to the lateral and posterior walls of the orbit. Intracranially, the concave surface of the greater wing is somewhat corrugated (reflecting the gyral impressions of the brain through the meninges). Superiorly, the surface of the greater wing may also bear the *groove for an anterior branch of the anterior division of the middle meningeal artery*. Particular to the sphenoid (and proceeding from top to bottom and from side to side) is the arcuate pattern created by the *superior orbital fissure, foramen rotundum, foramen ovale,* and *foramen spinosum*. The relatively large, subtriangular *sphenofrontal sutural* contact (*frontal margin*) the greater wing makes with the frontal bone above it is further distinguished by a pitted and spiky surface that is bounded laterally by the thin, almost vertical extension of the infratemporal surface of the greater wing; the latter overlaps externally the frontal and the parietal bones along the short sphenoparietal and sphenofrontal sutures (i.e., the lateral expression of the sphenofrontal suture).

Externally and laterally, the greater wing is relatively tall and thin; especially superiorly, its infratemporal/temporal fossa surface is concave along the vertical axis and bounded anteriorly by the lateral portion of the flat,

smooth *orbital surface*, which almost forms a right angle with the temporal surface. The relatively thin lateral extension of the orbital plate contributes to the *sphenozygomatic suture*; its sutural margin is deeply indented. The inner table of bone of the posterior and superior margins of the greater wing is extended beyond the outer table (mirroring the squamosal portion of the temporal, which overlaps the greater wing in the formation of the *zygomaticotemporal suture*). The inferior part of the greater wing's contribution to the zygomaticotemporal suture is thicker, and platelike extensions of inner and outer tables subtend a hollow that is pervaded by bony denticulations.

The descending greater wing curves medially along its vertical axis. At about the level of the inferior margin of the orbital plate (corresponding to the sphenoidal margin along the inferior orbital fissure), a variably modestly to markedly developed horizontal ridge courses across the external wing's external surface. Inferior to this ridge, the sphenoid is markedly displaced medially, and its lateral surface (*lateral pterygoid plate*) is concave. Just below this ridge, the sphenoid extends posteriorly, particularly along the basicranial portion of the sphenosquamosal suture, and terminates in a spikelike *sphenoidal spine*, which, in the articulated skeleton, nestles between the mandibular fossa and petrosal bone. The *foramen spinosum* and *foramen ovale* lie increasingly medial to the sphenoidal spine; the sometimes thin to incomplete bony wall between these foramina appears to emanate from the spine. (Even if the region is damaged, the sphenoidal spine and indented anterior margins of the sphenoidal foramina often remain intact.)

Because the *medial* and *lateral pterygoid plates* are thin and their inferiormost extents often peaked or pulled out into spikelike, posteriorly pointing projections, they are easily damaged. Inferiorly, the medial plate bears a fragile *pterygoid hamulus*. (Intact pterygoid plates are separated by an anteriorly invasive *pterygoid fossa*.) The bony strut enclosing the foramen ovale posteriorly is more often confluent with the superior "base" of the lateral pterygoid plate than with the base of the medial plate. The margin of the medial pterygoid plate is distinguished superiorly by its termination in a flattened, posteriorly directed *pterygoid tubercle*. (Damage to the body of the sphenoid, in the region between the medial pterygoid plates and inferior to the hypophyseal fossa and the area of the jugum exposes the vacuous, variably partitioned *sphenoidal sinuses*).

Development and Ossification

The sphenoid ossifies from multiple intramembranous and endochondral centers. The first to appear, during the eighth to ninth intrauterine weeks and in the presumptive region of the *foramina rotundum* and *ovale*, is the intramembranous center from which the *greater wing of the sphenoid* develops. These foramina and the *foramen spinosum* ossify around the structures they will surround. The foramen rotundum is completely formed by the middle of the second trimester and the foramen ovale, by late in the third trimester. The foramen spinosum does not ossify fully at least until the first postnatal year. The *medial* and *lateral pterygoid plates* may arise intramembranously as early as the ninth to tenth postnatal weeks but more frequently during the fourteenth to fifteenth weeks. The medial and lateral pterygoid plates may coalesce in the sixth fetal month, and the medial pterygoid plate unites with the greater wing about a month later, at which time the *vaginal process* develops from the medial and superior margin of the medial pterygoid plate. The *hamulus* ossifies from a separate center, which appears a year or two after birth.

By 4 to 4.5 fetal months, part of the *body of the sphenoid* has arisen endochondrally from either a pair of ossification centers or (less frequently) a single center; the depression representing the presumptive *hypophyseal fossa*

is also visible. At about the beginning of the fifth month, a pair of centers (or sometimes a single center) that will give rise to the *dorsum sellae* arise endochondrally in the posterior part of the presumptive body of the sphenoid; the sellae is morphologically identifiable within 2 months. Thereafter, its shape and that of the hypophyseal fossa change little. The dorsum sellae, hypophyseal fossa, greater wing, and pterygoid plates constitute the *postsphenoid*. Only during the seventh to eighth fetal months does the *posterior part of the body of the sphenoid* (the dorsum sellae and hypophyseal fossa) fuse with the *anterior part of the body* (from the region of the tuberculum sellae anteriorly). The anterior part of the body of the sphenoid together with the lesser wings is the *presphenoid*. The *lesser wings of the sphenoid* begin to ossify toward the end of the fourth fetal month, each from an endochondral center situated just lateral to the presumptive *optic canal*.

Although the lesser wings continue to enlarge in all directions, it is not until the sixth month that growth accelerates; then, the lateral ends become pointed and the lesser wings appear winglike. The anterior part of the body of the sphenoid arises endochondrally from a pair of centers that appear between the end of the fifth and beginning of the sixth fetal months. The optic canal does not become complete until (perhaps) the sixth fetal month—but definitely during the seventh to eighth fetal months—when the lesser wings unite with the anterior part of the body of the sphenoid. This "unit" then fuses with the posterior part of the body of the sphenoid. Although now surrounded by bone, the optic canal is triangular and does not assume its characteristic circular shape until after birth. Incomplete union of the presphenoid and postsphenoid results in the persistence of a *craniopharyngeal canal*, along which *craniopharyngeomas* (epithelium-lined cysts) may develop.

At birth, the lesser wings (although forming a unit with the body of the sphenoid) are unconnected across the midline, right and left *greater wings* and *pterygoid plates* are separate elements, and the sphenoidal sinuses are, at best, tiny. During the first postnatal year, the lesser wings unite at the midline and the greater wing/pterygoid plate unit fuses with the rest of the sphenoid (forming, for example, the *pterygoid canal*). The smooth-boned medial region posterior to the *ethmoid spine* (which abuts the *cribriform plate*) and between the optic canals represents the area of union between right and left lesser wings (although not clearly delineated in the adult, it is identified as the *jugum* or *jugum sphenoidale*). As mentioned above, the body of the sphenoid and the inferior angle of the basilar part of the occipital bone do not coalesce across the spheno-occipital synchondrosis until the twenty-first or twenty-second year or even as late as the twenty-fifth year.

Hyoid

The relatively thin adult hyoid (Figures 3–23 and 3–24) resembles an edentulous mandible. The *body of the hyoid* arcs forward or anteriorly. Posteriorly, its right and left branches turn upward, each terminating in a *greater cornu*. Above and more or less in front of the region of the "angle" between the body and a cornu, a smaller projection (*lesser cornu*) extends superiorly. A hyoid's anterior surface is vaguely reminiscent of a mandible's mental region: its inferior margin may be somewhat "cornered" (and perhaps distended or swollen

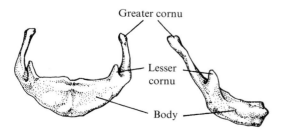

Figure 3–23   Hyoid bone: (*left*) anterior and (*right*) lateral views.

inferiorly) and its midline region somewhat raised. The posterior surface of a hyoid body's midline region is variably concave.

The hyoid is derived from the cartilaginous branchial arch skeleton, and each cornu (see Figure 3–8) as well as each side of the hyoid's body arises from a separate center of ossification. Ossification begins late in fetal development in the body and greater cornua but not in the lesser cornua until 1 or 2 years postnatally.

A hyoid is distinguished not only by not looking like any other bone but also because it bears no signs of contact with other bones.

Inner Ear Bones

The three inner ear bones (Figure 3–25) are often "discovered" by the careful excavator or "appear" during, or as a consequence of, preparation of a dirt-plugged, matrix-encased cranium. Sifting with extremely fine mesh (e.g., carburetor or window screen) will capture these tiny bones during archaeological, paleontological, or forensic excavation.

The inner ear bones (*auditory ossicles*) develop from the cartilaginous branchial arch skeleton; the *incus* and *malleus* arise from the first and the *stapes* from the second arch (see Figure 3–8). The incus and malleus appear during the latter half of the fourth fetal month, approximately a month or so earlier than the stapes. All three ossicles continue to grow postnatally. The malleus attaches to the tympanic membrane and the stapes to the fenestra vestibuli. The incus articulates between them.

The *malleus* (which resembles a hammer) presents a long, inferiorly tapering portion, on top of which sits a semiglobular "head." The *head of the malleus* is rounded anteriorly and concave posteriorly (as *facet for the incus*) and is tilted medially at its juncture with the handle or *manubrium*. Viewed medially, the manubrium appears relatively straight; viewed from behind, it is oriented medially except for its downwardly curving tip. A

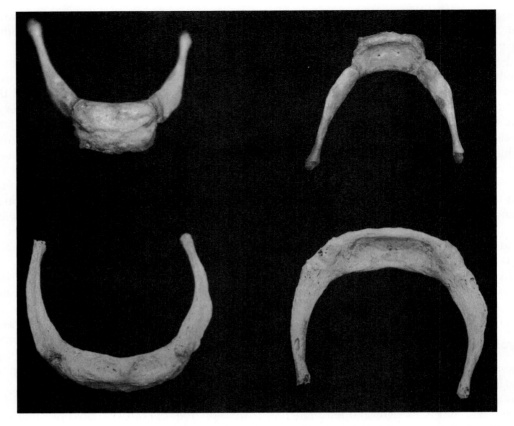

Figure 3–24   Hyoid bone: (*top row*) unfused body and cornua of a subadult and (*bottom row*) the fused hyoid of an adult; (*left column*), anterior views and (*right column*), inferior views.

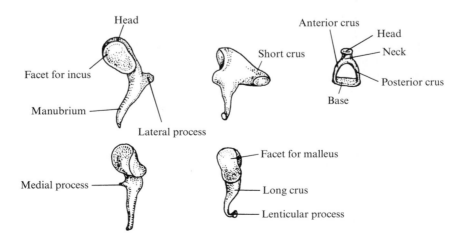

Figure 3–25   Right inner ear ossicles: malleus (*top left*), medial and (*bottom left*) posterior views; incus (*top middle*) medial and (*bottom middle*) anterior views; stapes, superior view (*right*).

small, variably blunt to pointed *lateral process* juts out laterally at the juncture of the manubrium and head of the malleus. (One can side the bone by orienting the head upward, the facet for the incus posteriorly, and the lateral process laterally.)

The *incus* is reminiscent of a tiny anvil. Like the malleus, it bears inferiorly a long, tapering extension (*long crus*), which looks somewhat straight when viewed medially but when viewed posteriorly is shaped like an elongate, shallow "S," with its tip arcing tightly in a medial direction. The tip flares out a bit into the *lenticular process*, which articulates with the stapes. The upper part of the incus is elongate and wedge-shaped and tapers posteriorly into the *short crus* (the part resembling an anvil). The taller anterior end is concave and bears the *facet for the malleus*.

The stirrup-shaped *stapes* is composed of a narrow end (*head* and *neck*) and a more or less flat *base*. The "U"-shaped section of the bone that bifurcates from the neck and meets the base comprises two parts, an *anterior* and a *posterior crus*. In vivo, the base of the stapes is oriented medially and slightly above the level of the laterally positioned head. Viewed from below, the base (medial surface) of the stirrup is slightly convex along its superior margin and, thus, slightly concave along its inferior margin.

For clues to siding inner ear bones, see Table 3–9.

Table 3–9   Clues to Siding the Inner Ear Bones

Feature	Position
Incus (anvil)	
Facet for malleus (saddle-shaped)	Anterior
Lenticular process	Medioinferior
Long crus (S-shaped)	Inferior
Short crus (thick and conical)	Posterior
Malleus (hammer, largest of the auditory ossicles)	
Anterior process	Anterior
Facet for incus (saddle-shaped)	Posteromedial
Head (club-shaped)	Superior
Lateral process	Lateral
Manubrium or handle	Inferior
Stapes (stirrup, smallest of the auditory ossicles)	
Anterior crus (shorter and straight)	Anterior
Base or footplate (footprint-shaped)	Medial
Head	Lateral
Posterior crus (longer and curved)	Posterior

Chapter 4

The Postcranial Axial Skeleton

Vertebral Column, Sacrum, Sternum, and Ribs

Along with the skull, the **vertebral column** (backbone), **sternum** (breastbone), and **ribs** constitute the **axial skeleton**. The latter three form the **thoracic cage** (**thorax**, **chest**), which is narrowest superiorly and broadens throughout most of its length. All 12 ribs articulate posteriorly with the thoracic vertebrae, but only the first 10 attach anteriorly either directly or indirectly via costal cartilages to the sternum. As in other hominoids (small- and large-bodied apes), but no other primates, the human thorax is not only laterally broad but also anteroposteriorly (dorsoventrally) compressed.

Vertebral Column: Overview

Morphology

The **vertebral column** (Figure 4–1), which extends from the base of the skull to the level of the pelvis, provides a somewhat rigid support for the upper body and protects the spinal cord it surrounds. Although there are details specific to each type of vertebra, most vertebrae (the coccygeal vertebrae excepted) have various features in common.

At birth, each "vertebra" consists of three components that eventually coalesce into individual **vertebrae**. In general, in humans, five vertebrae fuse to form the roughly triangular **sacrum** (sacral segments are identified as S1–S5). The upper portions of the sacrum's lateral sides—*auricular surfaces*—articulate with the posteriorly placed and internally facing auricular surfaces of right and left ilia. Extending downward from the inferior "apex" of the sacrum may be three to five but typically four rudimentary, sequentially tapering vertebrae; they constitute the **coccyx**, which may come to fuse with the sacrum. Sacral and coccygeal vertebrae are the *false* or *fixed vertebrae*.

The vertebrae between the skull and sacrum are the *true* or *movable vertebrae*. Three groups are recognized: the first seven are the **cervical vertebrae** (abbr. C), the next 12 are the **thoracic vertebrae** (T), and the remaining five are the **lumbar vertebrae** (L). One can identify a specific vertebra by its numerical position within its group (e.g., C1, T6, L3). All vertebrae possess a *body* or *centrum* and a *vertebral*

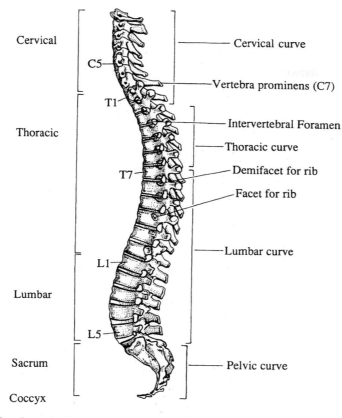

Cervical

C5

Thoracic

T1

T7

Lumbar

L1

L5

Sacrum

Coccyx

Cervical curve

Vertebra prominens (C7)

Intervertebral Foramen

Thoracic curve

Demifacet for rib

Facet for rib

Lumbar curve

Pelvic curve

Figure 4–1   Articulated vertebral column (*left lateral view*).

or *neural arch*. With the exception of C1–C2, elastic and fibrocartilaginous *intervertebral discs* (that do not survive burial) intervene between the bodies of all other neighboring vertebrae, including L5 and S1.

The posteriorly projecting vertebral arch delineates between it and the vertebral body the *vertebral foramen*, through which the spinal cord passes. The *spinous process* projects posteriorly from the midline of the vertebral arch. C1–C6 generally have short processes with bifid tips, although C2 may develop a slightly longer, nonbifid spine. C7–T12 have quite elongate spines. The noticeably protruding, virtually horizontally oriented C7 spinous process is identified as the *vertebra prominens*. The lumbar spinous processes are short and superoinferiorly tall. From T2 to T11, the

spines become increasingly deflected downward. In contrast, the spines of the lumbar vertebrae and, on occasion, those of the cervical region are essentially horizontal. Sacral spinous processes are usually vestigial. Coccygeal vertebrae lack them altogether.

In adult humans with normal vertebral articulation and positioning of intervertebral discs (normal posture), the vertebral column assumes a series of *spinal curvatures*. Proceeding inferiorly, the cervical vertebrae gently arc first inward (anterior, ventral) and then outward (posterior, dorsal). The posterior arc continues more aggressively in the thoracic vertebrae, peaking around T7, below which the vertebral arc curves inward. An exaggeration of the outward thoracic curvature (*kyphosis*) may result from degeneration of intervertebral discs (*primary kyphosis*) or vertebral collapse (*secondary kyphosis*) as a result, for example, of osteoporosis, osteomalacia, parathyroidism, or Paget's disease (see Chapter 1). The inward arc continues through L4. At L5 the curve again turns posteriorly, peaking in the midregion of the sacrum. Then, the curve reverses, sometimes so severely that the coccyx points straight forward. The sacral–coccygeal arc is the *pelvic curve*; the arc from T7 to the sacrum is the *lumbar curve* (an exaggerated lumbar curve produces *lordosis*). Because lumbar and pelvic curves are present in the newborn, they are referred to as *primary curvatures*. The curves from C1 to T2 and from T2 to T7 are, respectively, the *cervical* and *thoracic curves*; they are *secondary* or *compensatory curves* because they develop in tandem with increasing proficiency in bipedalism.

A vertebral body (**centrum**) is basically columnar, with roughened cranial and caudal tablelike surfaces whose perimeters are ringed by variably thin, slightly raised bands of smooth bone. The lateralmost borders of the cranial side of the bodies of C3–C7 are distended upward; this creates a saddle-shaped surface that cups the caudal side of the vertebral body above it. The lateralmost borders of the caudal side of the bodies of C2–C6 are truncated, and the band of smooth, circumferential bone appears "pushed up."

In general, vertebral body size increases in all dimensions as one proceeds inferiorly along the vertebral column. The outlines (viewed cranially or caudally) of the bodies of C3–T2 and T12–L5 are essentially similar: wider laterally than deep dorsoventrally, ventral surface arced somewhat convexly in its transverse plane but concave superoinferiorly, and dorsal surface ranging from relatively straight to gently concave (indented ventrally). The breadth of these bodies is reflected in the vertebral foramina being wider than anteroposteriorly deep and roughly triangular (the bases are delineated along the body's dorsal border, and the apices are located internal to the origins of the spinous processes). Pro-

ceeding from T3 to about T8 or T9, vertebral body shape changes to being anteroposteriorly deeper than wide; in addition, the ventral surface becomes more acutely convex and arced and the dorsal surface more noticeably concave or indented ventrally. The vertebral foramen thus becomes narrower and more circular in outline. From T8 or T9 through the lower thoracic vertebra, vertebral body shape becomes wider than deep and the vertebral foramen outline "reverts" from being circular to broad and shallowly triangular.

A vertebral body's sides are covered with compact bone, which may be perforated ventrally and ventrolaterally by a variable number of small nutrient foramina and dorsally by one or even several usually large and oddly shaped *vascular foramina* (through which the basivertebral veins course).

The bodies of thoracic vertebrae distinctively bear somewhat concave *facets for articulation with ribs*, which lie bilaterally at the dorsalmost extent of the body (just in front of the pedicle). A facet may be complete (representing total contact between a rib and a vertebra) or partial (*demifacet*, indicating a rib's contact with two adjacent vertebrae). A complete facet is variably ovoid in outline.

The body of T1 bears on each side a complete facet just below the edge of the cranial surface and, inferiorly, a demifacet that is confluent with the edge of its caudal surface (Figure 4–2). The body of T9 bears bilaterally only a single, but often large, demifacet that descends from the edge of its cranial surface; sometimes it may appear to be a complete facet that is separated from the cranial edge. T10–T12 are characterized by their single complete facets, which lie bilaterally just below the edge of the cranial surface; facet size and definition varies most in T12. The bodies of T2–T8 bear two demifacets bilaterally: one appears to be "dripping" from the edge of the cranial surface onto the side and the other, often smaller facet looks as if it was "pushed up" from the edge of the caudal surface.

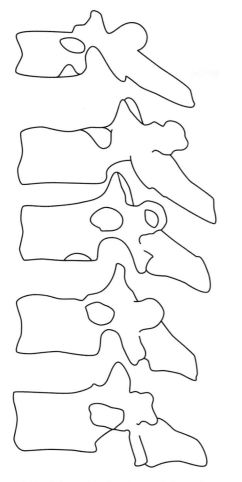

Figure 4–2 Schematic drawings of thoracic vertebrae to illustrate differences in facets/demifacets for ribs as well as in orientation and shape of spine: (*from top to bottom*) T1, T9, T10, T11, and T12.

Development and Ossification

The vertebral column develops from the mesenchymal segments—*sclerotomes*—that differentiate around the notochord and neural tube (see Figure 1–9). There is not, however, a correlation between one vertebra and one sclerotome. Rather, each sclerotome eventually divides in half. Coalescence of adjacent sclerotomal halves produces a vertebra. *Myotomes* —which give rise to associated musculature and initially matched the sclerotomes one-to-one—do not divide. Therefore, myotomes and "reorganized" sclerotomes are arranged half a segment out of sync with one another so that one myotome serves two vertebrae.

Vertebral anlagen differentiate craniocaudally, reaching the caudal end during the second fetal month. At this time, the cartilaginous vertebrae are complete in the sense that the vertebral arch and body have fused together and the various processes are identifiable. The distinctive upward projection from the body of C2 results from the fusion with it of the body of C1 (which therefore lacks a body). *Spina bifida*, which affects the lumbar region most and the cervical region least frequently, results from incomplete fusion of right and left sides of the vertebral arch. *Spondylolisthesis* results from the displacement of a vertebra due to incomplete coalescence of the vertebral arch with the body.

True Vertebrae

Morphology

Overall morphological similarity makes it possible to discuss true vertebrae, C3–L5, collectively (Figures 4–1 to 4–3). Each has a *body* (*centrum*), a *vertebral arch* that circumscribes a *vertebral foramen*, and a *spinous process* that emanates from the midline dorsally of the vertebral arch.

Approximately level with the posterior extent of the vertebral foramen (depending on the specific vertebra), and on each side of it, are the platform-like surfaces or processes at which neighboring vertebrae articulate with one another: a pair of *superior articular processes* and a pair of *inferior articular processes*, whose surfaces are aligned roughly parallel to each other. The superior articular processes of C3 are oriented slightly dorsally (tilted

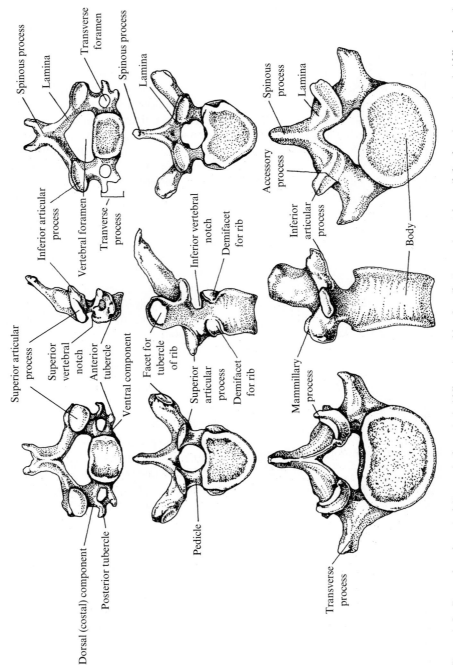

Figure 4–3 Examples of cervical (*top row*), thoracic (*middle row*), and lumbar (*bottom row*) vertebrae: (*left column*) superior, (*middle column*) left lateral, and (*right column*) inferior views.

up somewhat), while the inferior articular processes are oriented slightly ventrally (tilted down somewhat). The orientation of these articular processes becomes increasingly vertical inferiorly such that by T1 and definitely by T2 the superior articular processes face directly dorsally and the inferior articular processes, directly ventrally. At and below T12, the processes remain vertical, but the orientations of the articular surfaces change. With regard to T12, the articular surfaces of the inferior articular processes are rotated laterally almost 90°; in some cases, the surface of the articular surface is rounded or rolled. In concert with the inferior articular processes of T12, the superior articular processes of L1 are rotated medially and may be concave. The medial rotation of the superior articular processes and the lateral rotation of the inferior articular processes (with interlocking surfaces mirroring each other's contours) are maintained from L1 to L5. The articular processes of S1 that receive the inferior articular processes of L5 are also rotated medially, and their inner surfaces reflect the specific configuration of L5's processes. The posterior (dorsal) margins of the superior articular processes of L1–L5 distinctively bear variably thickened swellings (*mammillary processes*).

In C3–L5, the part of the vertebral arch on each side that proceeds dorsally to the base of the spinous process is the *lamina*. The portion of the vertebral arch on each side that proceeds ventrally from the articular process to the body of the vertebra is the *pedicle* or *root of the vertebral arch*. Each pedicle is constricted such that, when adjacent vertebrae are in articulation, the inferiorly "pinched" border of one pedicle and the superiorly "pinched" border of the neighboring pedicle create a laterally directed *intervertebral foramen* through which spinal nerves and vessels pass. On cervical vertebrae, the pedicle's superior border is at least as deeply notched as the inferior border (the *superior vertebral notch* is at least as deeply excavated as the *inferior vertebral notch*). In T1–L5, however, the inferior verte-

bral notch is greatly enlarged, which produces larger intervertebral foramina.

A *transverse process* projects laterally from the general region of each pedicle on C3–C6, jutting out level with the vertebral body. On C7, however, and thereafter, transverse processes emerge at the juncture of the pedicle and lamina (the root of the transverse process lies between superior and inferior articular processes). A specific feature of C1–C7 is a variably large, superoinferiorly and mediolaterally oriented *transverse foramen (foramen transversarium)* that perforates each transverse process. In C1–C6, the vertebral artery and vein and sympathetic nerves course through the transverse foramina. The lateral ends of the transverse foramina of C3–C6 bear two tubercles.

The *anterior tubercle of the transverse process* projects upward and slightly laterally away from the slip of bone that subtends the transverse foramen on its ventral side. This slip of bone is called the *ventral* or *costal component* because it (plus its tubercle) is a vestige of the *costal process* or *element* associated with a thoracic vertebra and from which a rib does develop. The spit of bone—*dorsal component*—that subtends the other side of the foramen actually borders only the dorsolateral aspect of the transverse foramen, taking origin from the columnar stretch of bone connecting the superior and inferior articular processes. The dorsal component is oriented laterally and ventrally and terminates in a noticeable projection, the *posterior tubercle of the transverse process*. This tubercle, which tends to point directly laterally, extends farther outward than the anterior tubercle. C7 typically possesses a ventral or costal component that closes off the transverse foramen, but it usually lacks an anterior tubercle. C7 also tends to develop a stout dorsal component that extends prominently and laterally as a true transverse process.

In summary, the dorsal component and its posterior tubercle together constitute the homologue of the transverse process of a thoracic vertebra. *Variations* in the expression of transverse foramina are common and take different forms, such as varying degrees of closure of the foramen by the costal component, asymmetry in foramen size between right and left sides of the same vertebra, and multiple or accessory foramina bi- or unilaterally.

The transverse processes of T1–T10, and occasionally T11, are distinguished by their length, shape (thick with a swollen, knoblike end), and orientation (dorsolaterally and even slightly upward), all of which give the vertebra the appearance of an odd-looking bird in flight. In T1–T10, the ventral side of the swollen end of each transverse process bears (or may even appear to "cup") a variably ovoid *facet for articulation with the tubercle of the rib.* The transverse processes of T12 and often T11 are not as prominent, projecting, or terminally knoblike; their typically knobby surfaces lack articular facets for ribs. The transverse processes of T12 are short, folded dorsally back, and often distended into three recognizable tubercles: the *superior* and *inferior tubercles* dorsally and the variably low-peaked and swollen, laterally oriented *lateral tubercle* just in front of the inferior tubercle.

Lumbar transverse processes are usually fairly horizontal and gently angled dorsally. At the root of each process, just to the side of the inferior vertebral notch, is a roughened, variably small, and elevated *accessory process.* Although they may project laterally quite prominently, the processes of especially L1–L4 are typically thinner, more compressed dorsoventrally, and generally more gracile than on a thoracic vertebra. The transverse processes of L5, which are often the thickest, may be the longest and may appear twisted, with their somewhat flat dorsal surfaces oriented toward the sacrum. There may also at times be signs of articulation between the transverse processes of L5 and the sacrum. If an L5 actually fuses with a sacrum, the transverse processes of the "new" last lumbar vertebra (L4) may develop some of the features of an L5.

Development and Ossification

Ossification of the true vertebrae (Figures 4–4 to 4–6) begins in the cervical region as early as the seventh and as late as the tenth week when a center appears in each side of the vertebral arch, near the bilateral junctures between the arch and cartilaginous vertebral body. Ossification of neural arches proceeds along the presumptive vertebral column in a caudal direction as ossification within individual vertebral arches spreading out laterally and ventrally (anteriorly) into processes and dorsally (posteriorly) toward the spine. The vertebral arches do not fully unite posteriorly until the first postnatal year as the process begins in the lumbar region and proceeds cranially.

Between puberty and 16 years of age, epiphyseal *secondary centers of ossification*

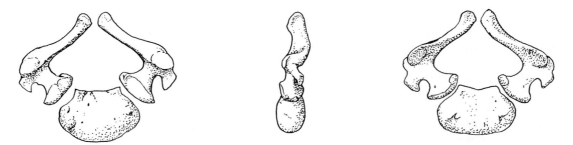

Figure 4–4 Vertebral arch components and body of typical cervical vertebra (C5) of third-trimester fetus: (*left*) superior, (*middle*) left lateral, and (*right*) inferior views.

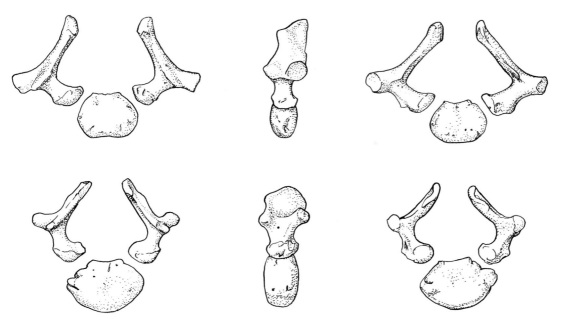

Figure 4–5   Examples of upper (T1) and lower (T11) thoracic vertebral arch components and bodies of third-trimester fetus: (*left*) superior, (*middle*) left lateral, and (*right*) inferior views.

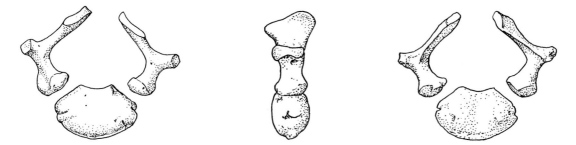

Figure 4–6   Vertebral arch components and body of typical lumbar vertebra (L2) of third-trimester fetus: (*left*) superior, (*middle*) left lateral, and (*right*) inferior views.

emerge in the tips of transverse and spinous processes. Usually by 20 years of age, processes and vertebral arches are completely united.

Within a week of the onset of ossification in the vertebral arches, ossification of the vertebral bodies begins, with centers appearing first in the lower thoracic and upper lumbar regions and then spreading cranially and caudally. Because the dorsolateral "corners" of a vertebral body are actually derived from

its vertebral arch, an unfused vertebral body looks somewhat scallop-shaped, with the narrow end pinched between the two ends of the arch and the rest of the body fanning out anteriorly. The cartilaginous junctures between the end of a vertebral arch and a body are *neurocentral synchondroses*; their ossification begins during the third year in the upper vertebrae (either cervical or thoracic) and extends to the lower lumbar region

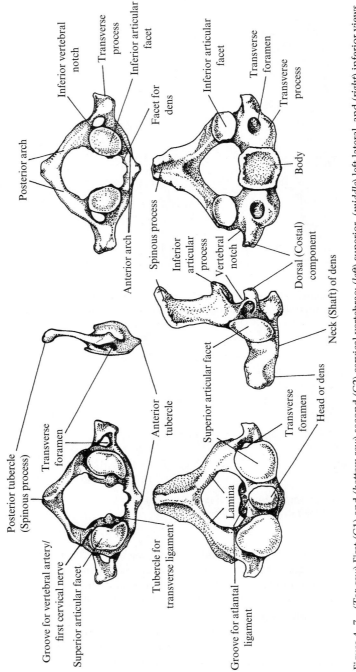

Figure 4–7 (*Top row*) First (C1) and (*bottom row*) second (C2) cervical vertebrae: (*left*) superior, (*middle*) left lateral, and (*right*) inferior views.

by the sixth year. Within the next 2 years, the lumbar accessory and mammillary processes ossify and the articular processes, which were originally oriented as in thoracic vertebrae, rotate medially away from the sagittal plane such that the superior articular processes face medially and the inferior ones laterally.

On the basis of their similar ontogenies, it appears that the transverse process of a lumbar vertebra is homologous with the costal process of a thoracic vertebra (which may thus be conceived of as a vestigial rib fused to the body of the vertebra; see Ribs, below). Morphologically, the thinner transverse process of a lumbar vertebra is more riblike than it is thick; it is also not terminally swollen, as a thoracic transverse process is. However, rather than the homologue of a thoracic transverse process being inhibited altogether in lumbar vertebrae, the former may be represented by a lumbar "superior articular process" that coalesced with the true superior articular process. This is further indicated by the fact that the distinctive orientations of the inferior articular processes of T12 and all lumbar articular processes are achieved during childhood.

During the sixteenth or seventeenth year, secondary centers appear in the cranial and caudal "surfaces" of the cartilaginous, disclike vertebral bodies. Continued ossification from these centers produces thin bony plates that fuse to the bodies of the vertebrae between the twentieth and twenty-fifth years.

First and Second Cervical Vertebrae

Morphology

[The distinctive morphologies of C1 and C2 (Figure 4–7) warrant separate discussion.]

Because the **first cervical vertebra** or **atlas** (named after the Greek god who held the earth on his shoulders) lacks a body, it is essentially an elliptical ring of bone that surrounds a large *vertebral foramen*. It bears (1) small, laterally projecting *transverse processes*;

(2) large, often teardrop-shaped *superior articular facets* that gently cups the skull's occipital condyles (Figures 2–4 and 2–5); and (3) a reduced to vestigial and vaguely bifid spinous process (*posterior tubercle*). The strut of bone in the region where a vertebral body might have been is the *anterior arch*, which bears, in the midline of its ventral surface, a variably swollen to peaked *anterior tubercle* and, on its dorsal surface (facing the vertebral foramen), a smooth, variably ovoid facet for articulation with C2's vertical process or *dens* (the "lost" vertebral body of C1). C1's vertebral or *posterior arch* broadens on each side, becoming superoinferiorly flattened toward the superior articular facets and transverse processes. The *groove for the vertebral artery* and *first cervical nerve* courses across the posterior arch's expanded cranial surface. The somewhat concave and obliquely inclined *superior articular facets* slope from the ventral border toward the vertebral foramen. The typically narrower, ventral end of each facet lies more medially than the swollen dorsal end; however, the dorsal end rises more noticeably above the posterior arch. *Variation* is sometimes seen in the delineation of the facet into two subequal moieties: as a crease coursing across the facet, with slight to marked pinching of the facet, or as two distinct facets. Such variation may mirror patterns of variation seen in the occipital condyles and may be expressed bi- or unilaterally, but variation may be differently expressed on each facet.

C1 transverse processes emerge from below the swollen ends of the superior articular facets; each is perforated by a relatively large *transverse foramen* and, unlike other cervical vertebrae, is not bifid. A small, swollen, facetlike *tubercle for the transverse ligament* lies on the other side of each superior articular facet, along the inner surface of the vertebral foramen and toward the ventral end of the facet; it is usually offset anteriorly and posteriorly by small depressions of variably porous bone. *Inferior articular surfaces* may be roughly ovoid, elliptical, or circular in outline. If there

is a narrower end, it faces the anterior tubercle. The anterior end of the articular surface
is set off from the anterior arch by a thin
rim. From here, the essentially planar or only
slightly concave articular surface becomes
increasingly elevated toward its posterior
margin, which projects over the transverse
foramen. The variably broad and shallow *inferior vertebral notch* lies at the juncture of the
posterior arch with the base of the inferior
articular surface; it courses laterally away
from the vertebral foramen. The inferior
articular surface, like the superior one, is
aligned obliquely along its long axis; it is also
oriented such that the side that borders the
vertebral foramen is much lower than the
opposite side. Since C1's inferior articular
surfaces articulate with C2's superior articular
surfaces, the shapes and orientations of these
opposing facets mirror one another. The
superior and inferior articular surfaces and
transverse process are referred to collectively
as the *lateral mass*.

The **second cervical vertebra** (**axis,
epistropheus**) is distinguished by the almost
phallic projection (*dens*) that rises out of its
vertebral *body*, the result of fusion with the
presumptive vertebral body of C1. The dens
has a *neck* or shaft on top of which sits a
blunt-tipped, superiorly tapering head that
swells out at its base beyond the width of the
shaft. It protrudes up into the vertebral foramen of C1—where C1's body would have
been—above the level of the anterior arch of
C1 and articulates against the inner surface
of C1's anterior arch via a smooth, roughly
ovoid articular facet on its ventral surface, just
above the midpoint of the shaft. A blunt midline ridge courses down the dorsal surface of
the dens' tip, terminating in a basal swelling,
below which lies the long and shallow *groove
for the transverse atlantal ligament* (which
stretches between the tubercles for the transverse ligament, essentially tethering C2 to C1).

C2's large *superior articular surfaces*, which
permit head rotation, mirror the morphology
and orientation of C1's. C2's *inferior articular*

surfaces sit on *inferior articular processes*; they
are much smaller than the superior articular surfaces and steeply inclined, with their
articular facets facing essentially ventrally.
Unlike in other vertebrae, C2's inferior articular surfaces do not lie immediately below
the superior articular surfaces; they lie more
posteriorly. C3's superior articular processes
and surfaces mirror the features of C2's inferior articular processes. The thick *costal* or
dorsal component of C2's *transverse process* is
deflected caudally such that it terminates
below the anterior half of the superior articular surface; it is separated posteriorly from
the inferior articular process by a variably
deep *vertebral notch*. The more or less laterally
directed *transverse foramen* lies directly above
or directly above and slightly anterior to this
notch. Caudally, the transverse foramina lie
well in front of the inferior articular processes
and face downward.

Although similar in shape to the arches
of lower cervical vertebrae, C2's vertebral arch
is thick and more rugose in all dimensions.
Cranially, C2's *lamina* is compressed into a
ridge that circumscribes most of the vertebral
foramen; its sides are deep. The *spinous process*
is thickened and roughened at its end, which
is quite variably bifid; more anteriorly, the
spine's sides are deeply indented. The spinous
process broadens and flares out caudally, and
its inferior surface is somewhat grooved or
roughly concave (as if to envelope and embrace
the spine of the vertebra below it). C2's spine
projects dorsally beyond C3's.

Development and Ossification

Ossification of C1's posterior arch and lateral
masses (Figure 4–8) begins as early as the
seventh and as late as the tenth fetal week.
The region of the posterior tubercle is still
cartilaginous at birth. The posterior arch does
not become continuously ossified until the
third to fourth years; a separate center of ossification may appear in its still-cartilaginous
region. The anterior arch, which arises from

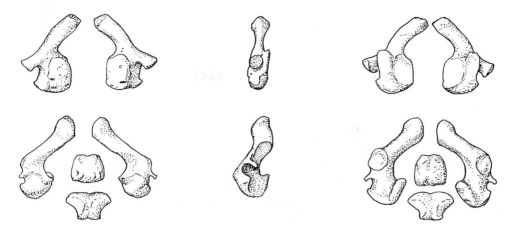

Figure 4–8   (*Top row*) First (C1) and (*bottom row*) second (C2) cervical vertebrae of third-trimester fetus: (*left*) superior, (*middle*) left lateral, and (*right*) inferior views.

the hypochordal plate (notochord), does not begin to ossify from its one or sometimes two centers until during or as late as the end of the first year. The anterior arch fuses with the lateral masses during the fifth to ninth years.

As in other vertebrae, the body and sides of C2's vertebral arch each ossify from a single center: that for the body appears at about the fourth fetal month and those for the halves of the vertebral arch during the seventh or eighth fetal week. The arch fuses with the body between the third and sixth postnatal years. During the fourth to fifth (or even sixth) fetal months, two centers appear at the base of the dens and soon coalesce, leaving a cleft at the top in which the still cartilaginous tip of the dens nestles. The ossified dens does not fuse with the body until the fourth to sixth years. Although the cartilaginous tip of the dens ossifies during the second to third years, it does not fuse with the rest of the dens until about the twelfth year. (The dens' "tip" corresponds to the pro-atlas or last occipital vertebra of primitive vertebrates.) Noncoalescence of this part of the dens produces an "atavistic" (or "hypostotic") variant. The caudal surface of C2's body bears an epiphyseal plate that ossifies around the seventeenth year.

Sacrum

Morphology

The adult **sacrum** (Figure 4–9) normally consists of five fused vertebral segments that form a solid, superoinferiorly arcuate (dorsally convex and ventrally concave), inferiorly (caudally) tapering (from side to side), and dorsoventrally thinning unit. In general, a female's sacrum is shorter, broader, and less severely curved caudally into the pelvic canal than a male's, which is a longer, more tapering triangular unit, whose caudal end may protrude into the pelvic canal more prominently. Vertebral morphology is suggested by the variably protrusive series of midline *spinous processes*, which become less prominent caudally and collectively form the *median sacral crest*. Vertebral morphology in the sacrum is particularly obvious superiorly in the large, lumbarlike *body* and medially directed *articular processes* of S1, which articulates with L5. The *promontory*, which is the extended and liplike ventral margin of S1's body, protrudes into the region of the pelvic canal. The fused spinous processes and associated *lamina* of the sacral segments form a continuous but increasingly caudally

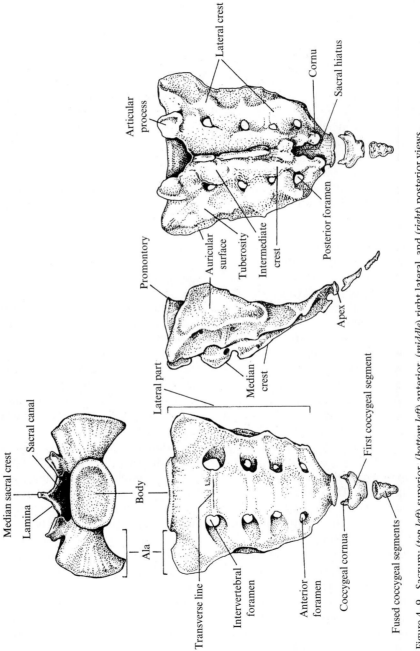

Figure 4–9   Sacrum: (*top left*) superior, (*bottom left*) anterior, (*middle*) right lateral, and (*right*) posterior views.

constricted *sacral canal* over the bodies of the fused segments.

Winglike expanses of bone extend bilaterally from S1's body. Because of their appearance, they are called *alae* (s. *ala*, means "winglike process"). A female's alae are typically quite large, distended, and markedly flared laterally and, thus, tend to appear large relative to S1's articular surface for L5 (or, vice versa, the articular surface appears smaller relative to the alae). A male's alae are usually smaller than, and thus often appear small relative to, S1's (relatively) larger articular surface. An ala is highest on its dorsal side; its superior (cranial) surface slopes increasingly caudally in the ventral direction. Sometimes it appears to bear a thickened, dorsally angled strut lying in front of and in parallel with the articular process. This strut represents a *transverse process* coalesced to an expanded *costal process*. The processes may not always be distinguishable, but the bone of the region of the transverse process seems to be rather consistently more roughened.

Caudally, S1's alae and body are fused to the respective parts of the segment below it, a pattern that continues throughout the sacrum. On each side, the fused alae form the *lateral parts of the sacrum*. Pairs of large, somewhat funnel-shaped *pelvic sacral foramina (anterior sacral foramina)* that diminish in size caudally occur between adjacent alae, just laterally to the lines of fusion (*transverse lines*) between segmental bodies. Ventrally, the foramina penetrate the bone lateromedially to become confluent with the sacral canal and at those places form *intervertebral foramina*, through which sacral nerves, arteries, and veins course. From the dorsal side, the corresponding pairs of *posterior sacral foramina*, which are slightly smaller than their ventral counterparts, open onto the intervertebral formina.

From the cranial edges of its alae and caudally for some part of its length, the sacrum is sandwiched between the posterior portions of the right and left ilia, whose granular auricular surfaces abut similarly granular *auricular sur-*faces on the sacrum. Ilial and sacral auricular surface shapes mirror each other. Although details of shape may vary, there is always an indentation dorsally and a protrusion ventrally (thus it is reminiscent of an ear, i.e., "auricular"). The male auricular surface typically extends from the edge of the ala to the region of S3; in females, it is usually restricted to S1–S2.

The sacrum's concave, relatively smooth-boned ventral or *pelvic surface* contrasts markedly with its roughened, morphologically highlighted, convex *dorsal surface*. In addition to features already mentioned, the dorsal surface bears (lateral to the series of posterior sacral foramina) a pair of wavy, at times peaked, thick or compressed, and rugose or smooth *lateral sacral crests*. These crests terminate caudally in variably distended swellings that define the sacrum's *inferior lateral angle*, which tends to be less pronounced in males than in females, in whom it adds to sacral breadth. The peaked elevations along each crest are the *sacral tuberosities*; there may be as many as five and as few as three of these, and their expression is not always bilaterally symmetrical. The portion of the sacrum that lies between these tuberosities and the dorsal margin of the auricular surface is typically rough-boned and pocked with variably large, craterlike pits.

On each side, medial to at least some of the posterior sacral foramina but at the same level as them, lie a variable number of smaller, less protrusive tuberosities (unnamed), which together constitute the *intermediate crest*. The most consistent feature of these crests is that they terminate caudally in variably minimally to prominently projecting *sacral cornua* (s. *cornu*). These cornua subtend the variably enlarged caudal opening of the sacral canal (*sacral hiatus*). The sacrum's ventral or pelvic surface extends caudally beyond the mouth of the *sacral hiatus* and terminates in the *apex*, which articulates with the coccyx. The morphologically detailed dorsal plate of bone that encloses the sacral canal may separate

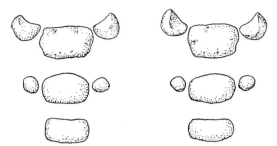

Figure 4–10 Sacral elements of third-trimester fetus: (*left*) superior and (*right*) inferior views of each segment.

from the body of the sacrum after burial, but the smooth surface of the sacral canal, the spinous dorsal surface, and the series of sacral foramina are good clues to its identity.

Development and Ossification

Information on sacral ossification is not consistent (Figure 4–10). For example, Clemente (1984) claims that ossification of S1's body begins at about the eighth to ninth fetal week, Netter and Crelin (1987) state 10 fetal weeks, but Fazekas and Kósa's (1978) analyses yield an age of 6 fetal months. Netter and Crelin put the onset of ossification of the vertebral arches at 10 weeks, but Clemente illustrates 6–8 months and Fazekas and Kósa 6–7 months. According to Netter and Crelin, ossification of the lateral part (costal element or process) begins during the sixth fetal month, but Clemente cites the sixth to eighth fetal months, while Fazekas and Kósa suggest birth or later.

There is more consensus on the fusion of vertebral arches and bodies, which begins caudally during the second year and reaches the upper segments during the fifth to sixth years. Coalescence of right and left vertebral arches occurs during the seventh to fifteenth years, with the sacral hiatus resulting from the lack of fusion in the lower segments. The costal processes fuse with the bulk of the sacrum at or about puberty. Between puberty

and 16 years, the cranial and caudal epiphyseal plates ossify and fuse with the sacral vertebral bodies. The lateral surfaces of the sacrum, which have remained cartilaginous, ossify during the eighteenth to twentieth years. At approximately the same time, ossification begins in the intervertebral discs between the caudalmost segments and proceeds cranially, reaching the uppermost disc between 25 and 30 years of age. The pelvic curve, which is a primary curvature of the vertebral column, is present at birth; but the sacral vertebrae only broaden and thicken in tandem with the development of bipedality.

Coccyx

Morphology

The *coccyx* (Figure 4–9) may be composed of three to five but typically four rudimentary, sequentially tapering vertebrae or segments, which, with the frequent exception of the first segment, typically coalesce into a single unit; fusion of the first segment is more typical of females than males. Only coccygeal segments 1–3 retain features that can be identified as vertebral—that is, vestigial bodies, transverse processes, and articular processes. As the segments taper caudally, their discernible morphology diminishes, such that the last segment is often only very rudimentary and variably pea-shaped.

The first coccygeal segment presents the most prominent transverse processes of the series. It also bears on either side of the surface that articulates with the sacrum's apex two hornlike projections (*coccygeal cornua*), which result from fusion of the presumptive articular, transverse, and costal processes. Each cornu points up toward, but is separate from, the neighboring sacral cornu.

Development and Ossification

Coccygeal vertebrae ossify from single centers that begin to appear after birth, in the first year in the first segment and variably between the fourth to tenth years in the second segment. The cornua of the first segment ossify from separate centers. The third segment ossifies between the tenth and fifteenth years and the fourth and frequently last segment, between the fourteenth and twentieth years (typically, the third segment ossifies prior to, and the fourth segment after, puberty). Coalescence of segments occurs between 25 and 30 years and proceeds caudocranially. The female coccyx may fuse to the sacrum.

Sternum

Morphology

The generally broader laterally than deep dorsoventrally adult **sternum** (Figure 4–11) is composed of three units: the *manubrium* superiorly, the *xiphoid process* inferiorly, and the *body* in between. Because the sternum is largely composed of spongy bone encased in a thin layer of compact bone, it feels unexpectedly light. In general outline, it looks spikelike: the typically pointed xiphoid process is the sharp end and the manubrium the broader, flatter head. The dorsal (posterior) surfaces of the manubrium and body tend to be slightly concave or depressed; the ventral surface is typically slightly swollen or convex, whereas the body's ventral surface is variably flat, shallowly depressed, or a combination of both. The sternal body's ventral surface is often elevated in the region of the *sternal angle*, which lies at the juncture of the manubrium and body. Even though it may be very shallow, this angle is discernible when viewed from the side. The xiphoid process is distinguished by often being narrower and rarely as thick as the inferior end of the body of the sternum, with which it articulates.

An isolated manubrium looks like a stubby "Y." The very shallow and broad superior *jugular* or *suprasternal notch* lies between the arms of the "Y." The ends of the Y's arms are subdivided along their lateral borders into a larger superior and a smaller inferior notch. The obliquely, superiorly, and laterally facing surface of the larger *clavicular notch* is typically composed of smooth articular bone; it receives the clavicle's sternal end. The costal cartilage of the first rib nestles into the more notchlike and essentially outwardly oriented

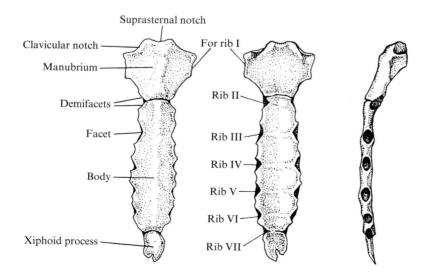

Figure 4–11   Sternum: (*left*) anterior, (*middle*) posterior, and (*right*) left lateral views.

smaller notch. The typically smooth-boned broad stem of the "Y" abuts the sternal body. In the rare case of a very old individual appearing to have fusion of the manubrium and sternal body, coalescence is usually only along the perimeter of these bones' contact while, interiorly, a cartilaginous disc persists between them. The lateral "corners" of the manubrium's base and of the superior end of the sternal body are truncated and indented into *demifacets*, which together on each side form a complete notchlike facet for attachment of the costal cartilage of the second rib. The line of contact between the manubrium and body delineates the midline of adjacent demifacets.

Proceeding inferiorly but at irregular intervals along the sternal body's lateral borders are four complete indentations or *facets*, which receive the costal cartilages of ribs III–VI. Variably discernible lines of union between ontogenetically once-separate sternal segments course across the ventral surface of the sternal body, between the midpoints of each right and left pair of costal cartilage facets. The sternal body's dorsal surface typically bears vertical lines or striations; its inferior end, which typically lies just below the facet for the sixth costal cartilage, is truncated and indented by a demifacet whose sister demifacet may be visible on the xiphoid process. Together these demifacets form the facet for the seventh costal cartilage. Sometimes, however, one may find only the one demifacet on the inferior end of the sternal body.

Development and Ossification

At or about the sixth intrauterine week, the sternum (Figure 4–12) emerges as two separate mesenchymal bars. Shortly thereafter, a suprasternal mesenchymal mass arises above each sternal mesenchymal bar; each suprasternal mass represents that portion of the future manubrium which will articulate with the clavicle. The presumptive upper ribs begin to coalesce with the sternal bars,

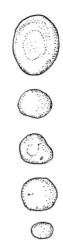

Figure 4–12 Sternal elements (anterior view) of third-trimester fetus.

and as they do, the sternal bars begin to fuse together (like a zipper). By 9 fetal weeks or so, the sternal bars have coalesced completely, the suprasternal masses have fused with the presumptive sternum, and the entire "unit" (ribs included) has become a cartilaginous cage. Incomplete union of the sternal bars anywhere along the sternum can result in a cleft or *sternal foramen* (a gap in the "zipper"). Incomplete fusion of the two inferior ends of the sternal bars results in a bifid xiphoid process.

The ribs, which ossify earlier than the sternum, cause the cartilaginous sternum to become segmented into (typically) six *sternebrae*. One *sternebra* corresponds to the manubrium and another to the xiphoid process, while the rest make up the sternal body. Ossification centers appear superiorly, in the manubrial sternebra, during the third to sixth fetal months. The second sternebra begins to ossify at the same time or at least by the seventh month, by which time all but the last two sternebrae have also begun to ossify. The last sternebra of the sternal body begins to ossify during the first postnatal year or so. The sternebra of the xiphoid process ossifies between the fifth and eighteenth years. Coalescence of the sternal body segments

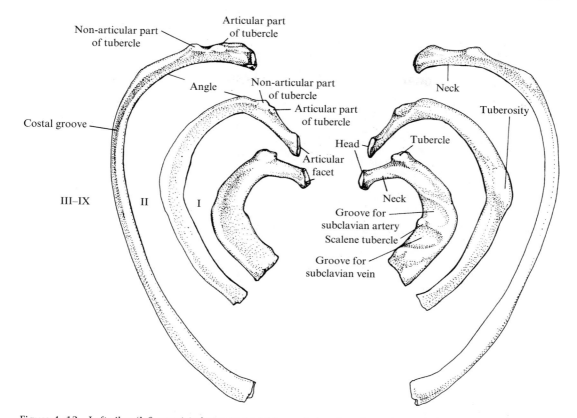

Figure 4–13   Left ribs: (*left group*) inferior and (*right group*) superior views; (*outermost*) example of ribs III–IX, (*middle*) rib II, and (*innermost*) rib I.

begins after puberty in the lowest portion and proceeds superiorly; the uppermost sternebrae fuse at or about the twenty-fifth year.

Ribs

Morphology

Ribs (Figures 4–13 and 4–14) are the most numerous elements of the thoracic cage. There are normally 12 pairs. All ribs attach posteriorly to the thoracic vertebrae, but only 10 pairs are anchored, directly or indirectly, to the sternum (via the costal cartilages). The seven pairs of facets on the sternum (the first of which lie on the manubrium) directly receive the costal cartilages of the first seven pairs of **true** or **vertebrosternal ribs**. The eighth rib attaches via its costal cartilage to the costal cartilage of the seventh rib; the ninth and tenth ribs attach via their costal cartilages to the costal cartilages of, respectively, the eighth and ninth ribs. Ribs VIII–X are the **vertebrochondral ribs**. The eleventh and twelfth, or **false**, **free**, or **floating**, ribs are unattached anteriorly. In addition to their relative slenderness, inferior declination away from their vertebral articulations (easily reconstructable), and other peculiar morphologies, the anterior ends of floating ribs taper, or at least do not flare out, as do ribs bearing thick costal cartilage.

All ribs present an identifiable *head* and *body*; ribs I–X also develop a *neck*. The head

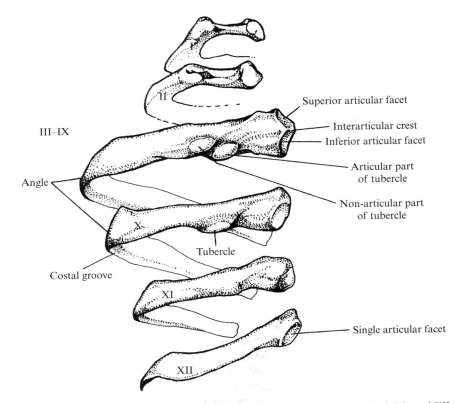

Figure 4–14   Posterior view of left ribs: (*from top to bottom*) I, II, example of III–IX, X, XI, and XII.

articulates directly with a vertebra or vertebrae. With the exception of the first rib's more rounded head, a rib head typically tapers or thins toward its end and is beveled, with the articular surface essentially facing medioventrally. The number of *articular facets* a rib head bears is correlated with the number of vertebrae with which it articulates: ribs I–XII articulate with a single vertebra and bear a single articular facet (as do T1 and T9–T12); other ribs articulate with two adjacent vertebrae, and their heads bear a somewhat cranially directed *superior articular facet* and a slightly caudally directed *inferior articular facet*, which are separated by a peaked, horizontally oriented *interarticular crest*. The neck is slenderer than the head. It is relatively straight and flattened dorsoventrally on most ribs; on ribs II and III, however, the flattened sides are twisted in a more craniocaudal direction. The dorsal surface of the neck of ribs III–XII is variably roughened, the internal surface of which is the smoothest stretch of bone on a rib. The caudal or inferior surface of the neck of the first two ribs is rougher-boned than the superior or cranial surface.

The terminus of the neck is delineated where a rib swells out laterally into a posterior *tubercle*. The first rib's tubercle bears an *articular part* that contacts T1's transverse process. On ribs II–X, an *articular part of the tubercle* (which contacts the transverse process of its associated thoracic vertebra) lies medial to a *nonarticular part of the tubercle*; these parts become increasingly separated from one another through the series. Floating ribs typically do not develop these tubercles; if present, they are poorly defined.

Either coincident with or a centimeter or so laterally beyond the tubercle, the rib's body begins to curve—sometimes subtly, sometimes abruptly. This region is the *angle*. On the first few ribs, the angle occurs at the tubercle. On most other ribs, the straightness of the neck extends beyond the tubercle; a low mound of roughened bone, from which the body then thickens, indicates the angle externally. Fractures can easily occur medial to the angle and isolate the occasionally amorphous tubercle/neck/head portion from the body. One might discern a slight angle on rib XI but never on rib XII.

Holding a rib in its anatomical position allows one to see the body arcing laterally. This arc is typically more severe near the rib's vertebral end and becomes gentler toward its anterior end (a rib's arc resembles the outline of half a heart). Ribs I and II are the most tightly and the floating ribs the most gently arced. When the ribs are in articulation with vertebrae, their dorsal (posterior) ends are higher than their ventral (anterior) ends; the anterior portion of the body thus assumes a downward twist, which is particularly noticeable on ribs III–XII. Except for rib XII, a rib's *superior border* is blunt and relatively thickened. For approximately half to two-thirds of its length (starting from the tubercle or angle), the body's *inferior border* is relatively sharp-edged and impressed internally by a *costal groove*, in which the intercostal vessels and nerves course. Most rib body fragments can be allocated at least to one side by determining the direction of the arc and then orienting the downward twist, inferior border, and costal groove properly. The floating ribs, especially the twelfth, are distinguished by their minimal morphology, while ribs I and II are individually identifiable because of their peculiar morphology.

The superior surface of rib I is highlighted along its dorsal half by two shallow, variably broad depressions or grooves that are separated by a variably triangular, slightly raised area of bone that emanates from the inner

Table 4–1   Clues to Siding Isolated Ribs

Feature	Position
Angle (I–X)	Posterior
Costal groove or sulcus (II–XI)	Inferior and internal
Groove for subclavian vein (I)	Superoanterior
Head	Medioposterior
Inferior border (sharp) (II–XII)	Inferior
Scalene tubercle (I)	Superomedial
Subclavian groove (I)	Superior
Superior border (blunt) (II–XII)	Superior
Tubercle (I–X)	Posterosuperior (I–II), posterior (III), posteroinferior (IV–X)
Tuberosity for serratus anterior (II)	Superolateral

margin of the bone (the *scalene tubercle*, where the anterior scalene muscle attaches). The *subclavian groove*, which lies closer to the middle of the rib's body, is often the smaller of the two depressions; the subclavian artery and part of the brachial plexus course across it. This groove may extend over and indent the rib's inner margin. The more sternally placed *groove for the subclavian vein* is typically the deeper of the two. Almost halfway around the curve of rib II, the outer margin of the superior or cranial surface is thickened and somewhat roughened into the *tuberosity (for the serratus anterior) of the second rib*.

See Table 4–1 for a summary of clues to siding isolated ribs.

Development and Ossification

The costal process of a thoracic vertebra, the anterior tubercle and ventral (costal) component of the transverse process of a cervical vertebra, and the transverse process of a lumbar vertebra are homologous structures. In thoracic vertebrae, the costal processes do not fuse to the vertebrae but remain separate and develop into ribs (Figure 4–15). Mesenchyme intervenes between the chondrifying rib and

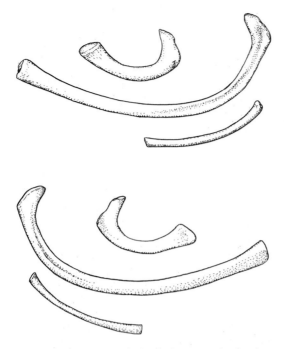

Figure 4–15 Ribs of third-trimester fetus: (*top group*) superior and (*bottom group*) inferior views of ribs I (*uppermost*), VII (*middle*), and XII (*lowermost*).

separate until later on. As early as the seventh but certainly by the eighth or ninth intrauterine week, the cartilaginous rib cage is defined; ossification begins with the appearance of a center in each of the angles of ribs V–VII (rib ossification begins slightly earlier than in vertebrae). Ossification centers appear rapidly thereafter in other ribs. Within each rib, ossification proceeds ventrally and dorsally away from the angle but does not continue throughout the costal cartilage, whose ventral end then remains as the costal cartilage; in old age, calcification may encroach on these cartilages. Ossification of the head and tubercle does not begin until 16–17 or even as late as 20 years; they fuse with the rest of the rib between the twentieth and twenty-fifth years. Ribs with articular and nonarticular components of the tubercle develop two centers of ossification in these areas.

Rickets may be indicated if the ventral ends of the ribs are overly enlarged and bear (at least on their inner surfaces) globular swellings of bone. Diagnosis of *scoliosis* (in which the vertebral column is deflected away from the vertical for some of its length) should be considered if "expected" rib curvature is noticeably deformed, with the posterior and anterior rib ends from one side of the body oriented more posteriorly and almost in parallel with one another, while ribs from the other side of the body are distorted on both ends in an anterior direction. The former set of ribs would be from the convex side of the lateral scoliotic curvature and the latter, from the concave side.

adjacent vertebrae and eventually develops into an articular cavity. Cervical ribs derive from separate centers of ossification, not from lack of fusion of the anterior tubercle and ventral component to the rest of the vertebra.

The presumptive ribs first become cartilaginous at their vertebral ends, with chondrification continuing into the presumptive sternum, from which the ribs do not become

Chapter 5

The Upper Limb

The upper limb is composed of the scapula (shoulder blade), which, with the clavicle (collar bone), forms the shoulder girdle; the humerus (upper arm bone), which articulates with the scapula; the radius and ulna, which together constitute the forearm; and the hand, which consists of the carpals (wrist bones), metacarpals (bones of the palm), and phalanges (bones of the fingers or digits of the hand).

Scapula

Morphology

The **scapula** (Figure 5–1) is a thin, internally mildly concave, and externally gently convex bone. It is fairly triangular in outline and adorned externally with a spine and laterally with various bony projections. The scapula's variably rugose and thickened long side (*medial* or *vertebral border* or *margin*) faces and lies roughly parallel with the vertebral column; it can be variably straight, undulating, gently arced inward, or slightly convex. The thickened and strutlike *axial* or *lateral border* or *margin* extends from the often-flattened *inferior angle* and slopes upward and laterally. Although the axial border is anything but straight, it always bears a slight inward dip (*groove for the circumflex scapular vessels*) approximately two-thirds along its length. Near its lateral- and uppermost extremity, the lateral border thickens into the *infraglenoid tubercle*, on which the long head of the triceps brachii muscle attaches. The infraglenoid tubercle is so named because it lies below the *glenoid process* or *head* of the scapula, which is the terminus of the lateral border. The humerus articulates in the gently depressed *glenoid fossa (cavity)* of the scapular head. From the scapula's thin *superior angle*, the equally thin *superior border* or *margin* descends steeply to a variably acute to obtuse *scapular* or *suprascapular notch*, which lies at the base of the projecting *coracoid process*. This process arises from behind the scapula's head and comes to overlie and then project beyond it.

The *scapular spine* emerges on the *dorsal* or *posterior side*, approximately one-fifth of the length of the medial border from the superior angle. As the spine

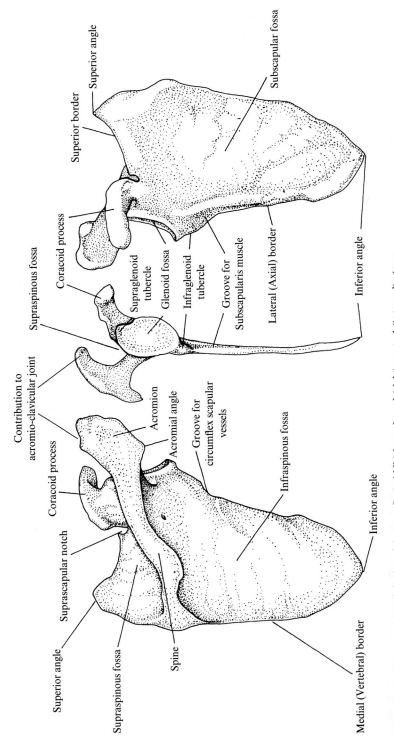

Figure 5–1 Right scapula: (*left*) dorsal (external), (*middle*) lateral, and (*right*) ventral (internal) views.

proceeds toward the scapular head, it increases in height and robustness laterally and somewhat superiorly, eventually broadening and projecting laterally beyond the glenoid process. The spine divides the scapula's dorsal surface into two areas of unequal size: the small superior *supraspinous* or *supraspinatus fossa* and the much larger *infraspinous* or *infraspinatus fossa*. Particularly in the region of the infraspinous fossa, the bone is quite thin and often translucent. With increasing age, the scapula becomes thinner, creating large patches of translucent bone within the two fossae.

The juncture between the spine and the medial border is flattened, relatively smooth, and somewhat triangular. As the spine becomes increasingly elevated, its surface becomes rougher and its course more sinuous; approximately one-third of its length away from the medial border, it typically becomes distended inferiorly. Just behind the scapular head, the spine is quite elevated dorsally. It continues over and laterally beyond the head as a spit of bone (*acromion* or *acromial process*) that broadens and becomes rather golf club-shaped, with the "heel" of the club inferiorly placed and the thick "toe" angled superolaterally. The platelike and slightly arced acromion, which contributes to an arch that protects the shoulder joint, is gently concave internally; as a whole, it is directed inwardly, or anteriorly, from its "heel" (*acromial angle*) to its "toe" (*acromioclavicular joint*). The internal surface of the acromion is somewhat smoother than its outer surface.

The scapula's *lateral border* bears a dorsally displaced edge that courses inferiorly from the infraglenoid tubercle; this edge may lose its discreteness approximately one-third to one-half the distance toward the inferior angle. The edge of this border subtends internally (ventrally) a shallow *groove for the attachment* (laterally) *of the subscapularis muscle*. The lateral border separates the subscapularis muscle from the dorsal muscles (the teres major below and the teres minor above).

The scapula's *head* or *glenoid process* is a rimmed, subovoid expansion of the lateral border of the bone superiorly. In outline, it is somewhat teardrop-shaped: narrow at the top, broad and round inferiorly, and arcuately expanded dorsally. The internal or ventral border of the glenoid process is expanded less fully, at least superiorly. The ventral border is often gently notched approximately one-third down from the top, beyond which the surface does swell out inferiorly. The *articular surface* or *glenoid fossa* is mildly concave and, in profile, may also have a slight inward curvature (from top to bottom).

The thick scapular *neck* lies behind the largely liplike rim of the head. Superiorly, the head's rim is less protrusive but is locally thickened into the *supraglenoid tubercle*, at or just behind which the deep base of the *coracoid process* is rooted. The medial edge of the platelike base of the coracoid process contributes to the formation of the *suprascapular notch*. Atop this base the coracoid process extends well above and laterally beyond the scapular head as a somewhat flattened, fingerlike projection, whose superior surface is quite roughened and often irregular; the inferior surface tends to be smooth and downwardly flexed. The base of the coracoid process typically arcs inward (like a "cocked hat" on the scapular head); because it may be skewed in that direction, the fingerlike projection may carry this orientation even farther.

The scapula's *internal* or *costal surface* is a fairly uninterrupted, slightly concave surface (*subscapular fossa*). A variable number of low, roughened ridges emanate from the medial border and proceed inward. The strutlike inner border of the groove subtended by the edge of the lateral border is the dominant feature of this side of the scapula.

Variation is noted not only in the variable delineation of the suprascapular notch but also in the presence or absence and number of *foramina* below and above the scapular spine. A single foramen below the spine and on the infraspinous portion is typical. Care should

Table 5–1   Clues to Siding the Scapula:
Its Features and Their Orientations

Feature	Position
Acromial angle	Posterolateral
Acromion or acromial process	Laterosuperior
Coracoid process	Superolateral
Glenoid fossa	Laterosuperior
Infraglenoid tubercle	Lateral
Infraspinous fossa	Posteroinferior
Scapular spine	Posterior
Supraglenoid tubercle	Laterosuperior
Suprascapular notch	Superior
Supraspinous fossa	Posterosuperior

be taken when encountering thin, platelike fragments of bone not to confuse scapula with os coxa (specifically ilium).

See Table 5–1 for clues to siding scapulae.

Development and Ossification

Scapular shape usually becomes recognizable by the end of the third or the beginning of the fourth fetal month (Figure 5–2). Mineralization up to this point is fairly rapid because the first center—near the scapular neck (lateral angle)—which gives rise to the body of the scapula, does not appear until the end of the second fetal month or perhaps even a few weeks later. At or about this time, the spine begins to differentiate and another center emerges near the base of the coracoid process that will contribute to the formation of the superior portion of the glenoid cavity.

Although the acromial end of the scapular spine is at first barely distinguishable from the glenoid cavity, its rate of growth is much faster than in the glenoid and neck regions, and it rapidly comes to extend laterally beyond the neck. The acromion and coracoid process are connected to the presumptive clavicle via the presumptive coracoclavicular ligament.

At birth, mineralization is restricted to the scapula's body and spine. The glenoid cavity, acromial and coracoid processes, and vertebral (medial) border are still cartilaginous. Each process ossifies from two centers, whereas the other cartilaginous regions ossify separately. The first postnatal center appears in the projecting portion (not the base) of the coracoid process. Fusion of the coracoid process to its base does not usually occur until 15 years or so; fusion of the coracoid's base to the scapular body does not usually occur until 17 or 18 years, when the glenoid fossa also ossifies and fuses to the scapular body. Ossification of the more medial acromial center occurs around 15 years, of the lateral acromial center and that of the inferior angle about a year later, and of the vertebral (medial) border coincident with the base of the coracoid process. Complete union of the acromion and inferior angle occurs at approximately 22 years and that of the vertebral (medial) angle a year or so later. There may, however, be a great deal of variation in the ages at which various epiphyseal unions occur.

Clavicle

Morphology

The elongate and shallow "S"-shaped **clavicle** (Figure 5–3) articulates between the scapula and the sternum. Viewed from above, it arcs gently outward (anteriorly) as it courses laterally from its *sternal end*, approximately one-third the distance from which it begins to curve inward (posteriorly) for slightly more than one-third of its length, and then it swings abruptly outward again, terminating in its

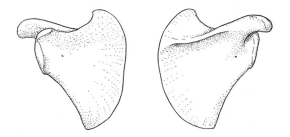

Figure 5–2   Right scapula of third-trimester fetus: (*left*) ventral (internal) and (*right*) dorsal (external) views.

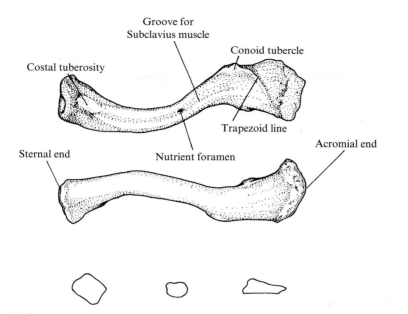

Figure 5–3   Right clavicle: (*top*) inferior and (*middle*) superior views, (*bottom*) cross sections through shaft.

acromial end. Although muscles do attach to portions of the *superior surface* of the clavicle (leaving roughened areas, particularly noticeable on the acromial end), the *inferior surface* is often the more scarred and even grooved surface.

Usually identifiable on the clavicle's inferior surface is a shallow *groove for the subclavius muscle* (along the second third of the bone), a *conoid tubercle* (where the acromial end begins to swing outward), and a *trapezoid line* (coursing from the conoid tubercle across the inferior surface to the lateral margin of the acromial end). Approximately halfway along the bone, in or sometimes above the groove for the subclavius muscle lies a small, slitlike *nutrient foramen* that pierces the shaft horizontally in a mediolateral direction; sometimes there is an additional nutrient foramen that may be well separated from the first. Another typically shallow groove lies along the anterior third of the bone inferiorly and toward its anterior margin; it bears a roughened surface for attachment of the pectoralis major muscle.

A triangular depression left by the costoclavicular ligament may also be visible just on the inside of this groove and more toward the sternal end.

The clavicle's *sternal end* differs from its *acromial end* in being bulkier, somewhat ovoid in outline, and more expanded. The acromial end is much flatter, wider anteroposteriorly, and compressed superoinferiorly. Regardless of its actual shape and outline (which can be quite variable), the sternal end appears to flare out from the shaft of the bone, particularly along its posterior side. Proceeding laterally from the sternal end, the bone's shaft is subtriangular in cross section (with the "triangle's" base facing upward and its apex inferiorly) as far as the conoid process, after which the superior and inferior surfaces flatten and widen in tandem. The lateral edge of the acromial end often bears a flattened area of contact with the scapula's acromion, and the sternal end's inferior border bears an area of contact (which may be indented) with the cartilage of the first rib

Table 5–2   Clues to Siding the Clavicle:
Its Features and Their Orientations

Feature	Position
Acromial end (flattened)	Lateral
Conoid tubercle	Inferoposterior
Costal tuberosity	Inferoposterior
Groove for subclavius muscle	Inferior
Nutrient foramen (mediolateral)	Posteroinferior
Sternal end (rounded)	Medial
Trapezoid line	Inferior (lateral)

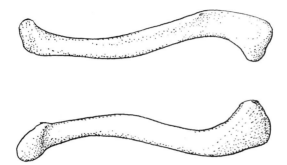

Figure 5–4   Right clavicle of third-trimester fetus: (*top*) inferior and (*bottom*) superior views.

(the first costal cartilage). The *costal tuberosity* is a raised area in the region of contact with the first rib's cartilage.

See Table 5–2 for clues to siding clavicles.

Development and Ossification

The clavicle is the first bone to begin to ossify and one of the last, if not the last, to become fully ossified (Figure 5–4). Starting at the acromial end, the presumptive mesenchymal clavicle elongates until, at about the sixth fetal week, it contacts and coalesces with the first rib. The clavicular shaft arises from two *primary centers*, which appear within the latter half of the second gestational month. By the end of the third fetal month, the body is ossified and its shape established. A secondary center of ossification does not appear in the cartilaginous sternal end until 18–20 years. The sternal end typically unites with the body of the clavicle around the twenty-fifth year, but union can occur as early as the twentieth or as late as the thirtieth year.

Humerus

Morphology

The **humerus** (Figure 5–5), the bone of the upper arm, is much longer and more robust than either of the two long lower arm bones (radius and ulna). The humerus' proximal and distal ends are morphologically quite different—the former is roughly globular, whereas the latter is wide from side to side and compressed from front to back. A long, relatively straight shaft separates them.

The anterior side of the humerus' *proximal end* bears two swellings (*greater and lesser tuberosities* or *tubercles*), which are separated from one another by a moderate to deep *bicipital* or *intertubercular groove* that courses only a short distance down the shaft just lateral to the midline of the bone. The *lesser tubercle* is the smaller of the two; it lies medial to the groove and is essentially confined to the bone's anterior surface. This tubercle is drawn out into an oblique crest, whose surface may be roughened; the face of the tubercle below this crest is longer and more vertically oriented than the shorter, more horizontal superior portion. The *greater tubercle* lies lateral to the intertubercular groove and expands out and around the lateral side of the bone proximally, encroaching upon the bone's posterior surface and underlying to some extent the humeral head. It is situated more proximally than the lesser tubercle and is separated from the head inferiorly by a shallow and broad depression. The greater tubercle bears a crest or thickened edge that distinguishes its long vertical side from a much shorter and horizontal superior plane. This crest continues for a short distance up the oblique surface established by the lesser tubercle but then

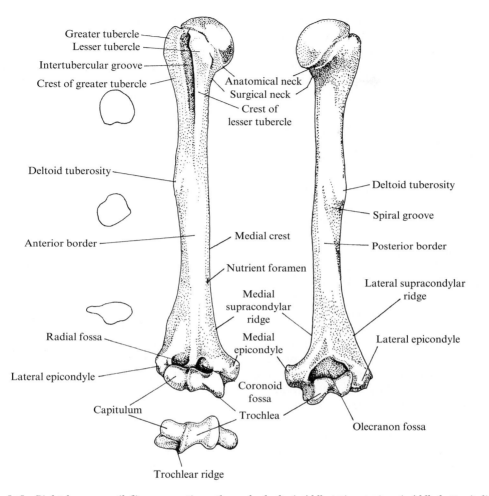

Figure 5–5   Right humerus: (*left*) cross sections through shaft; (*middle top*) anterior, (*middle bottom*) distal, and (*right*) posterior views.

turns downward to traverse the lateral aspect of the greater tubercle. A line drawn between the crests of the two tubercles would traverse an arc that roughly parallels the curvature of the spherical humeral head.

The large, relatively smooth, and posteromedially oriented humeral *head* represents only part of a sphere. Envisioning the humeral head as being sliced (well off center) from a sphere, the plane of this cut (which corresponds to the epiphyseal line and the *anatomical neck* it delineates) is obliquely oriented, with its upper margin lying anterior to the

midline axis and its inferior margin jutting out from the posteromedial side of the bone; this plane parallels the orientation of the crest on the lesser tubercle. The uppermost portion of the humeral head rises slightly above the level of the greater tubercle; the head's inferiormost portion descends below both tubercles. Just inferior to the humeral head's inferiormost extent, the bone's shaft narrows at the region of the *surgical neck*.

A crest of variable length and rugosity descends from each tubercle along the humeral *shaft*; a continuation of the intertubercular groove intervenes between these crests. The *crest of the lesser tubercle* tends to be shorter and less pronounced than the *crest of the greater tubercle*. The sometimes inferiorly grooved crest of the lesser tubercle is a blunt, truncated edge upon the shaft's medial surface. Below the inferior terminus of this crest, the more crisply defined and straighter *medial crest* proceeds down the medial side of the shaft, coursing to, if not even surmounting, the *medial epicondyle*. Not quite halfway up the bone from the distal end, a small, slitlike, inferiorly directed *nutrient foramen* that lies on or near the medial crest penetrates the shaft obliquely.

The usually roughened crest of the greater tubercle defines the stout, thick humeral *anterior border*, which extends almost halfway down the shaft. Toward the lower portion of this crest, which may be bounded inferiorly and laterally by a groove, the bone swells out into the *deltoid tuberosity*. The configuration of crest and tubercle together makes the shaft appear "twisted." The crest and tubercle also create a shallow and broad depression on the shaft that, from the posterior side, curves up and around the bone below the deltoid tuberosity. This is the *spiral groove* (for the radial nerve), which may be delineated further by a thin crest proceeding down the posterior side of the shaft. Occasionally, there is a tiny, inferiorly directed *nutrient foramen* penetrating the bone obliquely on or near this short posterior crest. The thin *posterior border* extends laterally from below the region of the spiral groove, forming an edge that continues inferiorly as the *lateral supracondylar ridge* to the *lateral epicondyle*. Coincident with or just below the emergence inferiorly of the lateral and medial supracondylar ridges, the humerus flattens posteriorly and widens markedly bilaterally.

The humerus' *distal end* bears a prominent, medially projecting *medial epicondyle*; a more truncated, flangelike, and laterally projecting *lateral epicondyle*; and a notably rounded, anteriorly tripartite *articular surface* that is delineated superiorly by two fossae in front and one large fossa behind.

(In visualizing the humerus' distal end, think of it as a "fist" in which the tripartite articular surface represents the "knuckles" and the medial epicondyle is a "thumb" sticking out; the large fossa on the posterior side of the distal end corresponds to the cupped "palm." The hand, of course, would be from the same side of the body as the bone itself.)

The large, medially projecting *medial epicondyle* is compressed from front to back and, in concert with the rest of the bone inferiorly, flatter posteriorly. (The unprotected ulnar nerve courses across the posterior surface of the medial epicondyle and produces the "funny bone" when hit inadvertently.) The *medial supracondylar ridge* is an extension of the medial crest; it may proceed far enough distally to surmount the medial epicondyle. In profile, the end of the medial epicondyle may be rounded, straight with "corners," or somewhat peaked. *Variation* is noted in the occasional development of a small *supracondyloid process* superior to the medial epicondyle.

The asymmetrical *articular region* of the humerus' distal end is expanded medially into a wheellike rim (*trochlea*) but is rounded and caplike laterally (forming the *capitulum*). The ulna articulates around the trochlea (in a parasagittal plane), and the radius articulates around and distally on the capitulum (parasagittally as well as rotationally or pivotally). The trochlea, which is typically more expansive and inferiorly distended than the capitulum and fully developed on all sides of the distal end of the bone, forms a fairly vertical wall against which the base of the medial epicondyle abuts. The capitulum, which becomes increasingly truncated lateromedially as it courses around the distal end, lacks a posterior component. The trochlea's lateral face and the capitulum's medial face slope inward, toward the midline of the articular region.

Anteriorly, a low, broad, vertical *trochlear ridge* wraps around the confluence of the trochlea and capitulum; it is more smoothly continuous with the trochlea than with the capitulum, from which it is often delineated by a narrow groove. Posteriorly, the trochlear ridge is visible only at the epiphyses' inferiormost margin. Viewed from behind, the "spindle-shaped" trochlea is smoothly and deeply concave with raised, ridgelike lateral and medial borders; the lateral border is elongate superiorly and the medial border (the edge of the trochlea's "wheel-like" end) is distended inferiorly. Posteriorly, the superior margin of the trochlea is inferiorly curved, descending from the higher lateral border and then back and up slightly to the top of the medial border. The superior margin of the trochlea bounds a large *olecranon fossa*, in which the olecranon process of the ulna nestles when the arm is extended. A nonmetric *variant* is sometimes noted in the perforation of the wall of bone in the olecranon fossa by a *septal aperture*, which may be as small as a pinhole or almost as large as the fossa itself. Septal aperture development seems to be more prevalent in females than in males and to occur on the left side more frequently than on the right.

Two small fossae, one above the trochlea and the other above the capitulum and separated by a ridge of bone, lie on the anterior side of the humerus' distal end. One is the *coronoid fossa* (because the coronoid process of the ulna encroaches upon it when the lower arm is flexed), which is bounded medially by a ridge of bone. When present, a septal aperture communicates between the olecranon fossa and the coronoid fossa.

The fossa superior to the capitulum is called the *radial fossa* because it receives the head of the radius when the lower arm is flexed. This fossa, which tends to be shallower and broader than the coronoid fossa, is bounded by, but may also extend onto, the lateral epicondyle.

The *lateral epicondyle* presents itself as a truncated, flangelike adornment on the lateral side of the humerus' distal end. It is

Table 5–3 Clues to Siding a Complete or Fragmentary Humerus: Its Features and Their Orientations

Feature	Position
Anatomical neck	Circumferential
Capitulum	Lateral (distal)
Coronoid fossa	Anterior
Crest of greater tubercle	Anterior
Crest of lesser tubercle	Medial
Deltoid tuberosity	Lateral (shaft)
Greater tubercle	Lateral (proximal)
Head	Medioposterior
Intertubercular (bicipital) groove	Anterolateral
Lateral epicondyle (smaller)	Lateral
Lateral supracondylar ridge	Lateral
Lesser tubercle	Anterior
Medial epicondyle (larger)	Medial
Medial supracondylar ridge	Medial
Nutrient foramen (superoinferior)	Medial
Olecranon fossa	Posterior
Radial fossa	Anterolateral
Spiral or radial groove	Posterolateral
Surgical neck	Circumferential
Trochlea	Anterior and posterior
Ulnar groove	Medial

traversed, at least in part, by the lateral border or crest, the inferior portion of which is the *lateral supracondylar ridge*. From the side, the lateral epicondyle appears to be bent forward, away from the lateral border of the trochlea, and tucked into the overlying capitulum. Posteriorly, the lateral epicondyle is relatively smooth. Laterally, nearer the capitulum, the lateral epicondyle's surface is more irregular and surmounted by at least one short, obliquely oriented crest or ridge.

See Table 5–3 for clues to siding humeri.

Development and Ossification

The humeral shaft begins to ossify from a single center by the end of the second fetal month (Figure 5–6). (The initially shorter

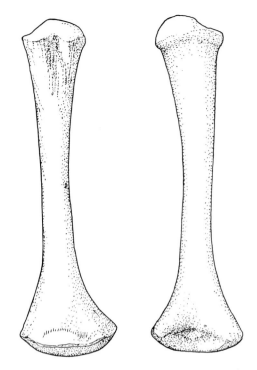

Figure 5–6   Right humerus of third-trimester fetus: (*left*) anterior and (*right*) posterior views.

femur also begins to ossify at this time; eventually, femoral growth speeds up so that, by the middle of the third month, the two bones are equal in length but, afterward, femoral length increases more rapidly.) The humeral shaft is almost completely ossified by birth, but the only other area of humeral ossification is a secondary center in the proximal epiphysis of the head. In most individuals, the humeral head begins to ossify during the first year of life. In general, the other cartilaginous parts of the humeral ends begin to ossify in the following sequence: capitulum (by the end of the second year), greater tubercle (by the third year), medial epicondyle (around the fifth year), lesser tubercle (by the fifth year), trochlea (during the tenth year), and lateral epicondyle (between the twelfth and thirteenth years). The head and tubercles begin to coalesce during year 5 and, within year 6, join to form a larger epiphysis, which

typically unites with the shaft during the twentieth year; complete union, however, may occur as late as the twenty-fifth year. Between the sixteenth and seventeenth years, the trochlea, capitulum, and lateral epicondyle begin to unite with one another and with the shaft; complete union may not occur until a year later. In general, then, the humerus' proximal end completes ossification somewhat later than its distal end. As a rule, the elbow joint (proximal radius and ulna and distal humerus) is completely ossified earlier than the shoulder (proximal humerus and scapula) and wrist joints (at least distal radius and ulna).

Radius

Morphology

The **radius** (Figure 5–7) articulates superiorly with the distal humerus' capitulum and laterally with the ulna; distally, the radius contributes to the wrist joint. The radius' *head* is subcircular in outline and presents itself as an upwardly directed, shallow, cup-shaped structure with a moderately thick rim circumscribing the cavity. The capitulum sits in this depression. A thick, bandlike *articular circumference*, which wraps around the perimeter of the radial head, rotates in the ulna's radial notch. In the anatomical position, the articular circumference is thinnest laterally and posteriorly and thickest anteromedially. Since the radial head is broader, especially laterally, than the shaft, the bone's *neck* is easily delineated.

A few centimeters below the radial neck and in the same anteromedial orientation as the thickest section of the articular circumference, a large ovoid to elliptical *radial tuberosity* swells markedly outward. The biceps muscle inserts on this surface, which is variably granular, pitted, grooved, and irregularly contoured overall. The radial tuberosity's anterior boundary tends to be less well defined than its posterior border, which is more edgelike. A low ridge courses inferiorly from the radial tuberosity, arcs toward the lateral side of

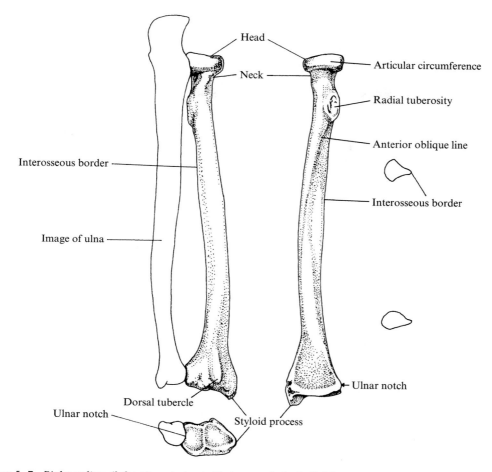

Figure 5–7   Right radius: (*left top*) posterior (with image of ulna), (*left bottom*) inferior, and (*middle*) anterior views; (*right*) cross sections through shaft.

the anterior surface (where it is identified as the *anterior oblique line*), and then extends the length of the shaft, veering outward with the lateral expansion of the bone's distal end. The anterior oblique line may become more crisply defined distally. Approximately at the level at which this low ridge assumes a more lateral position (a few centimeters below the radial tuberosity), the otherwise rather circular radial shaft becomes severely compressed medially into a sharp, crestlike border whose upper extent is variably distended and roughened. If present, a *nutrient foramen* will lie in the region between this variably distended and roughened section of the medial

crest and the oblique line opposite it; it will be minute and slitlike and penetrate the shaft obliquely in a superior direction. From this roughened medial region, inferiorly, the shaft arcs gently lateromedially.

The medial crest represents the bone's *interosseous border*. It extends as a sharp edge for much of the shaft's length, bifurcating inferiorly near the *ulnar notch*. For its last distal fifth or so, the radial shaft markedly broadens bilaterally toward its end. The bone of the anterior surface of this thin, subtriangular region is relatively smooth and flat. Near the epiphyseal region, this anterodistal plane arcs upward (becomes further elevated in the anterior dimension), especially in its mediodistal corner; this anterior expansion enlarges the articular surface of the distal epiphysis from front to back. Also on the anterior side distally, the mediodistal "corner," which is more rounded or "peaked" medially than squared up, projects medially to contribute to the side of the shallow *ulnar notch*. On the distolateral side, the thin crest described above courses down to the level of the mediodistal corner and, typically, swings or angles inward toward the medial side, sometimes continuing across the entire end of the bone. Just on the inside of this thin crest, the distally tapering *styloid process* arises, increasing the radius' length on its lateral side. The styloid process' anterior surface is usually slightly concave, smooth, and covered with articular-looking bone, which may be highlighted on its sides by thinly raised edges.

The radius' *posterior surface* is more rounded and swollen than its anterior side; it bears a variably low to ridgelike keel that arcs laterally as it descends partway down the shaft from its origin just above the superiorly roughened portion of the medial, interosseous border. In its most pronounced state, this keel delineates a shallow and moderately broad groove between it and the interosseous border. Otherwise, a thinner shallow groove courses just inside and along the interosseous medial edge. Sometimes the keel may extend weakly to the bone's distal end, where it becomes confluent with the ever-present *dorsal tubercle*, which is typically a prominently raised and vertically oriented feature at the midpoint of the distal end of the bone.

The dorsal tubercle may be subtended on its medial side by a groove of variable depth that is often bordered medially by a noticeable ridge; the dorsal tubercle thus appears "twinned." The slightly concave area between the tubercle and associated groove (and ridge, if present) eventually courses up to and terminates in the bone's variably right-angled to rounded mediodistal corner. Lateral to the dorsal tubercle, the bone's mildly concave

distal edge becomes distended into the *styloid process*, whose posterior surface may be variably irregular and faintly grooved. The groove or grooves medial to the dorsal tubercle are associated with various extensor muscles for the thumb (pollex) and other digits of the hand; muscles that extend the wrist lie lateral to the dorsal tubercle.

Typically, the *distal articular surface* is smoothly concave and roughly wedge-shaped in outline. The narrower, variably pointy to blunt end of the distal articular surface corresponds to the distension of the styloid process, medial to which the surface expands to the level of the dorsal tubercle. At this point, the anterior edge is indented slightly. The posterior edge (just medial to the dorsal tubercle) may also be indented such that it appears "pinched" or "waisted." A shallow but very faint groove or depression may course across the surface between the two indentations, creating the impression that one large facet is subdivided into two unequal portions, *demifacets*. The lateral demifacet articulates above and to the side of the scaphoid (of the proximal row of carpal or wrist bones). The smaller medial demifacet articulates with part of the lunate (another proximal-row carpal). Medial to the anterior indentations, the radius' distal articular surface arcuately expands posteriorly, terminating at the corner of the variably concave medial edge. The posterior edge proceeds rather directly to the medial edge. The articular surface does not terminate at the medial edge but wraps around and up into the shallow ulnar notch, in which the ulna's distal end sits.

See Table 5–4 for clues to siding radii.

Development and Ossification

A center of ossification appears in the radius' cartilaginous diaphysis (Figure 5–8) during the eighth fetal week. The shaft is well ossified at birth, but the proximal and distal ends are still cartilaginous. During the first, but sometimes not until the second year, the distal end

Table 5–4   Clues to Siding a Complete or Fragmentary Radius: Its Features and Their Orientations

Feature	Position
Anterior oblique line	Anterior
Dorsal tubercle	Posterior
Interosseous border	Medial
Nutrient foramen (inferosuperior)	Anterior
Radial or bicipital tuberosity	Anteromedial (proximal)
Rough area for pronator teres	Lateral (shaft)
Styloid process	Lateral (distal)
Ulnar notch	Medial

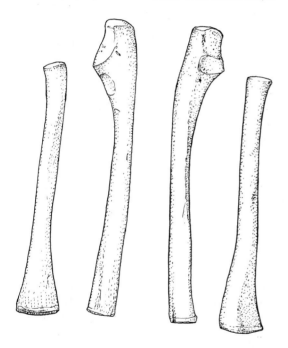

Figure 5–8   Radius and ulna of third-trimester fetus: (*far left*) anterior and (*far right*) posterior views of radius; (*middle left*) medial and (*middle right*) lateral views of ulna.

begins to ossify, but the proximal epiphysis does not show signs until the fifth year. The latter epiphysis begins to coalesce with the shaft coincident with puberty and may be completely fused between the ages of 15 and 18 or as late as 19 years. The distal end typically fuses with the diaphysis during years 17–20 (it may commence and complete union a year earlier in females than in males), although complete fusion may not occur until the twenty-third year. The radial tuberosity may sometimes arise during the fourteenth or fifteenth year from a separate center of ossification.

Ulna

Morphology

Whereas the radius is narrow on top and broad at its distal end, the ulna's shape is essentially the reverse (Figure 5–9). Its proximal end is enlarged, and the relatively thin shaft tapers to a small, globular distal end that is adorned posteriorly by a modestly pointed process. Because the ulna lies medial to the radius, features reflecting this association lie on the ulna's lateral side (on the same side of the bone as the side of the body from which it comes).

Anteriorly, the ulna's crescentic *proximal end* bears the *trochlear notch*, which cups the humerus' trochlea. The former rotates around the latter in a longitudinal or parasagittal plane. The posterior surface of the ulna's proximal end is essentially vertical and flat. When the lower arm is extended, the upper "horn" of the crescentic trochlear notch, which protrudes anteriorly as a ledgelike, relatively broad structure—the *olecranon process*—nestles into the olecranon fossa (on the posterior side of the distal humerus). The olecranon process' variably roughened superior surface is irregularly ovoid to subtriangular in outline. Its anterior rim slants downward toward the medial side and is variably distended by the terminus of a longitudinal, very low, ridgelike swelling that traverses the inner surface of the trochlear notch to its liplike, markedly protruding inferior horn, the *coronoid process*, which projects much farther anteriorly than the olecranon process. When the lower arm is flexed, the coronoid process lies in the coronoid fossa.

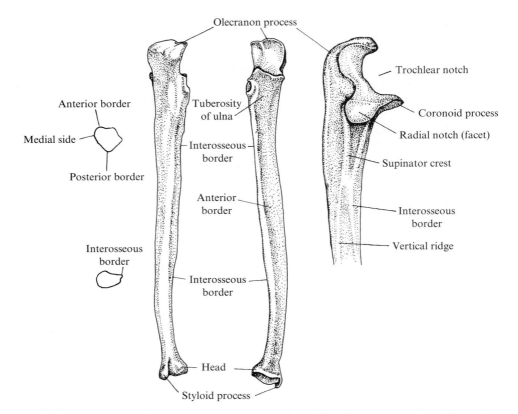

Figure 5–9   Right ulna: (*far left*) cross sections through shaft; (*middle left*) posterior and (*middle right*) anterior views; (*far right*) detail of proximal portion, lateral view.

The trochlear notch is subdivided unequally by a longitudinal ridge, from which, on each side, the articular surface slopes back. The bulk of the coronoid process lies on the medial side of this ridge and extends even farther medially as a ledgelike projection. Not too far lateral to this ridge, the coronoid process is severely truncated at an oblique angle and its facetlike articular surface appears creased and appressed to the side of the bone. This variably ovoid to elliptically shaped, modestly concave "facet" (*radial notch*) is pointier anteriorly; the articular circumference of the radial head rotates in it.

Variation in the proximal ulna is noted, for instance, in the degree to which the articular surface of the trochlear notch is indented by nonarticular bone where the coronoid process begins to expand medially; also from this point, a crease or groove may course across the articular surface, dividing it into upper and lower sections. Other variations are seen in the degree to which the posterior and inferior edges of the radial notch facet are distinct from the surrounding bone and in the presence of a secondary facet on the posterior side of this facet.

Two ridges, which subtend between them a variably excavated and roughened trough or groove, extend downward a short distance from the coronoid process' medial and lateral corners; the entire roughened area is the *tuberosity of the ulna*. The medial ridge is accentuated by a depression behind it and to some extent by the coronoid process' medial lip. The lateral ridge is enhanced by a longer depression that descends from the inferior edge of the radial notch facet. If the two ridges converge and enclose between them the trough and ulnar tuberosity, the rounded medial border of the shaft's anterior surface—the *anterior margin* or *border*—proceeds inferiorly from their point of confluence. But when the lateral ridge is fairly vertically oriented or fails to contact the medial ridge, the latter alone constitutes the anterior margin. In either case, the anterior margin becomes thickened and elevated approximately two-thirds to three-fourths down the shaft. Often, a small, slitlike, upwardly directed *nutrient foramen* penetrates the anterior margin just medial to, or a few or more centimeters below, the ulnar tuberosity.

The relatively sharp *supinator crest* descends from the posterior portion of the radial notch facet either as a single thick elevation or as a series of thickened elevations along the shaft's *lateral side*. At some variable point within 1 cm or more of the radial notch facet, the supinator crest appears to bifurcate into two crests or ridges (the posterior is usually fainter). The inferior or *vertical ridge* courses almost the entire length of the shaft. The much stouter and more pronounced anterior ridge constitutes the *interosseous margin* or *border*; it extends for much of the shaft's length, mirroring the crisp and distended radial interosseous border. An interosseous membrane tethers the radius and ulna together. (The ulnar interosseous border, like the radial notch superior to it, lies on the same side of the bone from which the bone comes.) The ulnar interosseous border is more anteriorly oriented than the anterior border; the bone between these two borders is variably concave or troughlike. The interosseous border and vertical ridge posterior to it subtend a much shallower and narrower groove or trough.

The raised *posterior border* of the ulna descends like a tail from the rather flattened, triangularly elongate surface atop the olecranon process. It is distinct, compressed, and crestlike for only two-thirds or so of the upper length of the shaft; for the distal third—all the way distally to the spikelike *styloid process* —it is more subtly expressed. The posterior border parallels the ulna's mildly sinuous configuration: it arcs laterally for approximately the proximal one-fourth to one-third of the bone, then medially, and laterally again for the last distal one-fourth or so.

The ulna's *medial side* is smooth and gently rounded for much of its length. It is flattened to varying degrees below the depression underneath the coronoid process' lip. In cross section and taking into account the various ridges, borders, grooves, and convexities, the ulnar shaft is more or less triangular for much or all of its length.

Although the ulna tapers rather markedly toward its *distal end* or *head*, it flares out somewhat above the epiphyseal region. The inferiorly projecting *styloid process* is appressed to the posteromedial side of the somewhat semicircular head, which expands out from the "apex" created by the styloid process, from which it may be separated by a shallow groove and/or pit(s). Even though it does not directly contact the triquetral (carpal) below it (an articular disc is interposed between the two), the ulnar head's inferior surface looks articular. Most of the head's circumference, which thickens vertically as it proceeds away from the styloid process, is, however, represented by articular bone. The distal radius' shallow ulnar notch rotates around this articular perimeter.

See Table 5–5 for clues to siding ulnae.

Development and Ossification

As in the radius and humerus, an ossification center appears in the ulnar shaft during the eighth fetal week (Figure 5–8). The shaft is well ossified by birth, but the distal epiphyseal region and most of the olecranon (which is an epiphyseal cap atop the process) remain cartilaginous. The ossification center in the distal end appears between years 5 and 6 or even as late as the seventh year; fusion occurs

Table 5–5   Clues to Siding a Complete or Fragmentary Ulna: Its Features and Their Orientations

Feature	Position
Articular surface of head	Anterolateral
Coronoid process	Anterior (proximal)
Groove for extensor carpi ulnaris	Posterolateral (distal)
Interosseous border	Lateral (shaft)
Nutrient foramen (inferosuperior)	Anterior
Olecranon process	Posterior
Pronator ridge	Anteromedial
Radial notch	Lateral
Styloid process	Posteromedial
Supinator crest	Lateral
Trochlear notch	Anterior
Ulnar or brachialis tuberosity	Anteromedial

between 21 and 25 years, much later than in the distal radius (which articulates directly with neighboring carpals, whereas the distal end of the ulna is separated by an articular capsule from the underlying triquetral). Ossification of the olecranon usually begins during the tenth year but may ensue as early as 7 or as late as 14 years; fusion to the shaft typically occurs during the sixteenth but sometimes as late as the nineteenth or even the twenty-third year.

The Hand

The hand is composed of eight carpals (short bones of the wrist), which make up the carpus; five metacarpals (lying just distal to the carpals), which constitute the metacarpus; and 14 phalanges (lying distal to the metacarpals), which are the bones of the digits or fingers. The general arrangement of bones of the hand is similar to that of the bones of the foot and ankle. Except for the distal phalanges, all cartilaginous, presumptive elements are present by the second fetal month.

Carpus

The eight small carpals of the **carpus** (Figures 5–10 and 5–11) are arranged in proximal and distal rows. Each consists of four bones. In the proximal row, from the lateral to the medial side (when the hand is in the anatomical position), are the **scaphoid** and **lunate** (which articulate with the distal radius), **triquetral** (**triquetrum**) (which is separated from the head of the ulna by an articular disc), and **pisiform**. The distal row, from the lateral to the medial side, is composed of the **trapezium** (between the scaphoid and first metacarpal), **trapezoid** (between the scaphoid and second metacarpal), **capitate** (which articulates proximally with parts of the scaphoid and lunate and distally with the third and part of the fourth metacarpals), and **hamate** (which articulates proximally with the triquetral and part of the lunate and, distally, with the fourth and fifth metacarpals).

Ossification of the Carpals

Elements other than the normal number of hand bones are detectable by the second fetal month. One corresponds to the *os centrale* (central carpal), which typically develops into a separate bone in monkeys and most prosimians, and the other to the hamate's hamulus (hook), which eventually coalesces with that carpal. Only postnatal humans, apes (variably), and a few species of prosimian lack a separate os centrale (Schwartz and Yamada, 1998). There has been much debate on the evolutionary significance and etiology of absence of the os centrale in humans (e.g., see Schultz, 1936). The common belief is that the cartilaginous precursor of this small bone fuses to the scaphoid (e.g., Cihák, 1972). But according to Fazekas and Kósa (1978), this cartilaginous element is merely resorbed during the third or fourth fetal month.

Each carpal ossifies postnatally from a separate center: the capitate and then hamate during the first year; the triquetral between

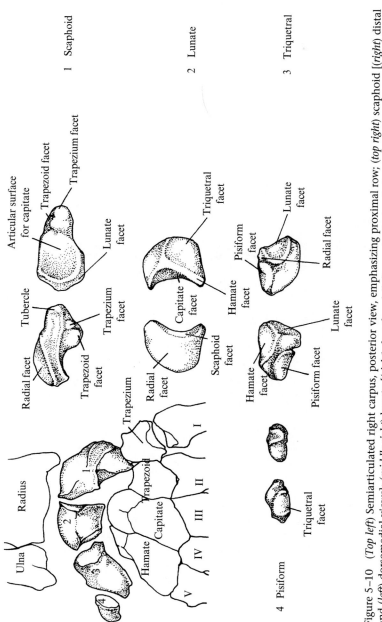

Figure 5–10 (*Top left*) Semiarticulated right carpus, posterior view, emphasizing proximal row; (*top right*) scaphoid [(*right*) distal and (*left*) dorsomedial views]; (*middle right*) lunate [(*right*) dorsal and (*left*) proximal views]; (*bottom right*) triquetral [(*left*) proximal and (*left*) distal views]; (*bottom left*) pisiform [(*right*) ventral (palmar) and (*left*) dorsal (posterior) views].

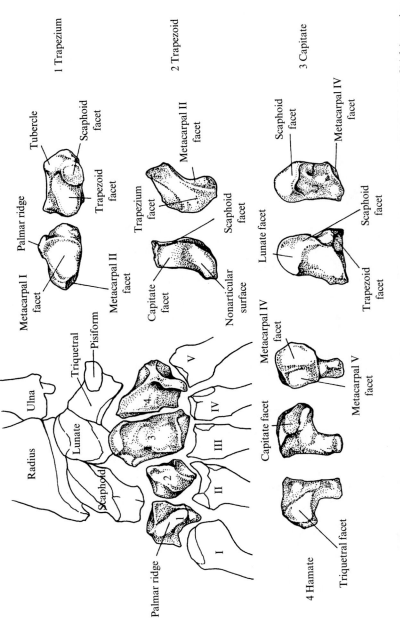

Figure 5–11 (*Top left*) Semiarticulated right carpus, anterior view, emphasizing distal row; (*top right*) trapezium [(*right*) proximal and (*left*) distal views]; (*middle right*) trapezoid [(*right*) lateral and (*left*) medial views]; (*bottom right*) capitate [(*right*) lateral and (*left*) dorsal (posterior) views]; (*bottom left*) hamate [(*right*) distal, (*middle*) lateral, and (*left*) proximomedial views].

the second and third years; the lunate in the fourth year in females and fifth year in males; the trapezium, trapezoid, and scaphoid during the fifth year in females and sixth to seventh years in males; and then the pisiform during the ninth year in females and as late as the eleventh year in males. These bones continue to enlarge as the individual grows.

Scaphoid

The **scaphoid** (Figure 5–10) was originally called the "navicular" after its presumed resemblance to an upside-down boat. It is, however, only "boat-shaped" in that its superior surface, in whole or in part, is arcuate and convex and its inferior surface may bear a circumscribed, concave articular region.

The scaphoid's *proximal* or *radial surface* is dominated by a roughly triangular to semi-circular convex *radial facet* that articulates with the lateral portion of the radius' distal epiphysis. This facet extends for two-thirds to three-fourths of the mediolateral width of the bone. The anterior and medial margins of the radial facet are "free" (they also form the border of the bone), whereas its posterior (dorsal) margin subtends a thin to moderately broad, horizontally oriented, variably pitted and grooved strip of nonarticular bone that subdivides the bone into proximal and distal parts; the latter lies more or less at a right angle to the former. The proximal part constitutes the radial facet. The narrower non-articular strip of bone extends beyond it medially and distally. Viewed from above, the nonarticular strip looks vaguely like the head of a snail and the radial facet like the snail's shell. The "snail's head" extends medially (to the side of the body from which the scaphoid comes) and swells somewhat proximally as the *tubercle* of the scaphoid. Although shifted medially, the distal component is almost a mirror image of the proximal part (recall, the distal component lies at a right angle to the radial facet and, thus, its predominant articular surface faces distally). The *trapezium-*

trapezoid facet of the distal component articulates with both the trapezoid and trapezium; it "hangs" below (distal to) the nonarticular strip of bone. The medial side of the trapezium-trapezoid facet is gently convex; its lateral side is variably mildly to markedly concave. The somewhat swollen proximal surface of this region is topographically irregular.

The scaphoid's *distal* or *inferior surface* of the *radial facet* is somewhat saddle-shaped, with the trapezium–trapezoid facet protruding above a larger articular region. The long axis of this vaulted, distal articular region is oriented mediolaterally; and because it articulates with the capitate's head, it is identified as the *articular surface for the capitate*. Arcing medioanteriorly around the circumference of the *articular surface for the capitate* is the obliquely offset semilunar-shaped *lunate facet*, which abuts the lateral side of the lunate.

Lunate

The **lunate** (Figure 5–10) is so named because its inferior or distal articular surface, when viewed laterally, is roughly lunate or crescentic in shape. The concave articular surface is mediolaterally narrow, anteroposteriorly elongate, and anteriorly broadening. Together, the lunate's *distal articular region* and the scaphoid's distal articular region cup the capitate's arced and convex proximal end. The articulated proximal surfaces of the scaphoid and lunate form a large convex surface. The bandlike articular region on the lunate's lateral side contacts the scaphoid's medial side; it parallels the concave lateral margin of the scaphoid's distal facet.

As the *radial facet*, the lunate's smoothly convex *proximal surface* nestles into the medial component of the articular surface of the distal radius. This facet is roughly triangular in outline, with rounded or blunted corners; the "triangle's" apex faces anteriorly (ventrally). Distally, the bone extends beyond the posterolateral "corner" of the articular surface (nesting a bit under the scaphoid).

The bone and its corner are obliquely truncated posteromedially. Viewed from its *distal* or *posterior aspect*, the lateromedial slant of the posterolateral corner of the lunate as well as its general lateral inclination are obvious. The somewhat paralleliform, nonarticular, posterior (dorsal) surface thus formed is irregular and deeply pitted.

The lunate's *lateral side* parallels the lateromedial slant of the bone's posteromedial corner. It is subdivided into (1) a large, flat facet for contact with the triquetral's medial side and (2) a narrow, sometimes slightly groove-like facet that follows the margin of the facet for the capitate and receives the compressed proximolateral edge of the hamate below it.

The lunate's anterior (*ventral, palmar*) *side* is narrow, rounded from side to side, and convex proximodistally; it contributes to the crescentic extension of the distal articular facet.

Triquetral

The **triquetral's** *proximal surface* (Figure 5–10), which is separated from the ulna's distal end by an articular disc, continues the configuration of the convex *radial surface* established by the scaphoid and lunate. The triquetral's large, slightly arced *distal facet* articulates laterally with the hamate. A distinct "corner" intrudes between two other articular facets that are at right angles to each another. The larger, variably crescentic to rhomboidally shaped, and almost vertically oriented facet articulates with the obliquely truncated posterolateral corner of the lunate. The triquetral's *anterior* (*ventral, palmar*) *side* extends only a short distance laterally from the edge of the facet for the lunate before the bone "corners" at the side of the anteriorly facing second (*pisiform*) facet. A short spit of bone proceeds away from the *triquetral–lunate facet*; its surface bears a deeply pitted depression.

The *pisiform facet* is generally ovoid in outline; irregularities on its perimeter are common. This facet's edges are usually delineated quite sharply and often emphasized further —especially proximally and laterally—by a constriction that pedestals the facet.

The triquetral's *proximal surface* bears a roughened and somewhat pitted groove that emerges from behind the pisiform facet and courses posterolaterally. This groove parallels the triquetral's long and posterolaterally oblique side; it may extend completely across the bone to its posterolateral corner (the posterior margin of the facet for the lunate) or fade out at any point. Regardless, two proximally situated raised areas are delineated posterior to the pisiform facet; the degree of separation of these two areas depends on the length of the groove between them.

The triquetral's *inferior* or *distolateral surface* bears a facet that articulates with the hamate. The fairly straight anterior (palmar) margin of this *hamate facet*, which is the largest of the three articular areas of the triquetral, courses laterally from the edge with the lunate facet and continues below the pisiform facet. The roughly parallel posterior margin of the straight hamate facet is much shorter (being truncated by the posteromedially oblique long side of the bone). The edge of the hamate facet also bears a shallow groove that is especially prominent along the posteromedially oblique long side of the bone.

The laterally facing hamate facet forms a crisply defined corner with the lunate facet but is separated from the pisiform facet. [After delineating the facet for the hamate, the facet for the lunate can be identified because it faces proximally (toward the side of the body from which the triquetral comes).]

Pisiform

The **pisiform** ("pea-shaped" carpal, Figure 5–10) is the fourth and last carpal of the proximal row. Since it articulates only with the triquetral, its sole, mildly concave *articular facet* faces posteromedially. Most of the pisiform rises from the articular facet like an amorphous mound (longer proximodistally than wide mediolaterally). Its *anterior* (*palmar*)

Table 5–6   Clues to Siding Carpals of the
Proximal Row: Their Features and Orientations

Feature	Position
Lunate	
Capitate articular surface (concave)	Distolateral
Radial facet	Proximal
Triquetral facet	Distomedial
Pisiform (difficult to side)	
Facet for triquetrum	Dorsal
Scaphoid	
Capitate articular surface (concave)	Medial
Radial facet	Proximal
Trapezium-trapezoid facet	Dorsal
Tubercle of scaphoid	Palmar-lateral
Triquetral	
Facet for hamate (largest)	Lateral
Facet for pisiform (ovoid)	Palmar
Lunar facet	Proximal

surface is less bulbous than its *posterior surface*. The articular facet for the triquetral may be offset by a narrow groove that constricts the perimeter of the bone to some extent. A broader groove typically lies on the bone's posterodistal side.

See Table 5–6 for clues to siding the carpals of the proximal row.

Trapezium

The **trapezium** (Figure 5–11) is the lateralmost carpal of the distal row. It articulates proximally with the scaphoid, distally primarily with metacarpal I, and distomedially with part of metacarpal II. A distinctive feature of the trapezium is the saddle-shaped *first metacarpal facet* that articulates with metacarpal I, which faces distally and laterally (toward the side of the body from which the bone comes). In outline, this saddle-shaped metacarpal I facet is subcrescentic; its "horns" point posteriorly or dorsally; its anterior peak lies just below the terminus of a stout crest on the *anterior (palmar) surface (palmar ridge)* that courses proximally in a mediolateral direction.

Where the palmar ridge juts out proximally, it is identified as the *tubercle of the trapezium*, which articulates beneath the scaphoid's distal articular process. Medial to, and coursing in parallel with, the stout anterior crest is a pronounced troughlike groove along which the tendon of the flexor carpi radialis courses. This groove is subtended medially by a peaked elevation of bone (corresponding to a distinct downward bend in the large, proximal articular surface). The consequent downward flexure subdivides the articular surface into two components; a smaller, essentially upwardly facing facet and a longer medially and posteriorly (ventrally) oriented facet.

The slightly concave proximal articular region for the scaphoid is roughly rhomboidal in outline. The longer, slightly concave to undulating facet, which articulates with the trapezoid, continues along and around the narrow distal end of the bone to contact between the trapezium and the proximolateral corner of metacarpal II. *Variation* in the trapezium is noted in a pinching off of the ventralmost portion of the long facet for the trapezoid and of the facet for metacarpal III.

Directly beneath the tubercle, the trapezium's *posterior* or *dorsal surface* bears a raised, heel-like eminence, from which emanates a low, variably distinct, medially directed, moundlike ridge that follows the flexure of the bone's medial surface. This low, moundlike ridge may terminate distally in a slight swelling just dorsal to the facet for metacarpal II.

Trapezoid

The **trapezoid** (Figure 5–11) nestles up against the lateral side of the trapezium and the medial side of the capitate. Proximally, it articulates with the scaphoid and distally with metacarpal II. While the trapezium's dorsal side tapers distally, its ventral side broadens distally. The narrow end of the trapezoid is directed anteriorly (ventrally) and its broad end posteriorly.

The trapezoid's *lateral side* mirrors the trapezium's medial side: the abutting facets are similar in outline and mirror images in contour. The variably convex to undulating *facet for the trapezium* typically expands toward the *palmar* (*anterior, ventral*) *side* of the bone; it may be pinched distoanteriorly in concert with *variation* in its sister facet on the trapezium. A shallow groove of pitted, nonarticular bone lies anterior (palmar) to the facet for the trapezium. It may course along the distal surface to separate the facet for the trapezium from the distally located *facet for the second metacarpal*; it may also be truncated, leaving the two facets confluent posteriorly across a bend in the bone.

The trapezoid's facet for metacarpal II is asymmetrical as it broadens posteriorly (it expands medially, following the shape of the bone). Thus, the facet's medial margin is arcuate and concave, while the lateral margin is mildly convex. A low "keel-like" ridge or narrow eminence may course down this facet, emphasizing its convexity mediolaterally. (Viewed from the side, the facet for metacarpal II is noticeably concave.)

The large facet for the scaphoid bone dominates the trapezoid distally. This facet's medial margin forms a "corner" and is confluent with the articular surface for the capitate; its lateral margin also forms a corner and is confluent with the articular surface for the trapezium. Together with the trapezium's scaphoid facet, the trapezoid's scaphoid facet creates a slightly concave surface in which the scaphoid's distal articular region nestles. The trapezoid's rough, nonarticular, posterior (dorsal) surface may extend onto the medial surface posteriorly.

Capitate

The **capitate** (Figure 5–11) articulates laterally with the trapezoid, medially with the hamate, proximally with the scaphoid and lunate, and distally primarily with metacarpal III and only slightly metacarpals II and IV. The bone's name reflects the "headlike" shape of its *proximal end*, which is relatively smooth and somewhat domelike and from which continuations of its articular surface may extend (like "earflaps" along its medial and/or lateral sides). Viewed from the palmar side, the proximal end is mildly convex superiorly (where it contacts the scaphoid) and laterally (where it meets the lunate); these two articular regions form a slight crease where they meet. The capitate's sometimes slightly concave lateral side (which forms a rather distinct corner with the superior articular prominence) is the more vertical. In palmar view, the proximal end's helmet-shaped appearance is enhanced further by a groove and/or pitting along the perimeter that delineates the roughened, obliquely oriented palmar (anterior) surface.

Below the capitate's head, the *palmar surface* becomes increasingly swollen distally as it comes to mirror the apex of the base of the third metacarpal. This palmar swelling is lopsided: the medial half of the surface is flatter than the lateral half.

The capitate's *distal end* is easily distinguished from its domelike proximal end. Viewed straight on, the distal end's articular surface resembles a right triangle. The short base courses along the posterior (dorsal) margin of the distal end and forms a right angle with the posteroanteriorly coursing side; metacarpal II articulates along this side's lateral edge. The right angle's shorter side elongates the distal end of the bone medially (away from the side of the body from which the bone comes); metacarpal IV overlaps this medial extension. In the anatomical position (lying on the shorter, posterior edge of its distal end) and viewed from either the medial or lateral side, the bone's distal surface is slightly concave and expands convexly from below the dome-shaped proximal end to the distal "peak" of the palmar surface. Viewed laterally, the posterior surface is pinched or indented, just below the proximal end.

The *trapezoid facet* occurs distally on the *lateral side* of the capitate. The trapezoid may articulate "high" or "low" relative to the capitate's distal surface. Roughly midway along its length, the capitate's lateral side is deformed outward (laterally), mirroring the concavity on the corresponding surface of the trapezoid; this "deformation" may take the form of a tiny flexure or elevation, a distinct swelling, or a markedly pronounced crest or ridge. The degree to which the capitate's lateral side is deformed outward appears to be correlated with the degree to which the capitate's trapezoid facet is (1) confluent with the more proximal facet for the scaphoid and (2) a single, broadly continuous facet or one that is divided into distinct areas of articulation.

For example, when the lateral deformation is minimally expressed, the capitate's trapezoid facet is smoothly expansive and broadly confluent with the capitate's scaphoid facet; a broad pit lies below or posterior to (and delineates posteriorly) the confluence of these two facets. When the lateral deformation is a pronounced crest, the capitate's trapezoid facet is pinched or "waisted" just below this crest, creating two semidistinct, trapezoidal articular regions. The facets for the trapezoid and the scaphoid are not confluent. Furthermore, it may appear as if the broad pit that otherwise lies posterior to the confluence of the capitate's scaphoid and trapezoid facets has interposed itself between these two facets and contributed to the subdivision of the trapezoid facet.

Distal to its smooth articular head, the capitate's *posterior surface* is roughened and irregular in contour. Its posteromedial margin may be gently concave or variably "edgelike," depending on the degree to which the facet on the medial surface for contact with the hamate is continuous or disjunct. If the (lateral) articular region for the trapezoid is markedly subdivided, the posterolateral margin may bear a few swellings.

The capitate's *medial surface* essentially consists of two different "zones": (1) a roughened, perhaps pitted, and somewhat swollen anterior (palmar) component that is a continuation of the palmar surface and (2) a *posterior articular region*. The latter articular region (for

the hamate) is continuous with the "head." Approximately midway down the medial side, the articular facet for the hamate narrows markedly. *Variation* occurs in this area of constriction in the extent to which the articular facet tapers or is pinched or subdivided into a large proximal and a much smaller subovoid distal facet. Just anterior to the point at which the facet may be disrupted lies a swelling whose size and pervasiveness covary with the degree to which the facet for the hamate is subdivided. The configuration of the capitate's hamate facet and accompanying swelling is reflected on the lateral surface of the hamate.

Hamate

"Hamate" means "shaped like a hook" or "bearing a hooklike process." Indeed, the hook is this bone's most distinctive feature (Figure 5–11). Because the hook protrudes anteriorly (toward the palmar side) and its end points laterally (toward the side of the body from which it comes), one can always side hamates.

Proximal to the *hook*, the bone tapers, reflecting the fact that it is wedged between the capitate (laterally) and triquetral (medially). At the hook's base, the hamate widens markedly toward the distal surface, which articulates with the bases of metacarpals IV and V. The hook is gently concave laterally throughout its length but convex medially only at its anteriormost portion. Posteriorly, it is buttressed medially and broadened by a strut of bone that forms the proximal border of the *facet for the fifth metacarpal*. The hook's base expands somewhat laterally and proximally as a tongue of slightly roughened, pitted bone, which may be circumscribed at its most proximal portion by the margin of an articular region that courses along the bone's perimeter. This margin is formed by the confluence of the *articular facets for the capitate* (laterally) and *triquetral* (medially). The convex and posteriorly oblique facet for the triquetral may either (1) continue up to and thus flare out in conjunction with the winglike extension of the distal surface (that contacts metacarpal V) or (2) be truncated in front by a depression that pervades the region behind the medial strut of the hook.

The hamate's *lateral surface* is relatively deep anteroposteriorly. In tandem with the contour of the distal *facet for the fourth metacarpal*, the bone's distolateral margin is slightly concave. The variably tapered proximal portion of the hamate's lateral side bears the smooth *articular facet for the capitate*. A depression along the bone's posterior border delineates the distal boundary of the large proximal portion of the facet for the capitate, which continues distally as a narrow articular strip. The latter contacts the capitate and may be separated completely from the larger articular region by a posterior invagination of the depression (groove) described above. The degrees to which the hamate's capitate facet is pinched or subdivided and the vertical groove extends posteriorly reflect, respectively, the subdivision of the corresponding facet on the capitate and the development of its midbody swelling.

The hamate's *posterior side* is roughly triangular in outline and its surface is irregular and pitted. Its *medial border* may be concave, reflecting the degrees to which (1) the winglike extension of the articular surface for metacarpal V protrudes medially and (2) its (proximal) apex is broadened by the confluence of the facets for the capitate and triquetral.

The hamate's *distal surface* is somewhat rectangular with rounded corners; one or both of the posterior "corners" may be swollen or expanded. A slight anteroposterior elevation of the distal surface divides it into two mildly concave articular planes that face slightly away from one another. The lateral plane (articular surface) receives the base of metacarpal IV; the slightly smaller medial surface (from which the hook protrudes) accommodates metacarpal V.

See Table 5–7 for clues to siding carpals of the distal row.

Table 5–7   Clues to Siding Carpals of the Distal Row: Their Features and Orientations

Feature	Position
Capitate	
Facet for hamate	Mediodorsal
Facet for lunate (domelike)	Proximal
Facet for metacarpal III	Distal
Facet for scaphoid	Lateroproximal
Facet for trapezoid	Laterodistal
Hamate	
Facet for capitate	Lateral
Facet for metacarpal IV	Distolateral
Facet for metacarpal V	Distomedial
Facet for triquetrum	Medial
Hamulus (points laterally)	Palmar
Trapezium	
Facet for metacarpal I (saddle-shaped)	Distolateral
Facet for metacarpal II	Distal
Facet for scaphoid	Medioproximal
Facet for trapezoid	Mediodistal
Groove for flexor carpi radialis	Palmar–medial
Tubercle of trapezium	Palmar–proximal
Trapezoid	
Facet for capitate	Medial
Facet for metacarpal II (crescentic and concave)	Distomedial
Facet for scaphoid	Proximal
Facet for trapezium	Lateral

Metacarpals

Morphology

The five **metacarpal bones** (metacarpals I–V, Figure 5–12) lie between the distal row of carpals and the proximal phalanges. Similar to the metatarsals of the foot, metacarpals have a *head* (distal end), *shaft*, and *base* (proximal end). Metacarpal I is the most lateral and metacarpal V the most medial of the set. (In the foot, metatarsal I is the most medial and metatarsal V the most lateral of the set.)

All metacarpal *heads* are wrapped anteroposteriorly by articular bone, which extends farther upon the palmar surface medially

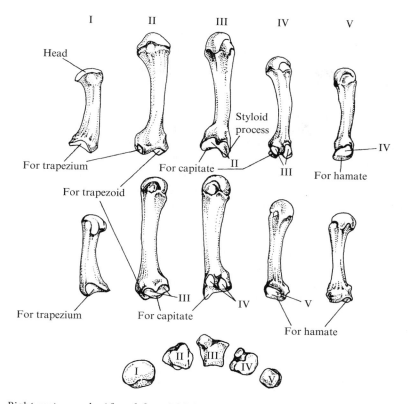

Figure 5–12 Right metacarpals: (*from left to right*) I–V; (*top row*) lateral (radial), (*middle row*) medial (ulnar), and (*bottom row*) proximal (basal) views.

and laterally (thus forming an inverted "U"-shaped border on this surface); sometimes, the lateral arm of the "U" is more pronounced or elongate. Posteriorly (dorsally), the head is narrower mediolaterally and bears shallow depressions on its medial and lateral sides that emphasize the medial and lateral "peaks" just behind the head (where the distal flare of the shaft terminates). Proceeding proximally from this distal flare, the *shaft* narrows toward, and then expands at, its juncture with the bone's base; this tapering is minimal on metacarpal I. The shaft's posterior (dorsal) surface is noticeably flat just behind the head. On metacarpals II–V, this flat surface tapers proximally (creating a long, flat triangular region); metacarpal I's posterior surface remains broadly flat for the shaft's length.

Metacarpal I is always distinguishable because it is the stoutest and shortest of the set. Metacarpals II and III are the longest and V is the shortest and most slender. Metacarpals I and V are distinctive because each articulates with only one other metacarpal. Thus, metacarpal I lacks a facet on its base's lateral side and metacarpal V lacks a facet on its base's medial side.

Each metacarpal's *base* is distinctive. On *metacarpal I*, the variable-in-outline proximal articular surface is consistently long and convex mediolaterally and short and gently concave anteroposteriorly. In conjunction with the trapezium, this surface forms the *saddle joint of the thumb.*

The *base of metacarpal II* is deep anteroposteriorly at its midline and elevated along its medial margin. This margin's medial side bears a moderately developed band of articular bone that contacts metacarpal III. Distally along the shaft, a variably deep pit or depression separates this articular strip (for metacarpal III) from a variably rugose buildup of bone. The outline of the base's articular surface is roughly triangular. Posteriorly (dorsally), the articular surface is indented. A groove circumscribing the bone's shaft separates the perimeter of the base from a variably rugose buildup of bone distal to it. Laterally, this metacarpal's base bears a small, somewhat ovoid, variably projecting facet against which metacarpal I's base articulates medially. This distolateral facet is oriented obliquely (generally faces palmad as well as laterally, i.e., toward the side of the body from which the bone comes). The *facet for the trapezium* lies proximal to this facet.

The *third metacarpal's* base mirrors the irregularly triangular to trapezoidal outline of the capitate's distal end. The long axis of this surface is oriented anteroposteriorly. The base of this triangle/trapezoid lies along the posterior (dorsal) aspect of the bone; it is oriented mediolaterally. Distention of this base's proximolateral corner gives a lateral twist to the articular surface (which is set obliquely, facing away from the side of the body from which the bone comes). The lateral side of metacarpal III's proximal articular surface bears a variably undulating to subdivided articular band that contacts the medial side of metacarpal II's base. Just distal to this articular region, the bone's surface may be pitted or depressed; farther distally, it may be variably rugose and raised. The medial side of metacarpal III's base (proximal articular surface) is articular and contacts metacarpal IV's base laterally. When this articular band is continuous, it is taller on its dorsal side and tapers to the palmar side; when subdivided into two facets, the dorsal facet is rather ovoid in shape. The dorsal surface around the base is roughened, irregular, and dominated proximolaterally by a spikelike projection.

In general, the *fourth metacarpal's* base is a reasonable match for metacarpal III's

base; however, its medial margin is concave (due to a slight expansion of its dorsomedial "corner") and its palmar portion is truncated. The lateral side of metacarpal IV's base mirrors the medial side of metacarpal III's base; it bears either a single articular strip or two separate facets. The longer and slightly concave medial side of metacarpal IV's base tends to be a continuously ovoid or elliptical facet against which a similarly shaped facet on the lateral side of metacarpal V's base articulates. Depressions and/or pitting may delineate the lateral, anterior, and medial sides of metacarpal IV's base from a variably raised and rugose buildup of bone farther distally along the shaft. The posterior (dorsal) margin of metacarpal IV's base may be indented or slightly grooved. At least the medial half, if not the whole, of the base of metacarpal IV may be angled slightly medially, paralleling the medial twist of metacarpal III's base. The articular surface of metacarpal IV's base tends to be gently convex.

The *fifth metacarpal's* base is short and concave mediolaterally, as well as broad and convex anteroposteriorly. A variably ovoid to elliptical facet for articulation with metacarpal IV's base lies along its lateral side. The variably swollen medial side of metacarpal V's base lacks a facet; this swelling contributes to a slight lateral twist in the base (metacarpal V's base angles toward the side of the body from which the bone comes). Just distal to the articular base, the posterior (dorsal) surface is variably depressed and irregular. Elsewhere, the bone may be roughened and somewhat swollen.

See Table 5–8 for clues to siding metacarpals.

Development and Ossification

Metacarpals II–V (Figure 5–13) each ossify from two centers: one in the shaft and one in the head (distal epiphysis). Ossification begins in the shaft typically during the ninth fetal week but not in the head until the second or third year. In a small percentage of individuals, a center may arise in metacarpal

Table 5–8   Clues to Siding Metacarpals: Their Features and Orientations

Feature	Position
Metacarpal I (Shortest of the metacarpals)	
Base (proximomedially distended)	Proximal
Facet for proximal phalanx (convex)	Distal
Facet for trapezium (saddle-shaped)	Proximal
Head (palmar–medially distended)	Distal
Nutrient foramen (proximodistal)	Medial (shaft; palmar concave)
Metacarpal II (longest of the metacarpals)	
Base (proximomedially distended)	Proximal
Facet for capitate	Proximomedial
Facet for metacarpal III	Medial
Facet for proximal phalanx (convex)	Distal
Facet for trapezium	Palmar–lateral (proximal)
Facet for trapezoid	Proximal
Head (palmar–laterally distended)	Distal
Nutrient foramen (distoproximal)	Medial (shaft; palmar concave)
Metacarpal III	
Base (dorsolaterally distended)	Proximal
Facet for capitate	Proximal
Facet for metacarpal II	Lateral (proximal)
Facet for metacarpal IV	Medial
Facet for proximal phalanx (convex)	Distal
Head	Distal
Nutrient foramen (distoproximal)	Medial or lateral (variable; shaft; palmar concave)
Styloid process	Dorsolateral
Metacarpal IV	
Base (medially distended)	Proximal
Facet for hamate (flattened)	Proximal
Facet for metacarpal III (double)	Lateral
Facet for metacarpal V	Medial
Facet for proximal phalanx (convex)	Distal
Head	Distal
Nutrient foramen (distoproximal)	Medial or lateral (variable; shaft; palmar concave)
Metacarpal V	
Base (medially distended)	Proximal
Facet for hamate	Proximal
Facet for metacarpal IV	Lateral
Facet for proximal phalanx (convex)	Distal
Head	Distal
Medial tubercle	Medial
Nutrient foramen (distoproximal)	Lateral (shaft; palmar concave)

II's base (proximal end). Although an ossification center arises in metacarpal I's shaft coincident with those in the other metacarpal shafts, 94% of the time its second center arises not in its head but in its base, as it does in the phalanges. The latter pattern of development has been interpreted as indicating that the first digit (thumb, pollex) lacks a metacarpal

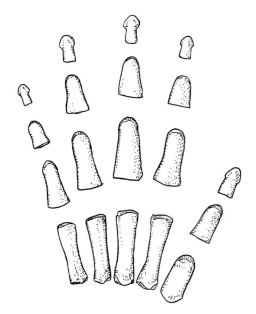

Figure 5–13   Bones of left hand (there are none in the wrist) of third-trimester fetus: metacarpals and phalanges, posterior view.

and possesses three, rather than two, rows of phalanges. Regardless, metacarpal I's base and the heads of metacarpals II–V unite with their respective diaphyses at approximately the same time as the phalangeal bases unite with their respective diaphyses (between 18 and 20 years).

Phalanges

Morphology

There are 24 **phalanges** (s. **phalanx**, depending on the interpretation of the bones of the pollex): three in each of digits II–V and two in the first digit (Figure 5–14). Like metacarpals, phalanges are miniature "long bones," replete with *shafts* and *proximal* and *distal ends*; the proximal end is the bone's base. Within each row, the phalanges of the *first* or *proximal* as well as of the *distal* or *terminal rows* are morphologically similar, the major differences between individual bones within a row being

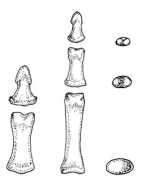

Figure 5–14   Right phalanges: (*left*) I, posterior view, (*top*) distal and (*bottom*) proximal phalanges; (*middle*) example of II–V, posterior views of (*from top to bottom*) distal, middle, and proximal phalanges; (*right*) example of II–V, (*from top to bottom*) proximal (basal) views of distal, middle, and proximal phalanges.

in size and robustness. On digits II–V, the phalanges of the *middle* or *second row* are, for the most part, similar to the proximal phalanges, from which they differ particularly in the articular morphology of their bases. As a set, the middle phalanges are smaller and slightly more gracile than the proximal ones. Within the set, the middle phalanges differ in size and degree of gracility and/or robustness.

In general, the *shafts* of the phalanges of the *first* and *second rows* are somewhat flattened on their anterior (palmar) surfaces and rounded posteriorly (dorsally). A cross section through a phalangeal shaft resembles a half-circle. These phalangeal shafts, while gently concave on their palmar sides, flare out on their sides toward the bone's base and head; the flare is broader and more marked proximally than distally because the proximal end is broader. The medial and lateral palmar margins of digits II–V's phalangeal shafts tend to be crisp and are sometimes raised and extend distally approximately two-thirds of the of shaft's length on the proximal phalanges but less than half the shaft's length on the middle phalanges; they are most distended at their distal termini. Proximally, these edges become more swollen, rounded, and expansive; this adds to the bulk of the bone's base and creates a variably shallow depression between them. The base of digit I's proximal phalanx is similar to other proximal phalanges.

The *distal end* or *head* of proximal and middle phalanges is covered largely by articular bone, which extends farther along the palmar than the dorsal side. The head is depressed somewhat toward its midline, which accentuates medial and lateral elevations that are more moundlike on the proximal phalanges (thus creating a somewhat undulating palmar surface). On the middle phalanges, these elevations are more crisply delineated (making the surface between them smoother and more broadly concave). In medial and lateral profiles, the bone's distal end is somewhat ovoid or subcircular (resembling a strangely shaped hubcap); the palmar portion rises above the level of the shaft. Typically, with the exception of its proximal portion, the distal end's perimeter is raised and subtends a shallow depression that may bear either a small crest or a few small, irregular bumps.

The proximal phalanx's *proximal end* or *base* bears a mildly depressed ovoid or elliptical articular facet for contact with the head of the metacarpal distal to it. This facet is buttressed, at least anteriorly, by the two (medial and lateral) swellings described above; the base may be distended elsewhere around the perimeter of the articular facet. A middle phalanx's base also bears an ovoid or elliptical articular facet, which articulates with, not a metacarpal's globular head, but a proximal phalanx's undulating head; thus, the surface of a middle phalanx's base bears a midline "keel," on each side of which lie two obliquely offset, shallowly concave surfaces. This keel nestles in the midline depression of a proximal phalanx's head, and the two surfaces on either side of the keel receive the two somewhat moundlike elevations of a proximal phalanx's head. A middle phalanx is distinguished further from a proximal phalanx in that, posteriorly (dorsally), its base may be distended into a swelling or tuberclelike feature (giving it a somewhat beaklike appearance).

Proximal and middle phalanges can be distinguished on the basis of differences in their bases (e.g., particularly on the middle phalanges of digit III or IV, with regard to the shaft's medial and lateral edges and the base's configuration; digit I's proximal phalanx is identified by the configuration of its base and distinguished from other proximal phalanges by lacking distinct medial and lateral edges on its shaft's palmar side). Digit I aside, identification of an isolated middle or proximal phalanx to a specific digit is basically impossible. Within the series, however, digit I's proximal phalanx is the shortest while digit V's is the most gracile; digit V's middle phalanx is usually the shortest and most gracile. One cannot "side" proximal and middle phalanges.

The *distal, third,* or *terminal phalanges* have very broad bases, short shafts, somewhat harpoon-shaped heads, and only one articular end; they are the smallest phalanges. Excluding the base's flare and expansion, a terminal phalanx is essentially flat on its palmar side and somewhat arced dorsally. Proceeding distally, the bone becomes more compressed anteroposteriorly, coincident with the head's fanning out.

In general, a terminal phalanx's *base* is similar to a middle phalanx, but its articular surface does not bear as distinctive a midline "keel" because a middle phalanx's head (with which a terminal phalanx articulates) is more broadly and smoothly arcuate than a proximal phalanx's head (with which a middle phalanx articulates). A terminal phalanx's shaft tapers drastically in from the base. The confluence of shaft and base on the palmar side tends to be quite roughened.

A terminal phalanx's harpoon-shaped *head* broadens beyond the sides of the shaft and sometimes sends two barblike extensions proximally, each projecting away from the side of the shaft. The head's margin is roughened and arcuate.

In contrast to the foot, in the hand, digit V's terminal phalanx may be the most gracile of its set, but it is not diminutive or

Table 5–9   Formulae for Estimating Stature (cm) from Maximum Length of Upper Limb Bones

White American males	White American females
3.08 (humerus) + 71.78 ± 3.37	4.74 (radius) + 54.93 ± 4.24
3.78 (radius) + 79.01 ± 4.32	4.27 (ulna) + 57.76 ± 4.30
3.70 (ulna) + 74.05 ± 4.32	3.36 (humerus) + 57.97 ± 4.45
African-American males	**African-American females**
3.42 (radius) + 81.56 ± 4.30	3.08 (humerus) + 64.67 ± 4.25
3.26 (ulna) + 79.29 ± 4.42	3.67 (radius) + 71.79 ± 4.59
3.26 (humerus) + 62.10 ± 4.43	3.31 (ulna) + 75.38 ± 4.83
Asian males	**Mexican males**
2.68 (humerus) + 83.19 ± 4.25	3.55 (radius) + 80.71 ± 4.04
3.54 (radius) + 82.00 ± 4.60	3.56 (ulna) + 74.56 ± 4.05
3.48 (ulna) + 77.45 ± 4.66	2.92 (humerus) + 73.94 ± 4.24

SOURCE: Modified from Trotter (1970) and Trotter and Gleser (1977).

morphologically truncated. Digit I's terminal phalanx is the broadest, most robust, and often longest bone of this set. Discriminating terminal phalanges, especially of digits II–IV, from one another and identifying "side" are essentially impossible.

Development and Ossification

Ossification of phalanges (Figure 5–13) generally begins at about the same time as that of the metacarpals and well ahead of the carpals. In some texts (e.g., *Gray's Anatomy*), the age of 8 fetal weeks is given for the onset of ossification in the hand's noncarpal bones, but there are differences in timing. The tips of the distal (third row) phalanges begin by intramembranous ossification, after which, during the seventh or eighth week, endochondral ossification spreads throughout the shaft. The diaphyses of the first row (proximal phalanges) begin to ossify during the ninth week and those of the second row (middle phalanges), as early as the eleventh and as late as the seventeenth fetal week. Ossification of phalangeal bases (proximal epiphyses) ensues after birth. In proximal phalanges, ossification of bases can occur between 1 and 3 years but in other phalanges, between 2 and 3 years or as late as the fourth year. Proximal ends may begin to coalesce with shafts as early as

14 years and fuse completely by 18 or as late as 25 years.

Determination of Stature from Upper Limb Bones

Trotter (1970; Trotter and Gleser, 1977) developed formulae for determining an individual's height (stature) from measurements of long bones (see Table 5–9). In all cases, these are maximum (not physiological) bone lengths that are taken using an osteometric board. **Maximum length** (in cm) is achieved by positioning the bone in whatever position yields the highest number; **physiological length** (in cm) is measured with the bone oriented as it would be in the body. The formulae are valid for individuals 18–30 years of age. For older individuals, one must adjust the results by subtracting 0.06 times (calculated age minus 30). The formulae are listed in order of decreasing accuracy (increasing standard error); all are less accurate than stature calculated from lower limb-bone measurements (see Chapter 6 for the latter). Since age-related changes after the age of 30 lead to a reduction in stature, Trotter (1970) developed the following equation to take this into account:

stature at death (cm)
= [maximum stature – 0.06(age – 30)] cm.

The Lower Limb

The major elements of the **lower limb** include the **pelvic girdle** [comprised of a right and a left hip bone or **os coxa** (still often, but not technically properly, referred to as the **innominate**), each of which articulates posteriorly (dorsally) with the upper segments of the sacrum and which together articulate anteriorly or ventrally at the pubic symphysis]; **femur** (thigh bone); **patella** (knee cap); **tibia** and **fibula**, which constitute the leg proper; and **foot**, which contains the ankle bones or **tarsals**, **metatarsals** ("long bones" of the body of the foot), and **phalanges** (toes). Only a few anthropometric landmarks are typically identified: *iliospinale* (the peak of the anterior superior iliac spine), *pubes* (a point in the midline of the superior border of the pubic symphysis), *tibiale* (the most proximal and medial point on the edge of the tibia's medial condyle), and *malleolare* (the distalmost extension of the medial malleolus).

Pelvic Girdle

Morphology

(See Chapter 4 for discussion of the sacrum and vertebral column.)

The **os coxa** (Figure 6–1, pl. **os coxae**) is a large, oddly shaped bone which, when viewed laterally, appears as a misshapen, top-heavy hourglass. Viewed from above, its superior and inferior portions lie at approximately right angles to one another, as if they had been rotated around the "waist" of the hourglass. Dominant in this constricted region is the socketed *acetabulum*, in which the femur's head articulates. In adults, a raised and thickened anteroinferiorly incomplete (and thus horseshoe-shaped) band—*lunate surface*—courses just within the rim of the acetabulum. Beneath the lunate surface's two ends lies the large *obturator foramen*, which is characteristically ovoid with blunted margins in males and triangular with crisp margins in females. Occasionally, the continuous part of the lunate surface may bear a small, roughened, triangular region or may be more extensively disrupted at its extreme expression by a distinct and superiorly elongate crease. These configurations may occur uni- or bilaterally and betray the fact that, developmentally, the acetabulum represents a region of

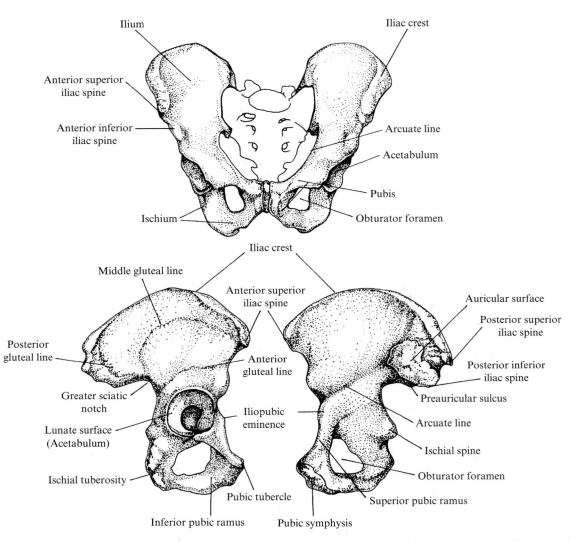

Figure 6–1 Articulated pelvis with image of sacrum, anterior view (*top*); (*bottom*) right os coxa [(*left*) lateral and (*right*) medial views].

fusion between three ontogenetically distinct bones: the **ilium** superiorly and, below it, the **pubis** and **ischium**.

The **ilium** is the largest of the os coxa's three bones. It is variably "bladelike" (i.e., somewhat thin, anteroposteriorly deep, and squat). Its bulk lies posteriorly; it is quite truncated anteriorly. A somewhat "U"-shaped *greater sciatic notch* is subtended, on its superior side, by the inferior border of the

posterior portion of the ilium and, on its anterior side, by the posterosuperior portion of the ischium. The greater sciatic notch tends to be narrower (like an upside-down "V") and appears deeply invaginated in males but more obtuse and shallower in females [this distinction generally pertains to children as well as to adults (Schutkowski, 1993)]. An ilium's outer surface is gently sinuous, bulging outward above the acetabulum and then curving smoothly inward until, at its most posterior extent, it again boldly swings outward. Internally (medially), an ilium is smoothly concave up to its area of articulation with the sacrum. An **arcuate line** highlights the curvature of the ilium from the sacroiliac articulation along the long axis of the superior pubic ramus and forms an inferior border to the internal aspect of the ilium. Arcuate lines tend to be more rugose in males than females.

The ilium's superior margin is the *iliac crest*, which is superiorly arcuate (more noticeably in males than females) and irregularly thickened. It is thickest around the highest part of the crest's arc and thicker and more pointed in males than females, in whom it is longer and less protrusive. The iliac crest terminates anteriorly in a bluntly pointed *anterior superior iliac spine*; the anthropometric landmark *iliospinale* is located at the peak of the anterior superior iliac spine. Posteriorly, the crest thickens broadly into the *posterior superior iliac spine*, which projects posteriorly beyond the sacrum and its vertebral spines. The bone beneath each superior iliac spine is indented to some degree, and below each indentation lies, respectively, an *anterior* and a *posterior inferior iliac spine.* The anterior inferior iliac spine is a knobby projection situated just above the level of the acetabulum's rim; the laterally more compressed and sometimes more spike-like posterior inferior iliac spine participates in articulation with the sacrum. A swollen to rugose *iliopubic eminence* lies below the anterior inferior iliac spine. Not to be confused with the posterior inferior iliac spine is another spinelike projection 1–2 cm away from it that subtends the ilial portion of the greater sciatic notch; this unnamed bony distension may protrude into the space of the notch, thus giving the impression that the notch is narrower than it really is. If the effect of the intrusive component on the shape of the greater sciatic notch is not taken into account, the angle of the notch may be incorrectly assessed, leading to misidentification of an individual's sex.

An ilium's outer surface bears muscle scars. Sometimes more markedly expressed in males than females, these scars correspond to the attachment sites of gluteal muscles that insert on the femur; they arc anteroposteriorly from top to bottom. The *posterior* or *superior gluteal line* is usually confined to the region just in front of the posterior superior iliac spine. The *middle gluteal line* courses from in front of and immediately below the midanterior thickening of the iliac crest. The *anterior* or *inferior gluteal line* originates above the anterior inferior iliac spine and fades out along the posterior rim of the acetabulum.

Except for its posterior, *auricular (sacral articular) surface*, an ilium's inner surface is relatively smooth. The auricular surface consists of distinctive superior and inferior components. The former is variably craggy in appearance and may bear accessory articular facets. The inferior articular region is club-shaped in outline (broadest where it abuts the arcuate line and tapering toward the end of the posterior inferior iliac spine); its surface is more granular than in the superior component.

A groove (*preauricular sulcus*) may lie immediately below the distinct inferior edge of the inferior articular component and course beneath the posterior inferior iliac spine. Preauricular sulci may be quite shallow and minimally expressed or well excavated and course along the entire length of the inferior edge of the auricular region. Some authorities consider the presence/absence of the preauricular sulcus a reflection of, respectively, female/male dimorphism (e.g., St. Hoyme and İşcan, 1989); others regard this as a non-sex-related nonmetric variant (e.g., Finnegan, 1978).

Right and left os coxae articulate anteriorly at the midline *pubic symphysis*; a cartilaginous interpubic disc separates adjacent pubic surfaces, which are roughly lozenge-shaped in outline. The shape, texture, and external morphology of this region change with increasing age (see Chapter 8).

The **pubis** consists of two branches (rami) that extend as arms of a "V" away from the *pubic symphysis*. The *superior pubic ramus* sub-tends the obturator foramen superiorly and includes the part of the acetabulum that lies anterior to the anterior inferior iliac spine. The *pubic tubercle* lies a few centimeters lateral to the pubic symphysis on the superior margin of the superior pubic ramus. The region between the tubercle and pubic symphysis is roughened and spanned by the superior pubic ligament. The inguinal ligament stretches between the tubercle and the anterior supe-rior iliac spine. The pubic tubercle tends to be pointier and closer to the symphysis in males and blunter and farther away from the symphysis in females. The anthropometric landmark pubes is located in the midline of the superior border of the pubic symphysis.

The *inferior pubic ramus* courses inferiorly and laterally away from the symphysis. It joins with an ascending ramus of the ischium to form the medially oblique boundary of the obturator foramen. Sometimes a slight pinch-ing of the bone occurs at the juncture of the inferior pubic ramus and ischium.

The *ischium* encompasses essentially the inferior half of the acetabular region. Pos-terior and slightly inferior to the acetabulum, the protrusive *ischial spine* delineates the infe-rior limit of the greater sciatic notch. Below the acetabulum, the ischium continues as a short, stout ramus, which is capped inferiorly by a roughened, subdivided, *ischial tuberosity*. The broad upper component of this tuber-osity is noticeably flatter than the narrower lower portion; a transverse ridge separates the two. The inferior portion of the ischial tuber-osity bears a longitudinal ridge. The ischium's thinner portion, which joins the inferior pubic ramus, courses up from the region of the ischial tuberosity.

See Table 6–1 for clues to siding complete or fragmentary os coxae.

Development and Ossification

Ossification of the os coxa (Figure 6–2) begins with the appearance of three centers in the

Table 6–1   Clues to Siding a Complete or Fragmentary Os Coxa: Its Features and Orientations

Feature	Position
Acetabular notch	Anteroinferior
Acetabulum	Lateral
Anterior superior iliac spine	Anterior
Arcuate line	Medial
Auricular surface	Medial
Greater sciatic notch	Posteroinferior
Iliac crest	Superior
Iliac tuberosity	Medioposterior
Iliopubic eminence	Anterolateral
Ilium	Posterior
Ischial spine	Posteroinferior
Ischial tuberosity	Inferior
Ischium	Inferior
Lesser sciatic notch	Posteroinferior
Obturator crest	Anterosuperior
Obturator groove	Inferoanterior
Pectineal line	Superoanterior
Posterior superior iliac spine	Posterior
Pubic symphysis	Anteromedial
Pubic tubercle	Anterosuperior
Pubis	Anterior
Tubercle of iliac crest	Laterosuperior

region of the eventual acetabulum. The first center, which corresponds to the presumptive ilium, appears above the area of the future greater sciatic notch by the end of the second or beginning of the third fetal month. By the middle of the fourth fetal month, the ilial contribution to the acetabulum and pos-terior inferior iliac spine are discernible (thus defining the greater sciatic notch). Within a few weeks, the presumptive anterior inferior iliac spine is distinct. The second center appears inferiorly along the ischium's body as early as the third or as late as the fifth fetal month. The ischium's characteristic shape is achieved prenatally. An isolated ischium looks somewhat like a comma or hook, with its thickened end contributing to the acet-abular region, its "stem" circumscribing the obturator foramen from behind, and its hooked

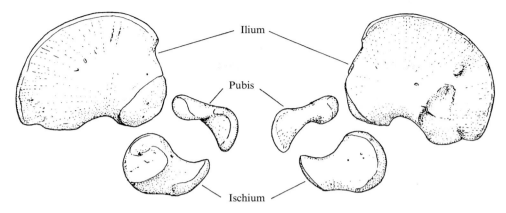

Figure 6–2   Right ilium, pubis, and ischium of third-trimester fetus: (*left*) lateral and (*right*) medial views.

end pointing anteriorly and superiorly. The third center arises in the presumptive superior pubic ramus, sometimes as early as the fourth but typically during the fifth or sixth fetal month. Throughout intrauterine life and until the os coxa unites as a whole, the pubis presents itself as a smaller mirror image of the ischium: its thickened end contributes to the acetabulum, its stem subtends the obturator foramen from above, and its shorter hooked end points inferiorly and somewhat posteriorly toward the hooked end of the ischium. (If the acetabular region is thought of as a hinge, the ischium and pubis vaguely resemble the arms of a spreading caliper.)

At birth, the ilium, pubis, and ischium are separated from each other in the acetabular region by a thick, "Y"-shaped *triradiate cartilage*; there are also four secondary centers of ossification, corresponding to the ischial tuberosity, pubic symphysis, anterior inferior iliac spine, and iliac crest. Between the second and sixth postnatal months, the small acetabulum becomes a shallow cup and the three ossifying bones extend farther into it. Coalescence of the pubis and ischium around the obturator foramen is usually well advanced by 6 years and virtually complete by 8. Between 9 and 12 years, ossification commences in the triradiate cartilage via one or more centers. Ossification and union in

the acetabulum occurs first between the pubis and ilium, then between the ilium and ischium, and, finally, between the pubis and ischium. These three bones may unify across the acetabulum between 14 and 16 or as late as 18 years. Onset of ossification in secondary elements usually coincides with the onset of puberty; their union to the bulk of the os coxa commences around 16 or 17 years and is completed between 23 and 25 years.

Femur

Morphology

The **femur** (Figure 6–3, upper leg) is the largest and sturdiest long bone. It articulates proximally with the acetabulum and distally with the proximal tibia; the patella rides anteriorly over its distal articular region. The long femoral *shaft* is essentially smooth along much of its anterior surface and more rugose and anatomically detailed posteriorly. The *greater trochanter* lies parallel to the long axis of the femoral shaft, cupping it proximally and anteroposteriorly; in lateral aspect, it broadens the proximal femur anteroposteriorly. A variably rugose and thickened bony line or band [*trochanteric (intertrochanteric) line*] courses across the anterior face of the proximal femur, from the raised medial aspect

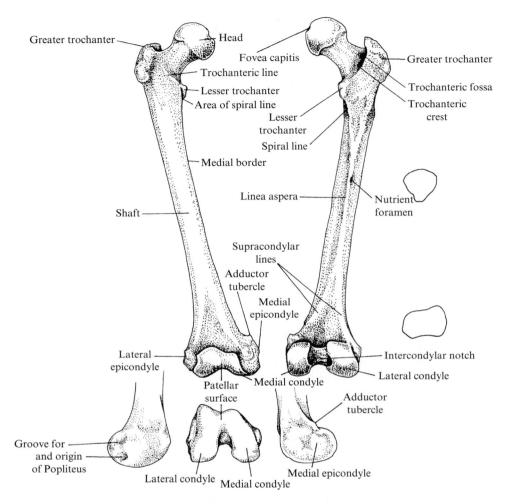

Figure 6–3 Right femur: (*top left*) anterior and (*top middle*) posterior views; (*top right*) cross sections through shaft; (*bottom*) distal end [(*left*) lateral, (*middle*) distal, and (*right*) medial views].

of the greater trochanter to the shaft's *medial border*. Posteriorly, the greater trochanter forms a vertical, liplike projection that is accentuated medially by a moderate to large pit (*trochanteric fossa*). Confluent with and continuing down and arcing medially away from the thick edge of the vertically projecting greater trochanter is the expanded *trochanteric (intertrochanteric) crest*, which terminates in a moderately pointed to broad *lesser trochanter*. When the femur is viewed

anteriorly, the lesser trochanter projects beyond the shaft's medial border. Extending superiorly and medially from the region sandwiched between the trochanteric line and trochanteric crest is the stout, short, slightly laterally compressed *neck* of the femur, which tapers slightly and then broadens in all directions. A semispherical *femoral head* caps the neck. Typically, a variably broadly shallow to deeply constricted pit (*fovea capitis*, in which the ligament of the femoral head attaches) lies posterior, if not also somewhat inferior, to the center of the femoral head.

At a level below the lesser trochanter, a muscle scar descending from the inferior region of the greater trochanter eventually thickens into an approximately 3 cm *gluteal tuberosity*, which appears more as a raised band than a projection and constitutes the scar left by the gluteus maximus muscle. An enlarged, medially projecting gluteal tuberosity occurs in humans as a nonmetric *variant*, when it is identified as a *third trochanter* (which is frequently found in mammals).

The *spiral line* may present itself as a continuation of the trochanteric line, or it may originate along the medial border of the femoral shaft, just inferior to the trochanteric line. It passes well below the lesser trochanter and courses to the midline of the shaft posteriorly to become confluent (less than halfway down the shaft) with the inferiormost extension of the gluteal tuberosity. Often, a moderately elongate and slitlike *nutrient foramen* penetrates the femoral shaft obliquely and in an upward direction (away from the knee) near this confluence. The merging of spiral line and gluteal tuberosity creates a single greatly thickened, raised, squared, and/or broad muscle scar (*linea aspera*). This scar may continue an additional one-quarter to one-third of the way down the shaft before it begins to fade and then bifurcate into *medial* and *lateral supracondylar lines*, which diverge as they course inferiorly to their respective sides of the distally widening bone; the medial line tends to be the more obscure. Between them, these lines subtend a somewhat posteriorly flattened, tall, triangular *popliteal surface*; the anterior surface is also somewhat flattened distally. A second nutrient foramen may lie variably near the apex of the triangular popliteal surface but more consistently closer to the medial supracondylar line; although sometimes larger than the primary nutrient foramen, it is similarly elongate and slitlike and perforates the shaft obliquely and in an upward direction.

The lateral supracondylar line terminates distally in a swelling (*lateral epicondyle*). The medial supracondylar line thickens into a variably crestlike *adductor tubercle* that lies atop the *medial epicondyle*. The bulk of a femur's distal end consists of large *medial* and *lateral condyles*, from the sides of which the two epicondyles bulge. These condyles are swollen posteriorly and separated by a deep *intercondylar notch*; each bears a characteristically smooth *articular surface* that originates just below the popliteal surface and wraps around its posteriorly swollen moiety. At about the midpoint of the inferior surface, the two articular areas merge around the intercondylar notch into one articular surface, which spreads anteriorly, eventually forming the *patellar surface*. The patellar surface is concave, due to the protrusion anteriorly of the medial and lateral condyles. But the patellar surface is also asymmetrical because the lateral condyle protrudes farther anteriorly than the medial one. The concaveness and asymmetry of the patellar surface mirrors the shape of the patella's posterior surface.

When viewed distally while on its posterior side, the condyles approximate, respectively, the perpendicular and long sides of an isosceles triangle: the lateral condyle is more or less vertical (in line with the lateral side of the femoral shaft), the apex of the triangle is the farthest point anteriorly of the articular surface, and the medial condyle appears to course (albeit arcuately) from this apex to the most posterior and medial corner of the distal articular region. Also apparent when the femur is in this position are (1) the difference between the medial and lateral condyles (the medial is arcuate, longer, and more swollen; the lateral appears more confined and ridgelike, even though it contributes anteriorly to much more than half of the patellar surface) and (2) the twist in the femoral shaft (the medial border of the shaft becomes more delineated in its upper portion and the femoral head and neck, which are oriented obliquely and anteriorly, are not in the same plane as the condyles). Severe femoral torsion may be an indication of repeated squatting behavior (personal observations).

In addition to being somewhat "bowed" or curved anteriorly, the femur does not assume the perpendicular when in the anatomical position. When it is resting flat on its condyles, one can see the outward or lateral orientation of the femur's long axis and appreciate the distinction between the medial and lateral condyles (the anteriorly protruding lateral condyle is shorter and oriented essentially anteroposteriorly; the medial is obliquely skewed outward and posteriorly), as well as the degree to which the femur is twisted slightly superiorly and the head and neck are oriented anteriorly.

See Table 6–2 for clues to siding complete or fragmentary femora.

Development and Ossification

The femur is the second bone of the body to begin to ossify, following quickly on the heels of the clavicle. Ossification in the shaft (Figure 6–4) usually begins by the end of the seventh fetal week. By birth, the only other sign of mineralization may be the presence of a secondary center in the middle of the distal epiphysis; in 7% of individuals, this center may arise 1 or perhaps even 2 months prior to term and in 12%, after birth). Ossification in the distal epiphysis gives rise to the condyles and epicondyles. The femoral head begins to ossify by the end of the first year, the greater trochanter during the fourth year, and the lesser trochanter between the thirteenth and fourteenth years. Fusion of these elements to the femoral shaft occurs after puberty,

Table 6–2 Clues to Siding a Complete or Fragmentary Femur: Its Features and Orientations

Feature	Position
Adducter tubercle	Medial
Gluteal tuberosity	Posterolateral
Greater trochanter	Lateral
Head	Medial
Intercondylar notch	Posterior
Intertrochanteric crest	Posterior
Intertrochanteric line	Anterior
Lateral condyle (smaller)	Lateral
Lateral epicondyle	Lateral
Lateral supracondylar line	Posterolateral
Lesser trochanter	Medial
Linea aspera	Posterior (shaft)
Medial condyle (larger)	Medial
Medial epicondyle	Medial (distal)
Medial supracondylar line	Posteromedial
Nutrient foramen (inferosuperior)	Posteromedial
Patellar surface	Anterolateral
Popliteal surface	Posterior
Spiral line	Posteromedial
Trochanteric fossa	Posterior (proximal)

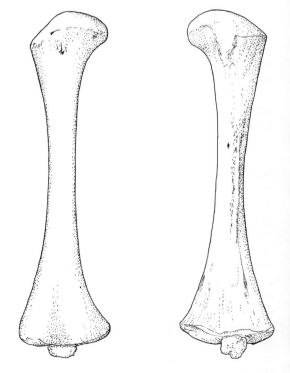

Figure 6–4 Right femur, with ossifying distal epiphysis, of third-trimester fetus: (*left*) anterior and (*right*) posterior views.

starting at about 15 years with the lesser trochanter and proceeding to the greater trochanter, head, and then distal epiphysis. The proximal femoral elements unite with the shaft between 18 and 20 years, prior to fusion of the distal epiphysis (at 20–23 years). The femoral and pelvic contributions to the hip region solidify before those of the knee joint. The typical shape of the femur can be recognized well before birth.

Patella

In basic outline, the **patella** (Figure 6–5) resembles a guitar pick. Its *base* is broadly arcuate superiorly and its *apex* points inferiorly. Its *anterior* and *posterior surfaces* are distinctly different from one another. The former is somewhat roughened and bears, to varying degrees, markings or even platelike scars left by the quadriceps femoris tendon, which encapsulates the bone. The posterior surface is dominated by a large, smooth *posterior articular facet*, below which the patella's apex extends. Since this facet articulates with the femoral patellar facet, it mirrors the anterior configuration of the femoral condyles: it is convex but asymmetrically so. The roundedly ridgelike, vertical "peak" of the surface separates a long, gently sloping *lateral facet* from a more steeply inclined *medial facet*. In turn, the medial facet is subdivided into an upper and a smaller and more steeply inclined lower section. The long lateral slope corresponds to the greater contribution of the lateral condyle to the femoral patellar facet. The shorter medial slope corresponds to the lesser contribution of the medial condyle to the femoral patellar facet. The smaller, lower portion of the medial facet contacts the medial condyle of the femur when the knee joint is in extreme flexion. Siding patellae is simple: when placed on its asymmetrical posterior surface, the bone comes to lie on the longer, larger, lateral portion of the facet (tilting toward the side of the body from which it came). The patella ossifies from several centers that may begin to appear as early as the second or as late as the sixth year.

Tibia

Morphology

The **tibia** (Figure 6–6) is the second longest bone of the leg and skeleton. It is easily distinguished from its smaller and slenderer partner in the lower leg, the fibula.

The tibia's *proximal end* is broad, especially laterally. It bears two subequal condyles whose superior articular surfaces correspond to, and reflect similar differences between, the femur's medial and lateral condyles, with which it articulates. Thus, the *medial condyle* of the tibia tends to be longer than the *lateral condyle*. The lateral is distinguished further from the medial condyle because, posterolaterally, it bears a somewhat inferiorly directed, variably ovoid or elliptical facet for its superior articulation with the fibula. The tibial condyles protrude posteriorly beyond both the neck and shaft. The anthropometric landmark *tibiale* is located at the most proximal and medial point on the edge of the medial condyle.

Both tibial condylar articular surfaces are somewhat lima bean–shaped, with the counterpart of the seed's scar being a raised, sometimes rather sharply delineated *intercondylar eminence*. There are thus two intercondylar eminences; the region between them may be

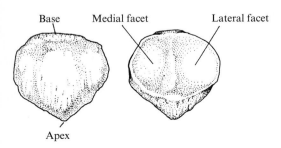

Base   Medial facet   Lateral facet

Apex

Figure 6–5   Right patella: (*left*) anterior and (*right*) posterior views.

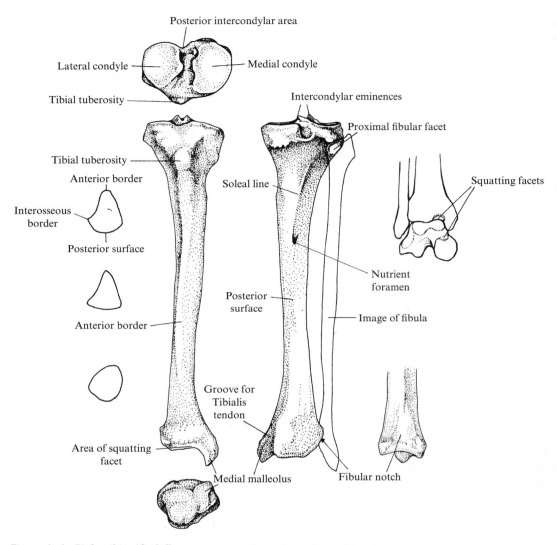

Figure 6–6   Right tibia: (*far left*) cross sections through shaft; (*middle left*) (*top*) proximal, (*middle*) anterior, and (*bottom*) distal views; (*middle right*) posterior view with image of fibula; (*far right*) (*top*) articulated distal tibia and fibula with talus, anterior view, and (*bottom*) distal end of tibia, medial view.

delineated further by a shallow depression. Two semitrapezoidally shaped, depressed areas expand outward—a large one anteriorly and a smaller one posteriorly—from the region of the intercondylar eminences. Variably impressed upon these depressed regions as well as upon the area between the intercondylar eminences are the attachment sites of the (semilunar and cruciate) ligaments that secure the knee joint.

The tibia's proximal end is expanded anteriorly by the *tibial tubercle*, to which the patellar ligament attaches; it may be delineated superiorly by a fairly horizontal groove. A variably distinct and distended crest may also lie 1–2 cm below this groove. The tibial tubercle is the proximal terminus of the *anterior border* of the tibial shaft.

The *anterior border* emanates from the region of the distal tibia's *medial malleolus* as a low, rounded margin. It typically becomes sharply and crisply defined as it swings laterally beyond the midshaft and then broadens toward the tibial tubercle. The lateral aspect of this border remains crisp as it arcs from the tibial tubercle to the lateralmost extremity of the lateral condyle. (The morphology and course along the shaft of the tibia's anterior border are relatively consistent clues to identifying bone fragments as tibial in origin and to siding them—i.e., the more sharply delineated border of the tibial tubercle swings laterally, or to the side of the body from which the bone comes.)

There are other distinctive aspects of a tibia's shaft. For instance, in cross section, it is roughly triangular throughout most of its length: the anterior border is the triangle's apex; the *posterior surface*, like a triangle's base, is somewhat flat; and the posterolateral edge, or *interosseous border*, is angular and "cornerlike" along much, if not all, of the shaft's length. The upper third or so of the shaft's posterior surface bears two major landmarks: (1) a rough muscle scar or *soleal line* (created by the attachment of fascia associated with the soleus muscle as well as with the popliteus and deep muscles of the leg), which courses obliquely from just below the lateral condyle to the bone's medial border, and (2) a typically very long and slitlike *nutrient foramen*, which enters the bone steeply from above and is located in the region between the soleal line and interosseous border.

The tibia's *distal end* differs markedly in shape and lies in a different plane from the proximal end. When the tibia is placed with its anterior surface facing up (lying on the posterior edges of both condyles), the medially situated, inferiorly distended "hook" or *medial malleolus* ("malleolus" meaning "little hammer") of the distal end is oriented upward approximately 45°; it expands the distal portion of the tibia medially. Viewed from the medial side, the malleolus' asymmetry (it is more inferiorly distended along its anterior portion) and narrowness (when compared to the lateral side of the distal end of the tibia) are obvious.

The medial malleolus is rather rugose and flat across most of its medial surface; it is delineated from the distended anterior region below by an oblique and crest- or ridgelike edge that is confluent with the lower extent of the tibia's anterior border. The medial malleolar surface also bears a posterior, somewhat vertical edge or crest (which extends a few centimeters superiorly along the shaft of the tibia) that subtends, just on the posterior surface, a *groove for the tibialis posterior tendon*. The posterior malleolar crest joints the more oblique anterior border to form an apex, which may lie to one side or the other of a variably excavated concavity that separates the malleolus' anteroinferior distension from the slightly swollen inferior extension of bone that bears the groove for the tibialis posterior tendon.

On the anterior side of the tibia's distal end, the medial malleolus is delineated by a variably pronounced concavity that is also variably confluent with one or more depressions along the region of the *epiphyseal line*. At times, a continuous, thin, undulating groove incises the epiphyseal line anteriorly; approximately midway along its course, this groove will be accentuated, if not also briefly interrupted, by a small swelling of bone. Just superior to the depression closest to the medial malleolus, the tibia's lower end tends to bulge. (The anterodistal margin, in conjunction with the medial malleolus, is helpful in identifying and siding tibial fragments.) The anthropometric landmark *malleolare* is located at the distalmost extremity of the medial malleolus.

The lateralmost limit of the anterior face of the tibia's distal end, whose surface tends to be rather rugose, often looks in profile like a corner whose point has been cut off obliquely. The lateral edge of this "corner" is typically quite crest- or ridgelike and extends inward and superiorly, typically becoming confluent with the interosseous line. This lateral crest or ridge also subtends on one side a large, somewhat triangular lateral depression (*fibular notch*) that broadens as it descends to the distal border; it receives part of the lower portion of the fibular shaft. Although the posterior and anterior surfaces of the tibia's distal end are similar in general outline, the former is distinguished, for example, by the groove for the tibialis posterior tendon and the presence of the shorter side of the medial malleolus. The lateralmost extent of the posterior surface is somewhat similar to its anterior counterpart in being a "corner without a corner." However, it differs in being less distended laterally and in bearing a variably developed crest, which may be *variably* weakly pronounced or sufficiently developed to become confluent with the interosseous line; this superiorly and inwardly coursing crest subtends the other side of the fibular notch.

The tibia's *distal surface* is represented by a large articular region, which partially cups the superior portion of the talus both from above and medially. A large, subtrapezoidal articular facet—broader laterally than medially—courses across the inferior surface of the distal end, "kinks" at approximately a 90° angle at the base of the medial malleolus, and then proceeds to cover most of the inner surface of the malleolus. Anteriorly, at the base of the medial malleolus, the articular surface might be disrupted either slightly by a minor "pinching" or more markedly by a definite crease or looplike invasion of nonarticular bone.

Occasionally, a distinctly delineated ovoid or elliptical surface may be present midway along the anterior edge of the tibia's distal margin; it represents an articular facet that developed as a result of constant contact with the talar head (which bears a counterpart to the tibial facet). Contact between the talar head and the tibia's anterodistal margin is achieved when the ankle is in extreme flexion (*hyperflexion*), as when squatting, thus creating tibial and talar "*squatting facets.*" Although squatting facets on the anterodistal margin of the tibia and on the talar head are the most common examples in osteology textbooks, facets may occur on the distal side of the proximal condylar surfaces as a result of extreme flexion at the knee joint (the femoral condyles will also exhibit facets at their points of contact with the tibia; see review by Kennedy, 1989). Additional tibial facets may include a rounded posterior margin of the lateral condyle, a twisting backward (retroversion) of the head, and a groove on the side of tibial tubercle created by the patellar ligament.

See Table 6–3 for clues to siding complete or fragmentary tibiae.

Development and Ossification

Ossification of the tibia (Figure 6–7) begins in the shaft during the eighth fetal week, a week or so after the femur. At birth, the only other

Table 6–3   Clues to Siding a Complete or Fragmentary Tibia: Its Features and Orientations

Feature	Position
Anterior border	Anterolateral
Fibular articular facet	Posterolateral
Fibular notch	Posterolateral
Groove for flexor hallucis longus	Posterior
Interosseous border	Posterolateral
Lateral condyle	Lateral (proximal)
Malleolar groove (for tibialis posterior)	Posteromedial
Medial condyle	Medial
Medial malleolus	Medioanterior (distal)
Nutrient foramen (superoinferior)	Posterior
Soleal line	Posterior (shaft)
Tibial tuberosity	Anterolateral

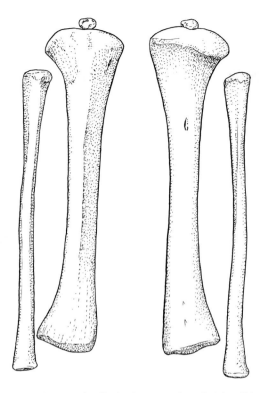

Figure 6–7   Right fibula (*outermost*) and right tibia with ossifying proximal epiphysis (*innermost*) of third-trimester fetus: (*left*) anterior and (*right*) posterior views.

mineralizing region is the proximal epiphysis, in which the secondary center usually arises shortly before term (although it may not appear until just after birth). The tibia's shape is recognizable early on, as is that of the proximal epiphysis (including the anterior projection corresponding to the tibial tuberosity). Ossification of the distal epiphysis begins during the second year; fusion to the shaft commences at about 16 years and is complete between 18 and 20 years. Coalescence of the proximal epiphysis to the shaft also begins at about 16 years, but fusion is not complete until between 20 and 23 years. Once the distal epiphysis begins to ossify, its adult shape is recognizable.

Fibula

Morphology

The long, slender **fibula** (Figures 6–6 and 6–8) articulates with the lateral side of the tibia at two points: (1) proximally, on the inferior and posterolateral aspect of the lateral condyle, and (2) distally, in a relatively long, shallow, somewhat triangular depression (whose base is formed by the distolateral margin). The fibula's distal portion, which extends below the level of the medial malleolus of the tibia, forms the lateral wall of the "cup" in which the superior portion of the talus nestles.

A fibula has distinctively sharp, crisp, longitudinal borders and edges. Although its proximal and distal ends are both pointed, the *proximal end* or *head* is much bulkier and more expanded three-dimensionally than the distal end, has a severely constricted *neck*, and in profile approximates the shape of an equilateral triangle. In marked contrast, the *distal end* or *lateral malleolus* is broader, flatter, and laterally compressed; in outline it resembles a right-angled triangle, with the apex being the most distal and posterior part. In addition, the terminus of the lateral malleolus tends to be blunter than the fibular head, which is often drawn out into a pointy *styloid process* that nestles up posteriorly against the tibia's lateral condyle. Thus, the fibula's styloid process bears an articular facet internally of variable size that mirrors its counterpart on the tibia. Below its process, the head swells somewhat into a "heel" that embraces the tibial condyle inferiorly and laterally. The fibula's thickened and flattened *anterior surface*, which is created medially by the sharp *interosseous border* and laterally by the equally distinct *anterior border*, may terminate on the internal surface of this heel.

The fibular head's less expanded side may bear an inferiorly coursing crest that may become continuous with the bone's medial border. The fibular head's posterior surface (defined when the fibula is articulated with

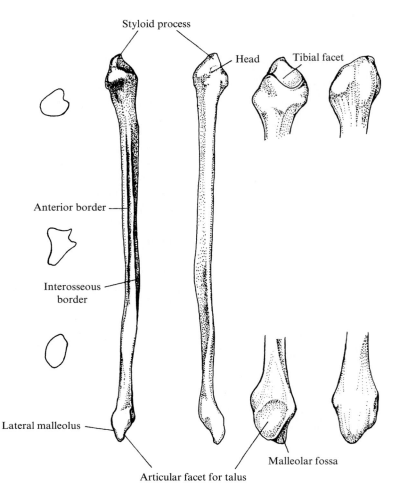

Figure 6–8   Right fibula: (*left*) cross sections through shaft; (*middle pair*) (*left*) anterior and (*right*) posterior views; (*right pair*) (*top*) detail of proximal end [(*left*) anteromedial and (*right*) lateral views] and (*bottom*) detail of distal end [(*left*) medial and (*right*) lateral views].

the tibia) may be rugose, but it is also the most planar part of the head. Taking the posterior surface as reflecting the orientation of the entire fibular head, it lies in a plane that is juxtaposed approximately 90° to the plane of the fibula's distal end (the major plane of the proximal end faces anteroposteriorly and the major plane of the distal end outward or laterally). This "twist" is also somewhat reflected along the shaft: a concomitant torque in cross-sectional shape (basically, an

equilateral triangle with concave sides) twists the courses of the crests and borders.

Internally, a large part of the posterodistal region of the fibular malleolus is demarcated by a pitted, depressed *malleolar fossa*, to which attach the various tibiofibular ligaments. Superior and anterior to the malleolar fossa is the edge of a large *articular facet for the talus*, which is shaped more or less like an (upside-down) equilateral triangle, whose base is usually delineated by the epiphyseal line. The

Table 6–4   Clues to Siding a Complete or Fragmentary Fibula: Its Features and Orientations

Feature	Position
Articular facet of head	Anteromedial (proximal)
Groove for peroneus brevis	Posterolateral (distal)
Interosseous border	Medial (shaft)
Lateral malleolus	Lateroposterior
Malleolar fossa	Posteromedial
Nutrient foramen (superoinferior)	Posterior
Styloid process	Posterolateral

lateral malleolus' posterior face is bisected by a variably excavated and vertically pervasive groove through which traverse the tendons of the peroneus brevis and peroneus longus muscles. [The peroneus brevis inserts onto the lateral tubercle on the base of metatarsal V and the peroneus longus onto (at least) the lateral sides of the base of metatarsal I and the medial cuneiform; both muscles participate, among other actions, in eversion of the foot.]

See Table 6–4 for siding complete or fragmentary fibulae.

Development and Ossification

Ossification of the fibula (Figure 6–7) begins in the shaft during the eighth fetal week, coincident with the onset of ossification in the tibia's shaft. Initially, the fibula and tibia are the same length. From about the middle of the third fetal month through term, the tibia is noticeably longer than the fibula. Subsequently, although the tibia remains the longer of the two bones, the difference in length is not as marked. Ossification of the distal fibular epiphysis begins during the second year and that of the proximal epiphysis between 3 and 4 years. Fusion of both ends to the shaft may begin at about 16 years; complete union of the distal end occurs at approximately 20 years and that of the proximal end between 23 and 25 years. The general adult shape of the fibula is recognizable early on.

The Foot

The **foot (pes)** is composed of seven **tarsal bones** (short bones of the ankle), which make up the **tarsus**; five **metatarsal bones** (situated just anterior to the tarsals), which constitute the **metatarsus**; and 14 **phalanges** (just anterior to the metatarsals), which are the bones of the **pedal digits** (toes). The general arrangement of the foot and ankle bones is similar, respectively, to that of the hand and wrist bones. Because of the orientation of the foot, the directional terms "inferior," "plantar," and "distal" are often used interchangeably, as are "superior" and "proximal." With regard to "anterior" and "posterior," for example, metatarsal heads lie anteriorly (ventrally) and their bases posteriorly (dorsally); phalanges lie anterior and tarsals posterior to metatarsals.

Tarsus

Morphology

The **tarsal bones** of the **tarsus** (Figures 6–9 and 6–10) are arranged in two rows. The **talus** and the **calcaneus** (also *calcaneum*) constitute the **first** or **proximal row**. The talus sits on top and to the medial side of center of the calcaneus and is the primary contact between the bones of the foot and leg. The **second** or **distal row** of tarsals comprises two subsets. The **navicular** (medially) and **cuboid** (laterally) articulate with the talar and calcaneal heads, respectively. However, while metatarsals IV and V articulate directly with the cuboid, the **medial**, **intermediate**, and **lateral cuneiforms** intervene between the navicular bone and metatarsals I, II, and III, respectively.

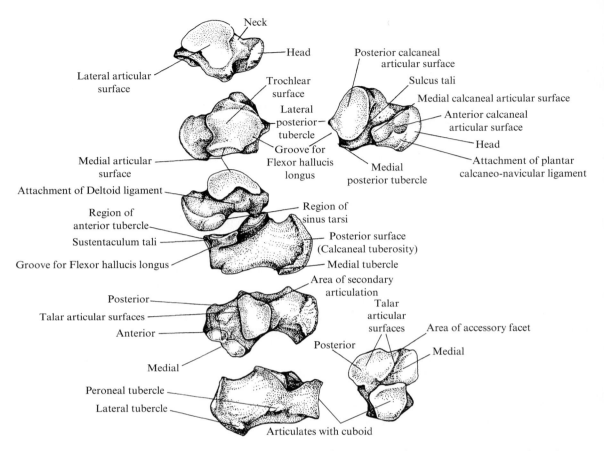

Figure 6–9 Right talus and calcaneus (proximal row of tarsus): (*left, from top to bottom*) talus, lateral view; talus, superior view; talus and calcaneus, medial view; calcaneus, superior view; and calcaneus, lateral view; (*top right*) talus, inferior view; (*bottom right*) calcaneus, anterior view.

Development and Ossification

Ossification of the tarsus (Figure 6–11) variably commences in the calcaneus and talus. A center may appear in the calcaneal body during the twelfth fetal week or be delayed until sometime during the seventh fetal month. Ossification in the talar body often begins during the latter part of the sixth fetal month. At birth, the only other area of mineralization in the tarsus may be in the cuboid: its ossification could begin close to term, but it could also be delayed until shortly after birth. After the cuboid, ossification ensues sequentially in the lateral cuneiform (year 1), medial cuneiform (2–4 years), and then in the intermediate cuneiform and navicular (3–5 years). Ossification of the calcaneal epiphysis begins as early as the seventh or as late as the tenth year; its fusion with the body begins at about 12 years and is complete between 20 and 22 years. Sometimes a separate center gives rise to the posterior process of the talus; if it does not coalesce with the body, it persists as a small, sesamoidlike bone (*os trigonum*).

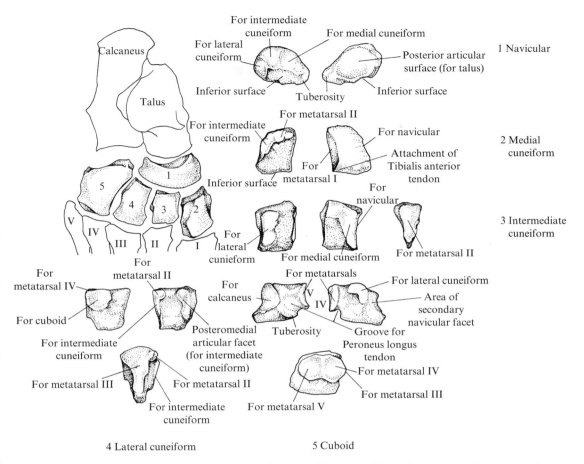

Figure 6–10 (*Top left*) Semiarticulated tarsus emphasizing distal row; (*right, from top to bottom*) navicular [(*left*) anterior and (*right*) posterior views], medial cuneiform [(*left*) lateral and (*right*) medial views], intermediate cuneiform [(*left*) lateral and (*right*) medial views], cuboid [(*left*) lateral and (*right*) medial views], cuboid, anterior view; (*far left, bottom*) lateral cuneiform, anterior view; (*left, bottom*) lateral cuneiform [(*left*) lateral and (*right*) medial views].

Talus

The stout and asymmetrically shape **talus** (Figure 6–9) looks somewhat like a subtubular object that was cut into top and bottom halves, with the bottom half then displaced anteriorly half a length. It articulates (1) with the distal tibial facet above it via its trochlear surface, (2) medially with the tibia's medial malleolus, and (3) laterally with the fibula's lateral malleolus. Anteriorly, the semi-ball-shaped talar head nestles into the shallow, cup-shaped posterior articular region of the navicular. The talus' superior articular or *trochlear surface*, which articulates with the distal tibia, is composed largely of an arced, semilunate articular surface with blunt, almost parallel medial and lateral edges; the latter edge is somewhat shorter. The region between these two edges is slightly depressed, thus giving the surface a shallow pulleylike appearance. The trochlear surface "drapes"

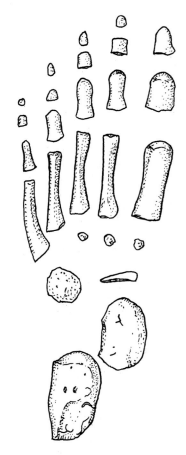

When the talus is viewed from above, (1) the lateral articular surface may be oriented rather vertically, with the inferior articular area for the fibula's lateral malleolus reflected outward; (2) in the anatomical position, the *head* and *neck* are anteriorly placed relative to the long axis of the bone's body (because the long axis of the body is oriented obliquely, it forms an obtuse angle with the neck's long axis); and (3) the "heel" is swollen in its midline into an elongate *lateral posterior tubercle*, which is separated from a bulkier *medial posterior tubercle* by an obliquely inclined, variably distinct *groove for the flexor hallucis longus*. In lateral aspect, the dominating triangular articular extension of the trochlear surface is circumscribed to some extent on its sides and below by nonarticular bone and adorned posteriorly by the variably tail-like protrusion of the posterior tubercle. Anteriorly, the talar neck is inset medially with an obvious downward angulation.

In medial aspect, the lateral edge of the trochlear surface rises above the level of the medial edge; the lateral posterior tubercle projects farther posteriorly than the medial posterior tubercle and is inclined superiorly (rising to a blunt end superiorly), whereas the truncated medial posterior tubercle comes to a blunt point inferiorly. The area of *attachment of the deltoid ligament* is marked by a swelling that lies just anterior and superior to the medial tubercle; the two may be confluent or delineated as separate entities. The *attachment site for the plantar calcaneonavicular ligament* lies near the most inferior and posterior limit of the articular region of the talar head; this site may be either indistinguishable from the rest of the head's articular surface or a distinct, flattened area that is further demarcated anterosuperiorly and inferoposteriorly by creases or shallow grooves.

Inferiorly, the talus presents two distinct articular regions, one corresponding to its body of and the other to its neck and head. A large, concave *posterior calcaneal articular surface (talar articular area for the calcaneus)*

Figure 6–11 Bones of left foot of third-trimester fetus, superior surface.

over the sides of the elevated eminence; the *lateral articular surface* (for the fibula's lateral malleolus) is roundedly triangular in shape and the *medial articular surface* (for the tibia's medial malleolus) is somewhat lunate. The general configurations of these medial and lateral articular areas conform to the shapes of the respective articular surfaces on the tibia and fibula. Viewed from above, the trochlear surface's anterior border is often straighter than the posterior border, which may be arced. Occasionally, the trochlear surface may be distended by a liplike articular projection that articulates with the tibia's medial malleolus.

that lies at approximately 45° relative to the long axis of the body articulates below with the calcaneus' large posterior articular surface. This former is bounded medially first by the thick *medial posterior tubercle* and then by the oblique *groove for the flexor hallucis longus*; the *lateral posterior tubercle* forms the posterior external corner of this groove. This surface descends rather drastically from the height of these tubercles but arcs more gently as it approaches its anterolateral extremity, where it is more tapered (or at least more rounded) than at its posteromedial edge. The anterolateral portion of this articular surface contributes to the variably developed inferior projection of the facet for the fibula's lateral malleolus.

The region of the talar *neck* and *head* is delineated sharply from the posterior calcaneal articular surface by a deep groove (*sulcus tali*), which may be variably smooth to deeply pitted. The sulcus tali, which emanates from the anterior face of the medial posterior tubercle and then follows its oblique orientation, widens and deepens as it proceeds laterally to the anteriormost extent of the talar head. The articular area on the neck and head's inferior surface is narrowest just below the region of the medial posterior tubercle. It then expands anteriorly, encompassing three variably distinct surfaces. The *medial calcaneal articular surface* lies immediately below the medial posterior tubercle and may be relatively flat and vaguely triangular in outline; it articulates with the middle articular surface of the calcaneus below. The relatively flat but thinner and broader *anterior calcaneal articular surface* is oriented horizontally and forms an obtuse angle with the middle calcaneal articular surface behind; it articulates with the calcaneus' anterior articular surface and is variably delineated from the rest of the talar head's articular surface, which articulates anteriorly with the navicular bone.

See Table 6–5 for siding tali.

Nonmetric variation in the talus is common and often takes the form of "pinching,"

Table 6–5 Clues to Siding a Talus: Its Features and Their Orientations

Feature	Position
Anterior calcaneal articular surface	Inferoanterior
Groove for flexor hallucis longus	Posteromedial
Lateral articular surface	Lateral
Medial articular surface	Medial
Medial tubercle	Medioposterior
Middle calcaneal articular surface	Inferomedial (distal)
Navicular articular surface	Anterior
Posterior calcaneal articular surface	Inferoposterior (proximal)
Posterior tubercle	Posterior
Sulcus tali	Inferior
Trochlear surface	Superior

"waisting," or complete subdivision of articular surfaces, especially of the posterior calcaneal articular surface and of the boundary between the middle and anterior calcaneal articular surfaces. As noted, the attachment sites for the deltoid and plantar calcaneonavicular ligaments are also variably delineated. There may be a squatting facet (corresponding to one that may develop on the anterodistal rim of the tibia) superiorly at the juncture of the neck and head.

Calcaneus

The **calcaneus** (Figure 6–9) is the largest and most robust tarsal. It is much longer anteroposteriorly and taller than bilaterally wide; its posteriormost portion constitutes the heel of the foot. It consists of three morphologically distinct components: (1) an *anterior component*, which provides the articular surfaces for the talus above and cuboid anteriorly; (2) a *posterior element*, which receives the insertions of the calf muscles and therefore provides the lever action of the bone; and (3) an *inferior* or *plantar surface*, which extends for the entire length of the bone. (These distinct components are invaluable clues to orienting the calcaneus properly.)

The articular surfaces of the calcaneus' *anterior component* mirror in general shape and contour their sister facets on the talus. In order to accommodate the talus, the anterior component is sharply concave in its midregion, with an oblique crease (*sulcus calcanei*) separating the *posterior talar articular facet* both from the *anterior talar articular facet* and medially projecting *middle talar facet*. The posterior facet is more than twice as expansive as the middle facet, which is typically much larger than the anterior articular facet. With the sulcus tali (on the talus), the sulcus calcanei encloses the *sinus tarsi*, which houses the interosseous talocalcaneal ligament.

The concave, obliquely oriented, and laterally tapering posterior calcaneal articular facet of the talus is mirrored to some extent in the calcaneus' large *posterior talar articular surface*. This region extends posteriorly over the calcaneus' superior surface, paralleling the talus' variable posterior extension and lateral posterior tubercle. The degree to which the posterior edge of the calcaneus' posterior talar articular surface is straight and squared up, rounded, or tapered is related to the development of the talus' lateral posterior tubercle as well as to the skewness of the plane created by the wall of bone that connects the talus' lateral posterior and medial posterior tubercles. Occasionally, a *secondary facet*, corresponding to the talus' lateral posterior tubercle, develops on the calcaneus, just behind the true border of its superior articular surface.

Proceeding anteriorly along the calcaneus, the *posterior talar articular surface* expands medially as it approaches the medially displaced *medial talar articular surface*. At this juncture, the posterior talar articular surface descends almost vertically. Coincidentally, the breadth of the posterior talar articular surface becomes truncated along its mediolateral surface. The sulcus calcanei separates the middle and posterior talar articular surfaces. A variably shallow to deep pit medially borders the laterally positioned inferior terminus of the posterior talar articular surface.

The elongate and variably but crudely ovoid to elliptically shaped *medial talar articular surface* is not always confluent with the smaller, ovoid *anterior talar articular surface* below it. Regardless, these two surfaces create a slightly concave surface that sits atop a bony strut (*sustentaculum tali*), which projects medially from the calcaneus' body and is most prominent below the region where the medial talar articular surface lies across from the posterior talar articular surface. The sustentaculum tali bears the anterior talar articular surface above it and projects anteriorly beyond the vertical facet (at the front of the calcaneus) for the cuboid.

Although the medial and anterior talar articular surfaces articulate, respectively, with the talus' middle and anterior calcaneal articular facets, there is no apparent correlation between the development of separate middle and anterior facets on the calcaneus and talus: these facets may occur as separate entities on one bone but be confluent on the other. The occurrence of separate medial and anterior facets on calcanei and tali, as well as the absence on the calcaneus of the anterior talar articular surface, is often noted as *nonmetric variation*.

The rest of the calcaneus anteriorly and superiorly is represented by a roughened hollow, which, in conjunction with the elevated concavity of the inferior surface of the talus above, creates the expanded *sinus tarsi*. This hollowed-out region of the calcaneus is typically devoid of much articular detail. However, an *accessory facet* may develop just on the inside of the medial talar articular surface as a secondary point of articulation with the medial rim of the posterior calcaneal articular surface of the talus. The pit described above (as lying just medial to the inferiormost portion of the posterior talar articular surface) may also become expanded anteriorly into a shallow basin, which may be subtended on both sides by elevated bony struts.

Viewed from above, the calcaneus' *posterior component* extends backward—at times with fairly straight and parallel sides but at others with a medial curvature—to terminate in a roughened *posterior surface*. The calcaneus' *lateral surface*, from the posterior talar articular surface to the posterior surface or "heel," is rather flat and vertical; it is concave superiorly between the posterior calcaneal articular surface for the talus and the calcaneus' posterior surface. The inferior border of the lateral surface is variably straight; it thickens posteriorly into a *lateral tubercle*. Proceeding anteriorly, the calcaneus' lateral surface decreases in height (following the descent of the posterior talar articular surface). Midway between the inferiormost extent of the calcaneus' posterior articular surface and its roughened inferior border, the lateral surface may *variably* bear a raised area (*peroneal tubercle*); it is typically minimally elevated, but it may be large enough to palpate in the living individual.

The calcaneus' *medial surface* is essentially divided into two components by a horizontal, thickened region that extends posteriorly from just below the sustentaculum tali and contributes to the apparent parallel-sidedness of the calcaneus when viewed from above. This thickening subtends the *groove for the flexor hallucis longus*, which lies on the underside of the sustentaculum tali. Beneath the sustentaculum tali, this thickened region arcs inferiorly, following the declivity of the bony projection, until it eventually fades out. Below this thickening, the calcaneus is variably concave, becoming more severely scooped out anteriorly.

Beneath the region where the medial and anterior talar articular surfaces might meet, the calcaneus swells into a modest *anterior tubercle*. Because of the increased concavity of the bone in that region, the anterior tubercle is situated quite medially. It is separated from the articular surface for the cuboid by a roughened area of bone, which is typically accentuated by a groove that delineates the medial border of the cuboid facet. A short plantar ligament attaches to both the anterior tubercle and the roughened area in front of it. Posteriorly and inferiorly, the calcaneus swells into the *medial tubercle*, which creates somewhat of a "corner" at the base of the posterior surface and is markedly larger than the *lateral tubercle*.

The calcaneus' *inferior surface*, on which the long plantar ligament attaches, looks generally like an elongate, right-angled triangle; the anterior tubercle is the apex, and its sides terminate in the medial and lateral tubercles. The vertical side of the triangle, which is a variably thickened band of coarse bone, extends between the anterior and medial tubercles, roughly paralleling the calcaneus' long axis. This band of roughened bone terminates at the base of the medial tubercle, which is distended downward and usually circumscribed medially, laterally, and anteriorly by a distinct edge or lip of bone. (The distension of the medial tubercle's edges is associated with the attachments of the abductor hallucis, the superficial part of the flexor retinaculum, the plantar aponeurosis, the flexor digitorum brevis, and part of the abductor digiti minimi; the latter originates on the lateral tubercle.) The borders of the lateral tubercle and larger medial tubercle may be either poorly delineated or separated by a crisply excavated groove.

Viewed from the side, the calcaneus' *posterior surface* (*calcaneal tuberosity*) is convex. Viewed straight on, it is broader inferiorly than superiorly. Frequently, the inferior surface of the small lateral tubercle lies well above the level of the inferior surface of the medial tubercle; at other times, it is distended inferiorly and creates a corner that squares up the base of the posterior surface. Three areas are usually distinguishable on the posterior surface. The smooth uppermost area is usually small and delineated inferiorly by a shallow, irregularly coursing, horizontally oriented groove. The middle area is bounded superiorly by this groove and inferiorly by another groove, which is accentuated by a raised edge of rugose bone, to which attaches the

Table 6–6   Clues to Siding a Calcaneus: Its Features and Their Orientations

Feature	Position
Anterior talar articular surface	Superoanterior
Anterior tubercle	Medioanterior
Calcaneal sulcus	Superomedial
Calcaneal tuberosity	Posterior
Cuboidal articular surface	Anterior
Groove for flexor hallucis longus	Mediosuperior (distal)
Medial tubercle	Inferomedial (proximal)
Middle talar articular surface	Superomedial
Lateral tubercle	Inferolateral
Peroneal tubercle	Lateroanterior
Posterior talar articular surface	Superior
Sustentaculum tali	Mediosuperior

calcaneal (Achilles) tendon. The third division of the posterior surface lies below this roughened elevation.

The calcaneus' entire *anterior surface* serves as the area of *articulation with the cuboid*. This vertical facet is oriented obliquely relative to the calcaneus' long axis (the cuboid articular facet's medial edge lies posterior to its lateral margin). This facet is slightly concave superiorly and gently convex inferiorly. Although the overall shape of the anterior surface is variable, it is often narrower inferiorly, even if minimally, than superiorly. [Viewing the posterior end straight on, the calcaneus' articular region, anteriorly and superiorly, looks like a mitten, with the medial talar and cuboid articular surfaces corresponding, respectively, to the thumb (which lies on the side of the body from which it comes) and ball of the hand. The posterior talar articular facet corresponds to the area of the fingers.]

See Table 6–6 for clues to siding calcanei.

Navicular

The **navicular** (Figure 6–10) articulates posteriorly primarily with the talar head; it may also contact slightly the calcaneus posteromedially. Anteriorly, the navicular articulates with the three somewhat wedge-shaped cuneiforms. The (vaguely) boat-shaped navicular is concave posteriorly and convex anteriorly and bears a "rudderlike" extension medioinferiorly.

The *posterior articular surface* mirrors in part the contour and configuration of the talus' head: it is teardrop-shaped, with its apex pointing medioinferiorly. The rudderlike extension (*tuberosity of the navicular*) is somewhat hooked inferiorly and projects beyond the apex of the posterior articular surface. The perimeter of the posterior articular surface, from the apex around the superior (proximal) and lateral sides to the midpoint of the inferior (distal) side, is elevated. The rest of the perimeter of the posterior articular surface is unbounded and "opens" inferiorly. Also at the midpoint of the inferior side, the outline of the posterior articular surface may become slightly angular or wavy. The body of the navicular extends beyond the perimeter of the posterior articular surface.

The navicular's roughened *inferior (plantar, distal) surface* may bear small, irregular, elevated patches of bone, some of which may also lie in the broad groove situated lateral to the navicular tuberosity, which, because it is thick anteriorly and tapers posteriorly, is obliquely oriented. The broad groove is usually more deeply incised just on the inside of the tuberosity; it transmits part of the tendon of the tibialis posterior (which inserts largely on the navicular tuberosity but also courses to the cuneiform bones as well as to the bases of metatarsals II–IV).

The *anterior surface*, although often described as convex, is actually composed of three vaguely triangular articular planes whose apices approach the midpoint of the inferior edge of the bone and whose bases fan out to their respective lateral and superior borders. These articular planes correspond to areas of contact with the three cuneiform bones. In general, the *medial articular area for the medial cuneiform* is the largest. The *middle* and *lateral articular areas* are subequal in size and associated with the *intermediate* and *lateral cuneiform bones*, respectively. The middle articular surface is somewhat concave, thus creating distinctly raised borders with the medial and lateral articular regions on either side of it. The "convexity" of the navicular's anterior surface results, therefore, more from the declination of the medial and lateral articular regions away from the middle articular surface than from a continuous arcing of the entire anterior articular surface. The broadened base of the navicular tuberosity extends beyond the limits of the articular region for the medial cuneiform.

The lateral side of the navicular bone is quite *variable*: it may be smoothly convex, rather vertical, or even angular in outline. Although the anteroposterior depth of the navicular seems to be consistently between 1 and 2 cm medially, lateral depth can vary markedly.

Medial Cuneiform

The relatively tall and thin **medial cuneiform** (Figure 6–10) articulates posteriorly with the medial articular region of the anterior surface of the navicular and anteriorly with metatarsal I. It makes contact laterally with the medial side of the intermediate cuneiform as well as with the base of metatarsal II. Of the three cuneiforms, the medial is the tallest, longest, and widest. It is roughly two and a half to three times taller (inferosuperiorly) and two times longer (anteroposteriorly) than it is wide (mediolaterally). Although extremely *variable* in the details of its shape and articular surfaces, certain features are relatively consistently presented.

In general, the medial cuneiform is broader inferiorly at its base than superiorly. The relatively flat or somewhat convex, typically rectangular or trapezoidally shaped *inferior* or *plantar surface* is usually quite roughened. Superiorly, the bone curves down from its distinct lateral margin (its contact with the base of metatarsal II) to its medial side. Viewed

from above, the medial cuneiform is broader anteriorly than posteriorly; a bony ridge may accentuate its lateral margin. Medially, the superior margin is convex, sloping downward as it proceeds posteriorly.

The upper and lower halves of the medial cuneiform's *medial surface* are gently convex, but the midpoint of this surface is noticeably concave. Anteriorly, the inferior portion of the medial cuneiform bears on its medial surface a slightly elevated, variably configured flattened area (the attachment site of the tibialis anterior tendon). Immediately posterior to this, there may be a shallow pit, a minor swelling, or a small, elevated secondary point of attachment.

The medial cuneiform's posterior *articular surface*, which mirrors the articular surface of the navicular behind, is slightly concave and variably teardrop-shaped. It may also be asymmetrical in outline, with the medial margin being more concave than the lateral (the medial margin contributes to the medial curvature of the bone and its articulation with the intermediate cuneiform). The bone's inferior portion swells out below the posterior articular surface; the medially cresting superior portion rises significantly above it.

Most of the medial cuneiform's *lateral surface* is variably straight to mildly concave. Its superiormost portion curves inward to some extent. Bandlike articular surfaces along the posterior and superior margins of the lateral surface are often connected and form an upside-down "L." (In some individuals, however, posterior and superior articular surfaces are disjunct.) The posterior articular strip forms a corner with the lateral margin of the posterior articular surface; it and most of the *superior articular surface* contact the *intermediate cuneiform*. Immediately in front of the superior articular surface lies a variably developed surface, which articulates with the posteromedial portion of metatarsal II; it may be quite large and distinctly separate from the longer articular region behind it or diminutive and barely distinguishable from the dominant superior articular area. The anterior margin of the medial cuneiform's lateral surface may bear a slight vertical groove just behind the edge of the anterior articular surface (which may encompass the bone's full height). Because this groove follows the contour of the anterior articular surface, it may curve in toward the bone's midregion. The bone's inferolateral "corner" is *variably* unexpanded to being markedly swollen and distended.

The medial cuneiform's *anterior articular surface* is taller than the posterior articular surface and its morphology more variable. (This articular surface looks like the footprint of a slipper or moccasin, with its "heel" inferiorly placed and the inward curvature of its arch located near the horizontal midline of the bone; the "heel" tends to be slightly concave and the "ball" somewhat convex.) Within this context, *variation* exists in the degrees to which the different components of the "foot" are narrow, broad, and/or curved. The anterior articular surface may also be noticeably "pinched" or "waisted" at its midline or bisected into separate articular facets.

Intermediate Cuneiform

The **intermediate cuneiform** (Figure 6–10) is the smallest and most wedgelike of the cuneiforms. Its *anterior* and *posterior articular surfaces* are essentially triangular in outline. The medial side may be in part convex, paralleling the slightly concave configuration of the lateral side of the medial cuneiform, against which it lies. The anterior articular surface may be gently convex, relatively flat, or mildly concave. Sometimes the anterior articular surface is longer than the posterior one, but the reverse may also occur. The posterior surface, which articulates with the middle articular region of the navicular's anterior surface, is often slightly concave. The *superior surface* is broad, variably bumpy to smooth, and relatively flat to gently convex. The *inferior* or *plantar surface* is more a thickened margin than a platform.

The intermediate cuneiform's *medial surface* bears an inverted, variably "L"-shaped articular surface. The base of the "L" courses along the superior margin; its side extends part or all of the way down the posterior margin. The configuration of this oddly shaped articular surface corresponds to the upside-down "L"-shaped articular surface on the lateral surface of the medial cuneiform. However, even if the articular "L" of the medial cuneiform is separated into distinct superior and posterior articular regions, the "L"-shaped articular surface on the medial face of the intermediate cuneiform may remain intact and vice versa. The area circumscribed by the "L"-shaped articular surface—usually just the bone's anteromedial corner—is a roughened surface within which secondary areas of articulation with the medial cuneiform may develop.

The intermediate cuneiform's *lateral surface* bears an articular surface that corresponds to its contact with the lateral cuneiform. The lateral articular surface may extend along the bone's superior and posterior margins, with the latter typically reaching and the former failing to reach the anterior margin; the lateral articular surface is, therefore, more truncated anteroposteriorly. Although forming more or less a right angle in its superoposterior corner, the lateral articular surface is not similarly indented internally. The rest of the intermediate cuneiform's lateral surface is roughened and may develop a *secondary articulation* with the lateral cuneiform.

Lateral Cuneiform

The **lateral cuneiform** (Figure 6–10) is also wedge-shaped. It is broad superiorly and narrow along its inferior or plantar surface. The *superior surface* is variably rough to smooth and flat to undulating. Its parallel sides are not straight: the medial edge bends inward and the longer lateral edge buckles outward at the midpoint of the superior surface. (The medial edge of the lateral cuneiform appears to wrap around the smaller intermediate cuneiform.) The longer lateral edge parallels the oblique orientation of the bone's posterior articular surface. The lateral cuneiform's inferior margin is typically straight and truncated posteriorly (does not extend the bone's full length).

The *anterior articular surface* primarily abuts metatarsal III's base; its anteromedial corner contacts metatarsal II's posterolateral margin. The variably flat to undulating anterior articular surface looks essentially like an elongate triangle but is swollen to some extent superolaterally and indented on its lateral margin.

The *posterior surface*, which contacts the lateral facet of the navicular's anterior articular surface, is small, subtriangular, and consistently, albeit slightly, concave overall. A raised rim may course around the perimeter of this surface's medial margin, from its superior to inferior extremities. The bone's thickened inferior margin extends markedly below the posterior surface. (The lateral cuneiform's posterior articular surface is oriented obliquely relative to the plane of the anterior articular surface.)

The lateral cuneiform's *medial side* (which contacts the intermediate cuneiform) bears a relatively large, variably ear- to inverted boot–shaped *posteromedial articular facet*, which is confluent around the bone's posteromedial edge with the small posterior articular facet. The somewhat convex posteromedial articular facet mirrors the slight concavity of the corresponding articular surface on the intermediate cuneiform. Most of the lateral cuneiform's medial side is roughened, nonarticular bone with irregular surface topography, which may bear as many as three small articular regions along its anteromedial border. One of these may be located in the bone's superior and anteromedial corner; the second at the inferior and anteromedial corner, where metatarsal II may make contact posterolaterally; and the third facet directly along the border, where a point of secondary articulation with the intermediate cuneiform may develop.

A flexure or "bend" in the lateral cuneiform's *lateral side*, just anterior to its midpoint,

subdivides it into two planes. The posterior plane is dominated by the large, vaguely wing-shaped *articular facet for the cuboid*, which is confined primarily to the upper two-thirds to one-half of the bone's posterolateral aspect. Sometimes, this articular region may extend as far as the bone's inferior margin. Just in front of the inferior portion of this articular region (regardless of its length) lies a variably small, shallow pit. An extremely small (sometimes barely visible) articular facet, which contacts the posteromedial edge of metatarsal IV, may occur with some frequency in the anterosuperior corner of the bone; it will be confluent at the bone's margin with the anterior articular surface and connect posteriorly to the larger posterosuperior articular facet (for the cuboid) via a ridge. The region below this ridge may be partially articular and roughened. As with the other cuneiforms, which maintain the foot's transverse arch, ligaments attach to the nonarticular surfaces of the bone's medial and lateral sides.

Cuboid

The **cuboid** (Figure 6–10) is the longest, tallest, and bulkiest of the distal row of tarsals. It articulates (1) posteriorly with the calcaneus' anterior articular surface, (2) medially with the lateral cuneiform and the navicular's anterolateral edge, and (3) anteriorly with metatarsals IV and V. Taking the positions of these bones relative to one another and the cuboidal–calcaneal articulation as reference points, the cuboid's body is skewed laterally while its articular surfaces for metatarsals IV and V are oriented obliquely. Viewed from above, the cuboid is broader posteriorly than anteriorly; it has a short, concave or notched lateral side and a markedly longer medial side that forms a corner near its posterior base.

The cuboid's *superior surface* is roughened and variably flat to undulating. Various dorsal ligaments—coursing between this tarsal and the calcaneus, navicular, lateral cuneiform, and metatarsals—attach to its superior surface. The cuboid's *inferior* or *plantar surface* is distinguished from its superior surface by a shallow to moderately excavated *groove* that courses between the bone's anterior margin and a modest to well-developed *ridge* situated approximately one-third the distance from this margin. The ridge parallels the orientation of the cuboid's anterior surface; the groove it subtends houses the peroneus longus tendon, which courses through the *notch* on the bone's lateral side. Occasionally, the ridge protrudes markedly laterally. When it does, it—in conjunction with a lateral lipping of the anterior articular surface—further accentuates the lateral notch. Laterally, the ridge may remain anterior to or become coincident with the corner of the posterior articular surface. The lateralmost portion of the ridge is the *tuberosity of the cuboid bone*. Its surface is variably smooth because it articulates with either a true sesamoid bone or an inclusion of unossified cartilage in the peroneus longus tendon. Posterior to the oblique plantar ridge, the remaining two-thirds to three-quarters of the cuboid is more or less triangular in outline and its surface mildly concave and roughened.

The cuboid's *posterior articular surface* roughly mirrors the triangular outline of the calcaneus' anterior articular surface. Its inferior margin is long and relatively straight (in the anatomical position, it is mediolaterally angled downward), and its superolateral side is somewhat longer than the superomedial side; the corners of this "triangle" are curved, not angular. Reflecting further the calcaneus' anterior articular surface, the cuboid's posterior articular surface is gently concave and slightly twisted (its inferomedial aspect is deflected partially upward; its superolateral border faces downward slightly).

The cuboid's long *medial side* is dominated superiorly and somewhat centrally by the large *articular facet for the lateral cuneiform*, which tends to mirror (or at least emphasize) the configuration of the corresponding facet on the lateral cuneiform (if the facet of the lateral cuneiform is elongate superiorly and somewhat truncated inferiorly or relatively short superiorly and thickened and extended inferiorly, so is the facet on the cuboid). Regardless of the range of potential variation, the anteroinferior edge of the cuboid's articular facet is delineated further by a moderately excavated and rather extensive depression, which may extend as far as the anterior articular surface's border. Posterior to this facet, the bone either angles toward the edge of the posterior articular surface or simply curves posteriorly. In either case, a smaller *secondary navicular articular facet* may be present near the cuboid's posterior boundary. This secondary facet reflects a point of articulation with the anteromedial margin of the navicular bone and may be accentuated by a pit or crease along its inferior border.

The cuboid's *anterior surface* is "pinched" or "waisted" just behind its rim and unequally subdivided into somewhat concave medial and lateral articular planes that differ in size and orientation. The medial articular plane faces more directly forward to articulate with the base of metatarsal IV; the most medial extent of this surface's rim may contact the posterolateral edge of metatarsal III. The lateral articular plane, which articulates with metatarsal V's base, is directed outward; it thus forms a very obtuse angle with the medial plane, from which it is sometimes further delineated by a vertical ridge. The (medial) border of the medial articular area is foreshortened; the lateral articular surface is more elongate.

See Table 6–7 for clues to siding the tarsals.

Metatarsals

Morphology

The **metatarsus** of the foot is composed of five **metatarsal** bones (I–V) (Figure 6–12) situated between the distal row of tarsals and the phalanges. The metatarsals are similar to long bones: each has a *head*, *shaft*, and *base*. Unlike a typical long bone, the metatarsal's head lies

Table 6–7 Clues to Siding Tarsals: Their
Features and Orientations

Feature	Position
Cuboid	
Cuboid tuberosity	Plantar–lateral
Facet for calcaneus	Posterolateral
Facet for lateral cuneiform	Mediodorsal
Facet for metatarsal IV	Anterior
Facet for metatarsal V	Anterolateral
Groove for peroneus longus	Plantar–anterior and lateroanterior
Intermediate (second) cuneiform (smallest of the cuneiforms)	
Facet for lateral cuneiform	Lateroposterior
Facet for metatarsal II	Anterior
Facet for medial cuneiform (L-shaped)	Medioposterior and mediodorsal
Facet for navicular	Posterior
Lateral (third) cuneiform	
Facet for cuboid	Lateroposterior
Facet for intermediate cuneiform	Medioposterior
Facet for metatarsal III	Anterior
Facet for metatarsal IV	Lateroanterior
Facet for navicular	Posterior
Medial (first) cuneiform (largest of the cuneiforms)	
Facet for intermediate cuneiform	Lateroposterior and laterodorsal
Facet for metatarsal I (footprint-shaped)	Anterior
Facet for metatarsal II	Lateroanterior
Facet for navicular (concave)	Posteroplantar
Navicular	
Facet for intermediate cuneiform	Anterior
Facet for lateral cuneiform	Anterolateral
Facet for medial cuneiform	Anteromedial
Facet for talus (concave)	Posterior
Groove for tibialis posterior	Inferomedial
Tuberosity of navicular	Medioinferior

at its distal or anterior end and the base at its proximal or posterior end.

The *first* (most medial) *metatarsal (metatarsal I)* is the shortest and, in all respects, the bulkiest metatarsal. It also bears an articular facet (for metatarsal II) only on its lateral side (the side of the body from which the bone comes).

Metatarsal I's *base* is vertically accentuated and vaguely ear-shaped in outline; its surface mirrors in part the shape of the medial cuneiform's anterior surface, with which it articulates. *Variation* in the *posterior surface* is noted in a pinching or notching approximately midway along the slightly concave lateral margin, below which a small pit might be found. Some individuals also develop a notch in the noticeably convex medial margin. Less frequently, the medial and lateral notches are more invasive and create two articular regions, between which a horizontal ridge may course. The somewhat concave posterior surface is circumscribed by a shallow groove and consists of a broader and more concave upper portion and a narrower, less concave, somewhat lateromedially facing lower portion.

In cross section, metatarsal I's *shaft* is essentially triangular, with its base oriented laterally and its apex medially. Consequently, the shaft's *superior surface* slants downward lateromedially, the *lateral side* is basically flat, and the *medial aspect* is highlighted by a longitudinal ridge, which may swell into a small tuberclelike structure just in front of the posterior articular surface's rim. Anteriorly, the *inferior* or *plantar margin* of the lateral surface swings laterally outward and becomes distended into a short, tonguelike projection, over which the articular surface of the bone's head spreads. Posteriorly, the inferior margin of metatarsal I's lateral surface terminates in a roughened, somewhat ovoid, flattened to gently concave attachment site for the peroneus longus tendon. Superiorly on the lateral side of the posterior articular surface's rim may be a variably developed *facet for* (contact with) *the second metatarsal* (which bears its corresponding facet anterior to the facet on the margin of its base for contact with the edge of the medial cuneiform). The inferior margin of metatarsal I's lateral side is somewhat concave. As this margin proceeds posteriorly from the head, it narrows slightly and then descends

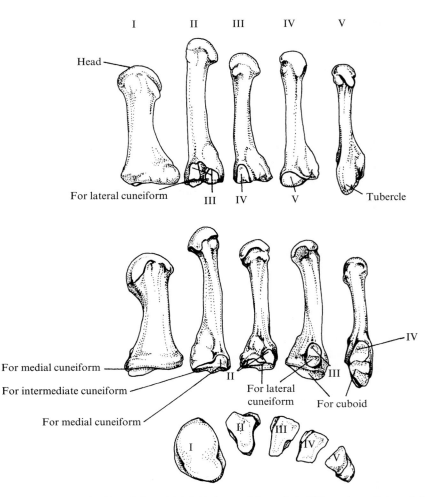

Figure 6–12 Right metatarsals: (*from left to right*) I–V; (*top row*) lateral, (*middle row*) medial, and (*bottom row*) proximal (basal) views.

more drastically as it nears the inferior limit of the bone's base. The superior margin remains relatively straight and terminates anteriorly in a small, low to spikelike projection.

Metatarsal I's *head* is broad. Its surface is largely articular and wraps around the bone's head, well onto the plantar surface. Viewed straight on, the superior margin of metatarsal I's head is slightly convex; its medial edge descends fairly vertically but, about halfway down, its lateral edge is deflected outward. The head's medial margin is quite thick. Just off center laterally is a ridge that thickens as it

proceeds downward. In concert with the head's inferomedial margin, the ridge's medial side bounds a relatively deep groove. The ridge's lateral side, in tandem with the lateral margin, circumscribes a shallower groove. These two grooves are most pronounced on the plantar area of the head's articular surface, which is where the off-center ridge is most exaggerated. The off-center ridge and its attendant grooves create a "W" shape; the lateral arm of the "W" is the more robust and elongate. Small *sesamoid bones*, which ossify between 8 and 14 years, ride in these grooves.

The **second metatarsal** is longer than, but similar in overall shape to, metatarsals III and IV. Although metatarsal II's *shaft* is somewhat more like that of metatarsal I (slanted and triangular in cross section), it is extremely thin, its inferior margin is more evenly concave, and its superior border is gently convex. Viewed from above, the bone's superior surface broadens slightly anteriorly and terminates in two marginal swellings, which are offset from the gently convex head by a circumferential groove. Posteriorly, the superior surface becomes somewhat flattened and roughened; it broadens dramatically toward the base.

Metatarsal II's *posterior surface*, which articulates with the intermediate cuneiform, is gently concave and roughly triangular in outline (its base is oriented superiorly and the blunt apex inferiorly). Confluent with the posterior surface is the variably developed *facet for the medial cuneiform*. This anterolaterally situated facet wraps around the medial and superior parts of the posterior rim; the area below is roughened. A *facet for the first metatarsal* may develop posterolaterally along the bone and anterior to this facet. A variably developed *facet for* (contact with the anteromedial edge of) *the lateral cuneiform* may be present laterally and superiorly, at the posterior surface's margin. Anterior to, if not also confluent with, it is (part of) the *facet for the third metatarsal*. The rest of the *facet for the third metatarsal* occurs laterally and in the bone's posteroinferior corner; it may be ovoid and localized to that corner of the bone, or it may extend upward to some degree. A roughened, variably excavated groove separates the inferior from the superior facet(s).

Metatarsal II's *head* is taller than wide (the reverse of metatarsal I's head); its medial side is taller (extends more prominently superoinferiorly) than the lateral side. (Viewed head-on, the superior border slopes upward and the inferior or plantar border slopes downward from the lateral to the medial side.) The head's lateral and medial margins continue along the bone's plantar side, giving the articular surface a somewhat "U"-shaped edge. The lateral arm of the "U" extends farther posteriorly than the medial arm (as in metatarsals III and IV; thus, the longer arm of the "U" is on the side of the body from which the bone comes). The medial border's swollen plantar extremity nestles superiorly in the "crook" of the metarsal I head's lateral projection.

Overall and in specific detail, the **third metatarsal** is similar to the second but variably shorter; its shaft is typically more slender, its *head* and *base* are narrower, and disparity in height between the head's medial and lateral sides is less pronounced. The gently concave *posterior surface* articulates with the lateral cuneiform. The ovoid, tongue-shaped *facet for the second metatarsal* is confluent with this surface's edge and superiorly, along the base's medial side; the area below it is roughened and may also be slightly concave. Somewhat inferiorly, an additional *secondary facet for* (articulation with) *the second metatarsal* may develop. The typically well-developed ovoid *facet for the fourth metatarsal*, which contacts the metatarsal at its edge, lies superiorly on the *lateral side*; roughened areas define it anteriorly and inferiorly. A shallow, horizontal groove may also be present inferiorly. In contrast to metatarsal II, the lateral side of metatarsal III's posterior portion does not develop a second, inferiorly placed point of articulation.

The **fourth metatarsal** is somewhat similar to the third and second. Metatarsal IV's *head* is virtually identical morphologically to metatarsals II and III (including emphasis on the lateral extension onto the plantar side of the "U"-shaped articular surface). But its head is more laterally and its base more medially deflected, and it bears a more distinct longitudinal ridge medially on its *superior surface* (which terminates posteriorly in an elevated, roughened region and gives the bone a longer, more slanted lateral side). Metatarsal IV's *posterior surface* is less consistently wedgelike (the base's medial and lateral sides are more parallel to one another). The posterior surface is variably concave to concavo-convex, mirroring variability in the cuboid's corresponding articular facet. A somewhat ovoid and elongate or elliptical *facet for* (articulation with) *the third metatarsal* lies superiorly on the base's *medial side*; a crease or groove just lateral to its margin delineates it superiorly and may also invade the raised and roughened area associated with the ridge of the superior surface. Its distance from the edge of the bone's base is variable. Below and behind this facet, the bone is quite roughened and irregular and circumscribed by a shallow to deep groove. A smaller *facet for* (contact with the anterolateral edge of) *the lateral cuneiform* occurs either just posterior to the facet for metatarsal III or along the lateral edge of metatarsal IV's base.

The *lateral side* of metatarsal IV's base bears a somewhat holster-shaped *facet for the fifth metatarsal* that courses down from the base's superior margin; its straighter side is folded along the bone's posterior surface and may form an edge. This distinct facet may be surrounded by roughened, irregular bone or set off by a sometimes deeply excavated groove. Anterior and perhaps somewhat inferior to this roughened to grooved region is a slightly elevated, ovoid to subtriangular area for the attachment of an interosseous muscle.

The **fifth metatarsal** is the most easily identified of metatarsals II–V: it bears only posterior and medial articular facets. Viewed from above, the bone broadens posteriorly, culminating in a large *tubercle* that lies laterally adjacent to the posterior articular surface. In tandem with the marked inward orientation of the *posterior surface* (which corresponds to the orientation of that surface on the cuboid), the tubercle extends metatarsal V posteriorly; sometimes the tubercle is a blunt, hooklike projection. The bone's *shaft* may be relatively straight or inwardly curved; in lateral aspect, it becomes increasingly compressed in a posterior direction (the *base* is wider than tall, the

reverse of the configuration of the bases of metatarsals II–IV). The broad *superior surface* slopes downward mediolaterally, elaborating further the pattern established on metatarsal IV (the slightly concave to saddle-shaped posterior surface is taller along its relatively vertical medial border and tapers laterally to a blunt or rounded terminus). A variably ovoid to lunate *facet for the fourth metatarsal* lies somewhat superiorly along the base's medial side; it is confluent with the posterior articular surface. The facet is surrounded inferiorly and somewhat anteriorly by roughened bone. Metarsal V's *head* may be more asymmetrical than on metatarsals II–IV, particularly in the extent to which the lateral articular margin on the plantar surface is enlarged.

See Table 6–8 for clues to siding metatarsals.

Development and Ossification

Mineralization of the metatarsals (Figure 6–11) begins during the eighth to tenth fetal weeks with the appearance of a center in each bone's shaft. In metatarsals II–V, ossification can occur in the heads (distal epiphyses) anytime between the third and eighth years. Secondary ossification begins during this period in metatarsal I's proximal epiphysis (at the bone's base). The pattern of ossification of metatarsal I is like that in the pedal phalanges, metacarpal I, and manual phalanges. Union of the relevant epiphyses to the metatarsal shafts may begin at about 12 years, with complete union occurring between 18 and 22 years.

Phalanges

Morphology

Although differing in size and robustness, foot **phalanges** (Figure 6–13, *phalanges digitorum pedis*) are similar in overall design. Each has a *head* (distally), a *shaft*, and a *base* (proximally). Their bases are broad and sublunate

Table 6–8   Clues to Siding Metatarsals: Their Features and Orientations

Feature	Position
Metatarsal I	
Base (proximolaterally distended)	Proximal
Facet for first cuneiform (concave)	Proximal
Facet for metatarsal II (if present)	Laterodorsal
Facet for proximal phalanx (convex)	Distal
Grooves for flexor hallucis brevis	Plantar (distal)
Head (plantar–laterally distended)	Distal
Nutrient foramen (proximodistal)	Laterodorsal (shaft; plantar concave)
Tuberosity for peroneus longus	Plantar–lateral
Metatarsal II (longest of the metatarsals)	
Base (proximolaterally distended)	Proximal
Facet for lateral cuneiform (if present)	Laterodorsal and lateroplantar
Facet for medial cuneiform	Mediodorsal (proximal)
Facet for metatarsal I (if present)	Mediodorsal
Facet for metatarsal III	Laterodorsal and lateroplantar
Facet for proximal phalanx (convex)	Distal
Facet for second cuneiform (concave)	Proximal
Head (plantar–laterally distended)	Distal
Nutrient foramen (distoproximal)	Lateral (shaft; plantar concave)
Metatarsal III	
Base (proximolaterally distended)	Proximal
Facet for metatarsal II	Mediodorsal and medioplantar
Facet for metatarsal IV	Laterodorsal
Facet for proximal phalanx (convex)	Distal
Facet for third cuneiform (flattened)	Proximal
Head (plantar–laterally distended)	Distal
Nutrient foramen (distoproximal)	Lateral (shaft; plantar concave)
Metatarsal IV	
Base	Proximal
Facet for cuboid	Proximal
Facet for lateral cuneiform	Mediodorsal
Facet for metatarsal III	Mediodorsal (proximal)
Facet for metatarsal V	Laterodorsal
Facet for proximal phalanx (convex)	Distal
Head (plantar–laterally distended)	Distal
Nutrient foramen (distoproximal)	Lateral (shaft; plantar concave)
Metatarsal V	
Base (proximolaterally distended)	Proximal
Facet for cuboid	Proximomedial (proximal)
Facet for metatarsal IV	Medial
Facet for proximal phalanx (convex)	Distal
Head (plantar–laterally distended)	Distal
Nutrient foramen (distoproximal)	Plantar–medial (shaft)
Tuberosity	Lateral

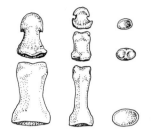

Figure 6–13 Right phalanges: (*left*) I, anterior view [(*top*) distal and (*bottom*) proximal phalanges]; (*middle*) example of II–V, anterior view [(*from top to bottom*) distal, middle, and proximal phalanges]; (*right*) example of II–V, proximal (basal) view [(*from top to bottom*) distal, middle, and proximal phalanges].

to elliptical in outline, and their shafts are narrower than their heads and bases (the sides of the bones have a slightly "scooped-out" appearance). Pedal digits II–V contain three phalanges each (a proximal or first row, a middle or second row, and a third distal or terminal row). Pedal digit I (*hallux*) comprises only two rows of phalanges.

The phalanges of the **proximal** or **first row** are similar: their *bases* are sublunate in outline (being flatter along the plantar or inferior edge) and uniformly concave (with a raised rim all around for articulation with the rounded metarsal heads). The first proximal phalanx is the largest of its set. A proximal **phalanx's** *head* is relatively compressed superoinferiorly. Viewed from above, the head of a proximal phalanx is somewhat concave centrally; this centrally placed indentation receives the central "keel" in the posterior articular surface of the phalanx in front. When viewed head-on, the proximal phalanx's superior margin is convex and its inferior margin concave. These curvatures create miniature "wings" on the phalangeal head that may be relatively flat on their sides or slightly concave or bear swellings (reminiscent of distal femoral epicondyles). The latter configuration is most characteristic of the first proximal phalanx, which is further

distinguished in that the curvatures of its head—above, in front, and below—are the most exaggerated.

The *superior surface* of a proximal phalanx's shaft, which is smoothly convex along much of its length, becomes flatter just behind the head. On the first proximal phalanx in particular, the region just behind the head is slightly roughened and may be irregular and even surmounted by a small elevation of bone. The shaft's *plantar* or *inferior surface* is flatter throughout most of its length; its medial and lateral borders may bear longitudinal ridges of some distinction. On the first proximal phalanx, a pit of variable depth usually develops just behind the bone's head; it delineates articular projections on either side of it.

Proximal phalanges are typically so symmetrical that clues to siding are virtually nonexistent. Sometimes, and especially on the first proximal phalanx, the head is deflected slightly laterally (to the side of the body from which it comes; the shaft's lateral side may be somewhat more concave than the medial side). Although the first phalanx can be identified with certainty, the slightly smaller and definitely more gracile second to fifth phalanges are usually indistinguishable from one another.

The **middle phalanges** of digits II–V are essentially short-shafted versions of the proximal phalanges, but their *posterior articular surfaces* are broader and sublunate in outline and bear gently raised, vertical, central "keels" that correspond to the indentations on proximal phalangeal heads. Pedal digit V's middle phalanx is most easily identified because it is often most asymmetrical, truncated in length, and prone to morphological alteration.

The **terminal** or **distal phalanges** have broad, flared-out *bases*; short, markedly inwardly curving *shafts*; and flattened, fanlike *heads*. A distal phalanx's base (sitting like a pedestal on which the rest of the bone sits) is somewhat elliptical in outline. Its *articular surface* is rimmed and elevated vertically in the midline (mirroring the indentation in the head of the phalanx behind it). The base's medial and lateral "corners" may be blunt and stout or curved and winglike. The *shaft* is essentially flat on its plantar or inferior surface and convex superiorly; a longitudinal, keel-like ridge may course along it. The *head* is convex superiorly and very flat and roughened inferiorly; its terminal edge is quite compressed, laterally extended, and variably smoothly arcuate to spearheadlike in outline. The terminal phalanx of the first pedal digit is the largest, while that of the fifth is typically the smallest and most amorphous in shape. Siding a terminal phalanx is nearly impossible, but the head of the terminal phalanx of the first pedal digit is sometimes tilted slightly medially (away from the side of the body from which it comes).

Development and Ossification

Mineralization of phalanges (Figure 6–11) starts with the appearance during the ninth or tenth week of centers in the shafts of the distal row of phalanges. At about the same time or only slightly later, centers arise in the

shafts of the phalanges of the proximal row. During the fourth month, ossification begins in the shafts of the middle row of phalanges. Centers usually appear in the bases (proximal epiphyses) between years 2 and 3, but this may be delayed until year 10. Union of bases to shafts begins at approximately 14 years and ends between 17 and 21 years.

Determination of Stature from Lower Limb Bones

Trotter (1970; Trotter and Gleser, 1977) developed formulae for determining an individual's height (stature) from measurements of the long bones (see Table 6–9). As with measurements on upper limb bones (Chapter 5), these are maximum (not physiological) bone lengths that are taken using an osteometric board. **Maximum length** (in cm) is achieved by positioning the bone in whatever position yields the highest number; **physiological length** (in cm) is measured with the bone oriented as it would be in the body. The formulae are valid for individuals 18–30 years of age. For older individuals, one must adjust the results by subtracting 0.06 times (calculated

Table 6–9   Formulae for Estimating Stature (cm) from Maximum Length of Upper Limb Bones

White American males	White American females
1.30 (femur + tibia) + 63.29 ± 2.99	1.39 (femur + tibia) + 53.20 ± 3.55
2.38 (femur) + 61.41 ± 3.27	2.93 (fibula) + 59.61 ± 3.57
2.68 (fibula) + 71.78 ± 3.29	2.90 (tibia) + 61.35 ± 3.66
2.52 (tibia) + 78.62 ± 4.32	2.47 (femur) + 54.10 ± 3.72
African-American males	**African-American females**
1.15 (femur + tibia) + 71.04 ± 3.53	1.26 (femur + tibia) + 59.72 ± 3.28
2.38 (tibia) + 86.02 ± 3.78	2.28 (femur) + 59.76 ± 3.41
2.11 (femur) + 70.35 ± 3.94	2.45 (tibia) + 72.65 ± 3.70
2.19 (fibula) + 85.65 ± 4.08	2.49 (fibula) + 70.90 ± 3.80
Asian males	**Mexican males**
1.22 (femur + tibia) + 70.37 ± 3.24	(femur + tibia) n.a.
2.40 (fibula) + 80.56 ± 3.24	2.50 (femur) + 58.67 ± 2.99
2.39 (tibia) + 81.45 ± 3.27	2.36 (fibula) + 75.44 ± 3.52
2.15 (femur) + 72.57 ± 4.66	2.36 (tibia) + 80.62 ± 3.73

SOURCE: Modified from Trotter (1970) and Trotter and Gleser (1977).

age minus 30) cm. The formulae are listed in order of decreasing accuracy (increasing standard error); all are more accurate than stature calculated from upper limb bone measurements (see Chapter 5 for the latter). As mentioned with regard to estimating stature using bones of the upper limb, one can adjust stature to account for age-related changes with the following equation (Trotter, 1970):

Stature at death (cm)
= [maximum stature − 0.06 (age − 30)] cm.

Chapter 7

Teeth

Although numerically representing a small component of an individual's skeleton, teeth are complex in development and morphology.

Basic Gross Dental Anatomy

A tooth (Figures 7–1 and 7–2) is composed of different tissues. The hardest is **enamel**, which encapsulates most, if not all, of the part of the tooth protruding from the gum (**crown**); humans have thick molar enamel, especially over the cusps, but enamel thickness varies by tooth and region of tooth. Mammalian teeth are secured to the jaw by one or more **roots**, which sit in sockets in the bone (**alveoli**, s. **alveolus**); polyphyodont animals (e.g., sharks) do not develop a root system. In humans, upper molars typically develop the greatest number of roots (three). Roots bear a thin layer of **cementum**, which is softer than enamel. The sometimes constricted or "waisted" enamel–cementum juncture is the **neck** or **cervix**. Deciduous teeth are typically distinguished from permanent teeth by their thinner, shorter, and, if multiple, more splayed roots, which accentuate the waisting of the tooth's cervical region.

A continuous tissue layer (**primary dentine**), which surrounds the **pulp cavity**, underlies enamel and cementum. Nerves and arteries penetrate the tooth via openings in root apices and course into the pulp cavity through root canals. Dentine exposed through the enamel—as a result of wear or breakage—wears away at a faster rate than the enamel around it. In time, severe abrasion can endanger the sensitive pulp cavity; if infection and decay are not complicating factors, the tooth will generate **secondary dentine**, which is somewhat harder than primary dentine and prolongs the life of the tooth. If infection invades the pulp cavity—because of carious infection or abscessing—nerve and thus tooth death will occur.

Teeth are not immovably anchored in the jaw. Each root is variably supported along its length by fibers of a **periodontal ligament** that stretches between the root and the wall of the alveolus, acting as a miniature shock absorber. [Some experimental evidence indicates that the periodontal ligament ultimately

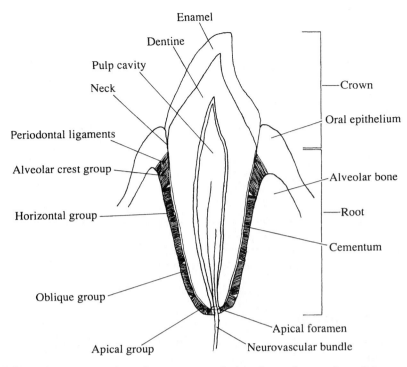

Figure 7–1 Schematic representation of gross anatomical regions of a tooth and its surrounding and supporting structures.

develops from the proliferating cell mass that also gives rise to the tooth, its different component tissues, and the surrounding alveolar bone (Ten Cate and Mills, 1972).] With bone in a state of constant flux—due to the differential interplay of osteoblasts and osteocytes—forward movement of teeth (*mesial drift*) is common (e.g., lower incisors of older individuals tend to be crowded together or even to overlap one another). Orthodontic prostheses to correct improper tooth position and/or orientation take advantage of the plasticity and responsiveness of bone relative to teeth.

Different Dentitions: An Overview

Most mammals—humans included—typically develop and erupt into their jaws two generations of teeth. Because some of the first generation of teeth are shed, they are referred to as **deciduous teeth**; deciduous teeth may also be called **milk teeth** because of the typically milky appearance of their enamel. Teeth that are not shed but retained throughout an individual's lifetime are identified as **permanent teeth**. Some permanent teeth replace deciduous teeth and thus represent a second generation of teeth developmentally. Other permanent teeth (e.g., human molars) do not replace teeth; they are thus defined as permanent teeth because of their fate. According to these definitions, humans and other mammals develop more permanent than deciduous teeth.

Animals that develop two "sets" of teeth during their lifetimes are **diphyodont** ("di" meaning "two," "phyo" meaning "families or generations," and "dont" meaning "teeth"; the condition is *diphyodonty*). Some extant mammals (e.g., manatees) develop only a single generation of teeth; they are **monophyodont**. Most reptiles and fish develop a lifetime of

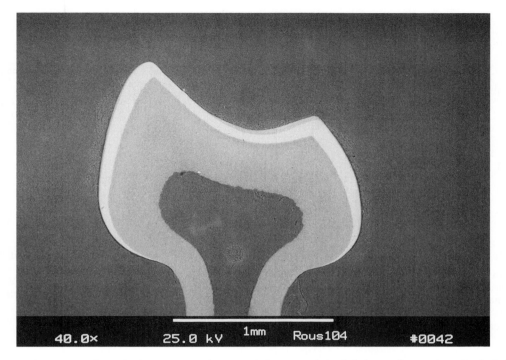

Figure 7–2 Section through the crown and some of the root of the left lower first molar of a fruit bat (*Rousettus amplexicaudatus*) illustrating (from outer layer to inner chamber) the layer of enamel (*white*), dentine (*light gray*), and pulp cavity (*dark gray*) (*courtesy of E. Dumont*).

generations of successional teeth: as if on a conveyor belt, wave upon wave of maturing teeth move toward the edges of the jaws as earlier rows of teeth are shed; these animals are **polyphyodont** ("poly" meaning "many"). In addition to a brief functional life, the rootless and poorly anchored teeth of polyphyodont animals are usually morphologically simple (e.g., conical or relatively flat and triangular in outline).

An animal with similarly shaped teeth is **homodont** ("same-toothed"); among mammals, seals and cetaceans, for example, are homodont. Most mammals, however, develop morphologically distinct groups of teeth; they are **heterodont** ("different-toothed"), and their morphologically relatable groups of teeth are **tooth classes.** According to received wisdom, the maximum number of tooth classes a mammal can

develop is three: **incisor, canine,** and **molar** (e.g., see Figures 7–3 and 7–4). Following traditional terminology, there are **deciduous** and **permanent incisors** and **canines** (e.g., see Figures 7–3 and 7–4). **Deciduous molars** (sometimes, but incorrectly, referred to as **deciduous premolars**) are replaced by **premolars;** by definition, however, "premolar" refers only to a permanent tooth that lies in front of the **permanent molars** (e.g., see Figures 7–3 and 7–4). Typically in heterodont mammals, successional teeth, although sometimes larger than the teeth they replace, are morphologically less complex (or, to put it another way, successional teeth tend to be no more morphologically complex then their predecessors, although, at least in some fruit-eating bats, it is the reverse—deciduous teeth are small, pointed, and morphologically simple).

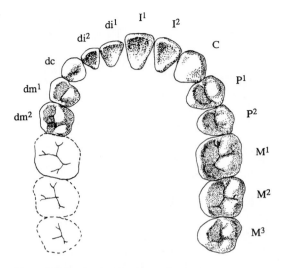

Figure 7–3 Occlusal view of upper permanent (*right*) and deciduous (*left*) dentitions; images of the permanent molars are drawn distal to dm² to illustrate the morphological continuity within the primary molar class.

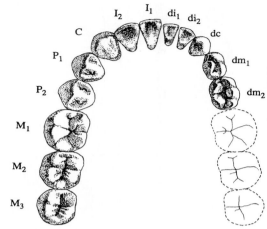

Figure 7–4 Occlusal view of lower permanent (*left*) and deciduous (*right*) dentitions; images of the permanent molars are drawn distal to dm₂ to illustrate the morphological continuity within the primary molar class.

Incisors are traditionally identified as the anteriormost teeth in the jaws. The stereotypical incisor is a somewhat spatulate or chisel-like, occlusally straight-edged tooth; teeth that look like an idealized incisor are described as *incisiform*. Many mammals, including humans, develop the same number of deciduous and permanent incisors in both upper and lower jaws.

The tooth traditionally identified as the canine lies behind the last incisor. Mammals are supposed to develop at most only one pair of canines in the upper and one pair in the lower jaw. Stereotypically (as in carnivores), the canine—especially the upper—is depicted as a relatively tall, pointed, single-cusped tooth; in humans, even the upper canine is only modestly elongate and pointed. A tooth that looks like a canine, even though it might not be in the same position in the jaw as the "canine," is described as *caniniform*. Although caniniform teeth can be found in various positions in the jaws of different mammals, the "true" upper canine is traditionally identified as the first tooth behind the premaxillary–maxillary suture; the "true" lower canine is

supposed to occlude with the front of the upper canine.

The teeth behind the canine belong to the molar class. Molars and at least the last deciduous molars are consistently the most morphologically complex teeth (the last deciduous molar of a heterodont mammal looks like, and is often misidentified as, a molar). When considered as members of the same tooth class, deciduous and permanent molars grade from one to another in both size and shape. When the typically morphologically less complex premolars are compared to permanent molars, this size–shape gradient is not apparent; thus, except when a mammal develops a *molariform* last premolar (e.g., horses), premolars are usually quite distinct morphologically from permanent molars. Although teeth identified as premolars may be single-cusped, they are typically lower-crowned and more morphologically embellished than canines. Human premolars are either uni- or bicuspid. Teeth that look like stereotypical premolars are described as *premolariform*.

The number of kinds of teeth in each jaw may be represented in a **dental formula**. The *dental formula of the normal human permanent dentition* (compare with Figures 7–3 and 7–4) is as follows:

$$\frac{2.1.2.3}{2.1.2.3}$$

The first column represents the number of permanent incisors; the second column, the canine; the third, premolars; and the fourth, permanent molars. The numbers above the line refer to upper teeth and those below the line, to lower teeth. Since mammals are bilaterally symmetrical animals, a dental formula is often constructed and read as representing one-half of the upper and one-half of the lower dentition. The total number of teeth in an upper or lower jaw is calculated by adding the upper or lower row of numbers and then multiplying by 2 (e.g., the number of human upper teeth is $2 + 1 + 2 + 3 = 8 \times 2 = 16$). Although humans and other mammals develop the same number of each kind of tooth in the upper and lower jaws, this is not true of all mammals (thus the numerator and denominator of a dental formula will not be identical), nor is the development of a tooth in each category (e.g., cows and deer lack permanent upper incisors altogether but develop them in the lower jaw).

The *dental formula of the normal human deciduous dentition* (compare with Figures 7–3 and 7–4) is as follows:

$$\frac{2.1.2}{2.1.2}$$

The first column denotes deciduous incisors; the second, deciduous canines; and the third, deciduous molars. Humans, like most other mammals, develop as many replacing teeth as deciduous predecessors, but this is not universal. For example, among prosimian primates, some slow lorises (genus *Nycticebus*) erupt two pairs of upper deciduous incisors but replace them with only one pair of upper permanent incisors, and species of *Lepilemur* (sportive lemur) develop one pair

of unreplaced upper deciduous incisors; some insectivores retain "deciduous" molars into adulthood (Schwartz, 1980).

In shorthand, a tooth is referred to by the first letter of its name. For permanent teeth, the letter is capitalized (I, incisor; C, canine; P, premolar; M, molar). Deciduous teeth are depicted in various ways (e.g., dI, di, i; dc, dC, c; dm, dM, m). A common approach to identifying individual teeth of a tooth class is to number them sequentially. Using the tooth abbreviations above, upper teeth are identified by a superscripted number and lower teeth by a subscripted number. Thus, the central or first permanent upper incisor is I^1 and the second lower deciduous molar is dm_2. Upper and lower permanent and deciduous canines are noted in one of two ways. The C or dc may be placed above (upper) or below (lower) a line that represents the line in a dental formula ($\underline{C}$ or $\overline{dc}$) or a super- or subscripted "1" is used (C^1 or dc_1). When referring to both upper and lower teeth, the number follows the letter.

Tooth Surfaces and Orientation

Teeth are anchored in the bony **alveolar ridge** or **process**; root sockets are **alveoli** (s. **alveolus**) (Figure 7–1). Alveolar bone apparently derives from cells involved in tooth development rather than from the basilar bone of the upper or lower jaw (e.g., Ten Cate and Mills, 1972). Sometimes, teeth and the alveolar process in which they sit are referred to collectively as the **dental arcade**. Mammalian dental arcades are commonly "U"- or "V"-shaped; thus, most, if not all, of one side of a jaw is straight, with many teeth oriented along an anteroposterior axis. The human dental arcade, however, conforms to a relatively squat, posteriorly broadening parabola, with many teeth oriented along different axes; for example, the central incisor's long axis forms an obtuse angle to that of the third molar. In order to accommodate different tooth orientations, a set of terms has been devised to refer to the four sides of a tooth, regardless of where in the dental arcade a tooth sits.

A tooth's **lingual** side faces the tongue; the opposite side is the **buccal** (referring to cheek) or **labial** (referring to lip) side. "Buccal" is used more frequently in referring to human teeth because our lips (at rest) do not extend beyond our anterior teeth. "Labial" is used more frequently for animals whose lips extend along much or all of the dentition. Here, we follow the convention for human teeth and use "buccal" and "lingual." The buccal–lingual (combined as **buccolingual**) dimension describes a tooth's **width** or **breadth** (Figure 7–5).

Mesial (anterior) and **distal (posterior)** refer to a tooth's other sides. "Mesial" and "distal" are more accurate for describing human teeth because, strictly speaking, "anterior"

means "forward-facing" and "posterior" means "backwardly directed." Given human dental arcade shape, no tooth is truly oriented anteroposteriorly. "Mesial" denotes the side of a tooth that is closer to, or faces toward, the midline of the jaw; "distal" is the side of a tooth farther or facing away from the midline of the jaw (e.g., a central incisor's mesial side faces directly toward and its distal side faces away from the midline; anterior on a human incisor is actually the buccal and posterior the lingual surface). The **mesiodistal** dimension is the tooth's **length**.

Two other tooth surfaces need to be discussed. One is the chewing or **occlusal surface**. Chewing or mastication [or grinding (*bruxating*) one's teeth while asleep] wears down the occlusal surfaces: the higher portions (e.g.,

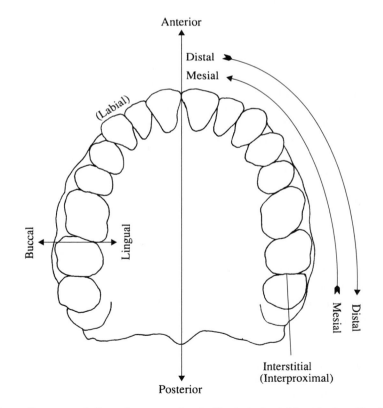

Figure 7–5   Schematic representation of an upper jaw to illustrate dental terminology for defining position and direction.

cusps, crests) are abraded to form flat surfaces (**facets**); eventually, the bulk of the enamel and even much of the tooth's crown may be worn down.

The surface between two neighboring teeth is the **interstitial (interproximal) surface**. For example, interstitial surfaces between central incisors are their mesial surfaces; interstitial surfaces between a first and second molar are the distal surface of the former and mesial surface of the latter. Since teeth move in the jaws, neighboring teeth will abrade against one another and eventually produce an *interstitial (interproximal) wear facet* on the mesial and distal surfaces of neighboring teeth. Increasing interstitial wear will shorten a tooth and add to the inaccuracy of measuring tooth length. The presence or absence of interstitial wear facets indicates the existence of neighboring teeth, which can be helpful, especially when dealing with isolated teeth or dentally incomplete jaws: for example, in determining whether a first molar had erupted behind a second deciduous molar or a second molar behind a first could affect aging or in identifying an erupted third molar which lacks a distal interstitial wear facet. Matching up interstitial wear facets by size and shape using digital technology could also assist in associating isolated teeth.

Basic Nomenclature of Crown Anatomy

There is a specific nomenclature for identifying the various cusps, crests, depressions, and other structures that might occur on a mammal's tooth (and humans are mammals), regardless of the specific tooth being studied (Figure 7–6). **Primary cusps** are represented by the word root "con"; *cone* refers to an upper-tooth cusp and *conid*, to a lower-tooth cusp (in general, the suffix "-id" identifies a feature on a lower tooth). **Secondary cusps** on upper teeth are *conules* and those on lower teeth, *conulids*. **Crests** or **ridges** are *cristae* (s. *crista*) on upper teeth and *cristids* on lower teeth. A three-dimensional band or ledge of enamel on a crown's side (not occlusal surface) is *cingulum* (pl. *cingula*) or *cingulid* (pl. *cingulids*) depending on whether the tooth is an upper or lower. A cingulum/cingulid may ring a crown or be restricted to one or a few regions; humans rarely develop cingula/cingulids, but, when present, they are weakly pronounced. A small and topographically restricted swelling or distension of enamel on a crown's side (usually buccal and/or lingual side) is a *style* or *stylid*. Prefixes (e.g., **proto-**, **para-**, **meta-**) are used in conjunction with *cone/conid*, *conule/conulid*, *style/stylid*, and *crista/cristid* to refer to specific structures in specific locations on a tooth.

Although it would be ontogenetically more appropriate to describe first primary and then secondary teeth, we follow convention by describing the "adult" or permanent teeth as a "set" and then the deciduous teeth. Since certain conventions used in identifying molar morphology are applied to morphologically simpler teeth, we begin with an overview of the permanent molars and proceed to the upper and then the lower teeth.

Molars: An Overview

Certain generalizations apply to human upper and lower molars (see Figures 7–3, 7–4, 7–6 to 7–9):

1. Typically, they decrease in size and morphological complexity from M1 to M2 to M3. That is, a cusp (or any other feature) is less fully expressed on M2 than M1 and less so on M3 than M2 (it could also be absent on M3). In short, whatever the degree of expression of a feature on M1, it will likely be less on M2 and even less on M3.

2. Molar roots also tend to conform to a gradient. Typically, M1 roots are the most splayed and separated and M3 roots, the most closely approximated. An extreme gradient would be where the M1 roots are separated only near their tips and not separated at all on M3 (although longitudinal

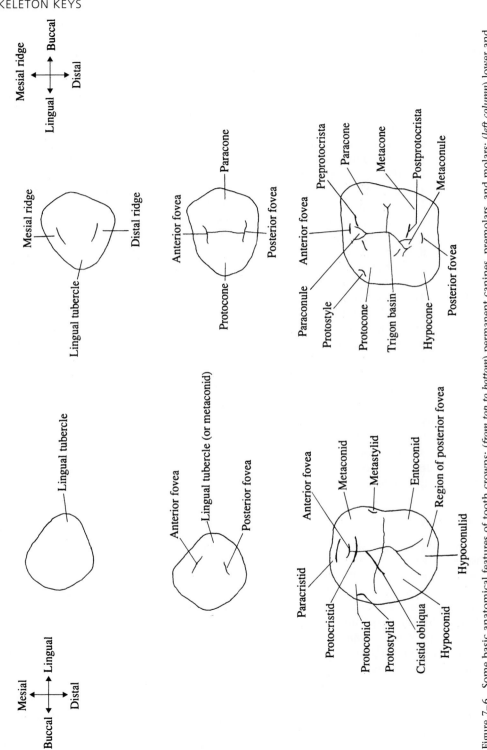

Figure 7–6 Some basic anatomical features of tooth crowns: (*from top to bottom*) permanent canines, premolars, and molars; (*left column*) lower and (*right column*) upper teeth.

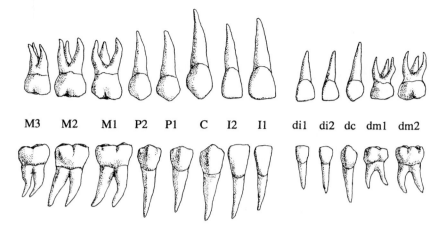

Figure 7–7   Buccal view of (*top*) upper and (*bottom*) lower dentitions; (*right*) left deciduous teeth; (*left*) right permanent teeth.

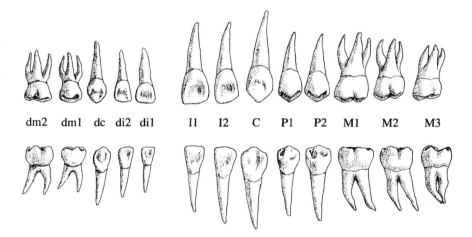

Figure 7–8   Lingual view of (*top*) upper and (*bottom*) lower dentitions; (*left*) left deciduous teeth; (*right*) right permanent teeth.

grooves or furrows would indicate root "boundaries").

3. Molar roots tend to arc distally (posteriorly) toward their tips (providing a secondary clue to tooth identification), which is a function of two factors: the path along which the tooth erupts and continuing growth of the jaws at the time of molar eruption. The intensity of root deflection distally increases from M1 to M3.

Upper Molars

M^1 is typically square (or the squarest of the three), M^2 is more rectangular (its mesiodistal length is shorter than its buccolingual breadth/width), and M^3 tends to be more triangular (Figures 7–3 and 7–6 to 7–9). Upper molars usually develop three roots. If M^3 roots are undivided (forming a pyramidal configuration), they are indicated by three

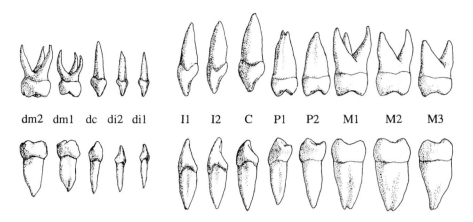

dm2   dm1   dc   di2   di1   I1   I2   C   P1   P2   M1   M2   M3

Figure 7–9   Mesial view of (*top*) upper and (*bottom*) lower dentitions; (*left*) left deciduous teeth; (*right*) right permanent teeth.

longitudinal grooves and three distinct root canal openings.

M¹, M², and even often M³ present a triangular arrangement of low, bulbous cusps; two cusps lie buccally and the third (the apex of the triangle) lingually and somewhat mesially. They form the *trigon*, the center of which bears a shallow *trigon basin*. The trigon's mesiobuccal cusp is the *paracone*, the distobuccal cusp the *metacone*, and the mesiolingual cusp the *protocone*. Since a root is associated ontogenetically with each trigon cusp, an upper molar has one lingual and two buccal roots. An upper molar often bears a fourth cusp (*hypocone*) in its distolingual corner that is delineated by an oblique groove or crease along the distal side of the trigon. Technically, a hypocone or any swelling in that region is the *talon* ("heel" in Greek). If an individual develops upper molar hypocones, they will be most fully expressed on M¹ and least developed or even absent on M³. If the upper molar size–shape gradient is quite steep, the metacone may also decrease in size. Human molar cusps may be lower and more bulbous than in many other mammals, but the buccal cusps are still taller than the lingual cusps and remain so even with wear (a clue to tooth identification).

Among other features that might occur on human upper molars are two stout, thick crests or cristae emanating from the protocone: the *preprotocrista* ("pre-" because it courses in front of, or "before," the protocone) connects the protocone and paracone and the *postprotocrista* ("post-" because it courses behind or posteriorly away from the protocone) connects the protocone and metacone. Collectively, a pre- and postprotocrista on the same tooth form a *protocrista* (pl. *protocristae*). In humans, the preprotocrista is frequently the more observable and dominant protocone crest.

Sometimes there is a swelling midway along the preprotocrista that may even be emphasized by thin longitudinal creases. This small, cusplike structure is a *paraconule*. Less frequently, a *metaconule* can be delineated on the postprotocrista. As with other features, conules tend to be expressed most on M1 and least M3.

Mesial to the preprotocrista and distal to the postprotocrista (if present) are variably developed depressions called *foveae* (s. *fovea*). The *anterior fovea* lies in front of the preprotocrista and is typically more excavated than the *distal* or *posterior fovea*. Human upper molar cingula are rare (e.g., see Swindler and Olshan, 1988) but, when present, usually occur lingually. More commonly, although still in low frequencies (e.g., see Swindler, 1976), a localized swelling of enamel may develop on distally alongside Carabelli's cusp or only a pit (e.g., Figure 7–10). Carabelli's cusp/pit expression conforms to the molar size–shape gradient.

Upper Premolars

Upper **anterior (first)** and **posterior (second) premolars** are morphologically rather similar (Figures 7–3 and 7–6 to 7–9). They develop two cusps—one buccally and the other lingually opposite it—and two roots, each above the region of a cusp. Premolar cusp names are based on the invalid assumption that molars are premolars with additional cusps. Thus, a premolar's buccal cusp is the *paracone* and the lingual cusp, the *protocone*. The paracone is somewhat larger than the protocone in height as well as in mass (the disparity is more obvious on P^1). Buccolingually oriented *anterior* and *posterior foveae* bound the bases and thus add to the delineation of paracone and protocone. A plane passing through the apices of these two cusps divides the crown asymmetrically (the mesial part of the crown is slightly smaller than the distal part). This asymmetry is also noted in the buccal outline of the paracone: the mesial edge (sloping from the cusp's apex to the mesial extremity of the crown) is shorter than the distal edge.

The P^1 cusp apices lie slightly more internally on the crown than on P^2. Thus the angle between the opposing faces of these cusps is steeper and more acute on P^1 than on P^2; when viewed from the mesial or distal side, the opposing faces of the paracone and protocone on P^1 form a narrower and more acute "V" than those on P^2. Also when viewed from the mesial or distal side, the buccal and lingual surfaces of P^1 are slightly more bulbous than those on P^2.

Upper premolar roots are typically bulkier and buccolingually deep (less delineated at the neck from the crown) than on lower premolars; they are not conical but typically flattened along their mesial and distal surfaces. The stereotype about upper premolars is that they have two roots while lower premolars have

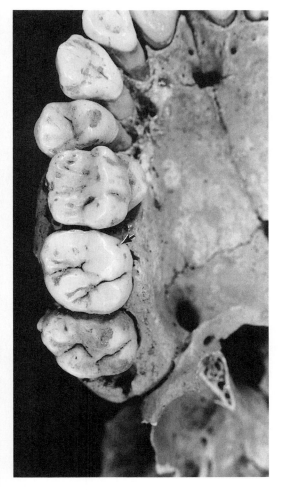

Figure 7–10 Upper jaw illustrating how premolar number could be reduced while still maintaining three molariform teeth: the second deciduous molar is retained as the first molariform tooth, the third molar is inhibited, and the first and second molars then become the second and third molariform teeth in the series [in reality, the second premolar of this individual (visible under the crown of dm^2) had started to erupt]; also note, for example, that the canine crown is circumscribed by a deep linear hypoplastic groove and M^1 bears a small Carabelli's pit (*arrow*) (prehistoric, Pennsylvania).

the protocone's lingual face. In nonhuman mammals, this structure is identified as the *protostyle*. In humans, it is called *Carabelli's cusp*. *Variations* may include a pit (*Carabelli's pit*) one. In practical terms, two roots are more frequently identifiable as somewhat distinct entities on P^1 than on P^2; however, fully separated P^1 roots are not that common. The ontogenetic potential for two roots is demonstrated on an upper premolar by longitudinal grooves that course, respectively, along the mesial and distal sides of the undivided root complex. Although root deflection as a clue to tooth identity is less obvious in premolars than in molars, the root tips may arc or bend distally.

Upper Canine

The upper canine is a stout, single-cusped, roughly conical, single-rooted tooth (Figures 7–3 and 7–6 to 7–9). The crown is subtriangular in buccal outline; its mesial edge (measured from the cusp apex to the most mesial extent of the crown) is shorter than the distal edge. An upper canine's root is quite robust and long; the tip may be deflected distally.

Aside from its prominent primary cusp (which shall remain unnamed), an upper canine usually lacks major additional morphological adornment. A *lingual swelling* (*lingual tubercle* or *tuberculum dentale*) or variable elevation may develop near the crown's base. Low, rounded lingual marginal ridges of enamel (*mesial* and *distal ridges, margocristae*) may emanate from the lingual tubercle, with one ridge proceeding mesially and the other distally. When surface topography is lost through wear and attrition, the prior existence of a ridge may be indicated by a thin groove, which would have bound the inner margin of the ridge.

Upper Incisors

Human upper incisors (Figures 7–3 and 7–7 to 7–9) are typically morphologically simple, single-rooted teeth. Although upper incisors are often described as spatulate, they are actually more spade-shaped: the buccal surface is typically gently concave and the lingual surface minimally convex with, on occasion, some lingual swelling near the base. Viewed buccally, the mesial edge is the more vertical and the distal edge the more arcuate or flared. There is individual variation in the degree of mesial verticality versus distal flaring, but the mesial edge can reliably be distinguished from the distal edge.

Typically, the **lateral (second) incisor** is smaller than the **central (first)**, although difference in size may be less marked in some individuals. Shape tends to be more variable in I^2 than in I^1 (e.g., somewhat conical, narrow, and/or even more markedly curved along the distal edge).

An adornment of the lingual surface of upper incisors most frequently attributed to Asian peoples and those of Asian origin (especially Native North Americans) (e.g., Turner, 1984) consists of the buildup of thickened mesial and distal marginal ridges (*margocristae*) that are confluent with an enlarged lingual tubercle. In cross section, such a tooth looks like a small scoop and is described as *shovel-shaped* (see Figure 9–28). Sometimes (but more frequently on I^2 than I^1) the marginal ridges are so thick, particularly toward the lingual swelling, they create a tubular configuration with a pitted bottom; the tooth is described as *barrel-shaped*. A tooth is *double shoveled* when mesial and distal marginal ridges are present lingually and buccally.

Lower Molars

Although lower molars decrease in size and morphological complexity from M_1 to M_2 to M_3, they all tend to be rectangular (mesiodistally longer than buccolingually wide) (Figures 7–4 and 7–6 to 7–9). Lower molars typically develop two buccolingually broad and somewhat mesiodistally compressed roots, which become increasingly appressed to each other if not also incompletely separated from M_1 to M_2 to M_3. Each root corresponds to roughly half of the tooth (a mesial half and a distal half). A third, distinctively spindly root may emanate from the cleft between the two primary roots (lingually near the neck); this is an isolated developmental phenomenon that can affect any lower molar (e.g., Schwartz, unpublished data; Turner, 1984).

Human lower molar crowns usually bear four major cusps that are generally of similar size and shape. The *metaconid* lies in the crown's mesiolingual corner; the *protoconid* lies opposite the metaconid in the mesio-buccal corner; the *entoconid* (the only cusp whose prefix provides a clue to its position) sits behind the metaconid in the distolingual corner; and the *hypoconid* lies opposite the entoconid in the distobuccal corner. The reverse of the condition in upper molars, lower molar lingual cusps are taller than buccal cusps (whose bases are more swollen). The mesial root lies under the metaconid–protoconid region and the distal root under the entoconid and hypoconid.

There may also be a fifth cusp (*hypoconulid*), which presents itself as a pie-shaped wedge between the entoconid and hypoconid near the midline of the tooth [a centrally placed hypoconulid actually characterizes Old World monkeys, apes, and hominids (Kay, 1977)]. The hypoconulid will be best expressed on M_1 and decrease in size to M_2 and then to M_3 (although only M_1 may have a hypoconulid).

Of other features one finds on primate lower molars, the most common in humans is a relatively well-defined *anterior fovea* in front of the metaconid and protoconid that is bounded mesially by a ledge of enamel (*paraconid shelf* or *paracristid*). A low ridge behind the anterior fovea and between the metaconid and protoconid is the *distal trigonid ridge, distal trigonid crest*, or just *protocristid*. Lower molars lacking a hypoconulid may bear a short *posterior fovea* just behind the entoconid–hypoconid pair. The ledge of enamel delineated distally by this fovea is usually not identified in humans; in other primates, it is a *hypocristid*. Like other primates, humans possess an oblique crest (*cristid obliqua*) that courses between hypoconid and

metaconid. Unlike most primates, in which this crest is longer and more clearly defined, it is stout and truncated in humans.

Humans rarely develop a lower molar *cingulid*. When present, it is typically found on the crown's buccal side, usually as a truncated structure (e.g., Swindler and Olshan, 1988). Human lower molars may also (but infrequently) bear a *protostylid* on the protoconid's buccal face. In general, however, protostylids are rare among primates [e.g., the slender loris (*Loris tardigradus*) of Sri Lanka, *Indraloris* from the Miocene of Indopakistan, and various australopiths (Schwartz, 1986; Schwartz and Tattersall, 2005)].

A feature often identified as a *metastylid* (cf. Scott and Dahlberg, 1980; Turner, 1984) may present itself as a small conulid nestled between the metaconid and hypoconid. In nonhuman primates, the metastylid is more stylidlike in position (buccal and on the posterior face of the metaconid) and morphology (elevated and peaked).

Lower Premolars

Contrary to received wisdom (e.g., Clark, 1966), human lower premolars (Figures 7–4 and 7–6 to 7–9) are not, as a set, *bicuspid*. Each is dominated by a buccal cusp—by convention, a protoconid—lingual to which the crown is somewhat swollen or distended. In no lower premolar is the buccal cusp as tall or well defined as in an upper premolar. In the lower **anterior (first) premolar**, lingual "cusp" development is often constrained to the level of what in other primates (e.g., orangutans) would be identified as a *lingual tubercle*, which is bound by small *anterior* and *posterior foveae* and sometimes connected to the protoconid's apex by a crest. Human P_1s are essentially *unicuspid*—as in most anthropoid primates with the exception, for example, of many australopiths—with *variation* in the degree of lingual expansion. Indeed, the unique morphology of the human P_1 makes it one of the easiest isolated teeth to identify.

Sometime Inuit/Aleut P_1s bear an *odontome*. This cusplike elevation of enamel lies closer to the protoconid than a metaconid would and results from the deposition of enamel directly atop an anomalous extension of the pulp cavity, not dentine (e.g., Cruwys, 1988).

Human lower **posterior (second) premolars** possess cusplike structures lingually that are often large enough to be identified as *metaconids*; these cusps do not approach the protoconid in size or height (in contrast to upper premolar protocones versus paracones). Typically, a P_2 metaconid, which may be delineated by weakly incised mesial and distal grooves, is lower than, and broad relative to, its protoconid's base. A P_2 also typically bears *anterior* and *posterior foveae*. Viewed buccally, because each premolar is somewhat expanded distally, the outline of P_1 and P_2 protoconids is characteristically asymmetrical: the cusp's mesial edge (as measured from its apex to its most mesial extent) is shorter than its distal edge (as measured from its apex to its most distal extent).

A lower premolar typically bears a single somewhat conical root. The P_1 root is quite slender and acutely tapered toward its apex. The P_2 root is somewhat thicker but does not approach the robustness or shape of a fused P^2 root. Occasionally, a lower premolar may bear a supernumerary root; like a lower molar "third" root, it is anomalous.

Lower Canine

A lower canine (Figures 7–4 and 7–6 to 7–9) is a semiconical, single-cusped, single-rooted tooth that is smaller and less robust than an upper canine; it is also narrower, more straight-sided, and, in cross section, less circular or ovoid at its base. A C_1 lingual surface is also less swollen basally and less frequently embellished (even with a small tubercle). The root is shorter and less massive. However, like a C^1, a C_1 crown is asymmetrical in buccal outline: its mesial edge is shorter than its distal edge.

Lower Incisors

Like lower canines and premolars, lower **central (first)** and **lateral (second) incisors** (Figures 7–4 and 7–7 to 7–9) are simpler morphologically than their upper counterparts. In general, lower incisor crowns are narrower, more straight-sided, and less buccolingually thick at their bases than upper incisors; their thinner and shorter roots are mesiodistally compressed. I_1 is smaller (even if minimally) than I_2 (the reverse of upper incisors). Lower and upper incisor crowns are similar in that distal and mesial edge orientations differ: in the lowers, the distal edge is slanted more laterally away from the vertical (even if subtly so) than the mesial edge. Newly erupted lower incisors typically bear more and better-defined *mamelons* (small mounds along the occlusal edge) than uppers. Lower incisors may occasionally be *shoveled*.

Summary of Permanent Tooth Identification

Incisors

"Spatulate"/spade-shaped (somewhat buccolingually flattened, occlusally broadening mesiodistally) crown, more or less straight occlusal edge, single-rooted.

Upper: Crown broader than lower; distal edge arcuate and somewhat laterally flared; some lingual swelling; root thicker, longer, and more ovoid/circular in cross section; I^1 > I^2.

   Side: Mesial edge straighter and more vertically oriented than arcuately flaring lateral edge, root tip may point mesially.

   I^1: Overall, larger and more robust than lateral incisor, crown broader, lingual swelling more pronounced, lateral flare more marked and arcuate, stouter root.

   I^2: Crown narrower; shape more variable (sometimes nearly conical);

lateral flare less pronounced, sometimes negligible; root slenderer.

Lower: Crown narrower than upper, sides straighter and divergent laterally, root slenderer and somewhat laterally (mesiodistally) compressed, mamelons (visible on unworn teeth) more developed, $I_1 < I_2$.

> **Side:** Distal edge more divergent laterally from vertical, root tip may point mesially.

> I_1: Crown narrower, root slenderer.

> I_2: Crown broader, lateral edge more noticeably divergent from the vertical, root slightly stouter.

Canines

Crown roughly (sub)conical; strong, single cusp; some lingual development, possibly with mesial and distal (accessory) ridges (margocristae/cristids); relatively stout single root, ovoid to circular in cross section.

Upper: Crown stouter and more circular near base in cross section; single cusp higher, with more widely divergent mesial and distal edges; lingual swelling/tubercle development more marked; margocristae more distinct; root stouter, longer, more circular in cross section.

> **Side:** Distal edge longer than mesial edge; buccal surface characteristically smooth and convex; lingual surface flatter, with topographic relief; root may curve gently mesially along most of its length (root tip direction unreliable).

Lower: Crown narrower, shorter, generally more gracile; mesial and distal edges shorter and less divergent from apex; lingual side flatter and less swollen at base; root more slender and variably more compressed laterally.

> **Side:** Distal edge longer than mesial edge; buccal surface characteristically smooth and convex; lingual surface flatter, with topographic relief; root orientation uninformative.

Premolars

Two or one cusp; if two cusps, lingual cusp not taller than buccal cusp; if one cusp, lower-crowned and more gracile than any canine; double- or single-rooted; if two roots undivided, distinguishable from large single root; if one root, noticeably gracile and slender.

Upper: Demonstrably double-cusped, with lingual cusp mirroring buccal cusp in shape and approaching (but not exceeding) it in height; demonstrably double-rooted, with roots being variably separate along their length or at least bifid at their tips, especially on P^1; when fused, mesial and distal longitudinal grooves present; root complex deep buccolingually.

> **Side:** Buccal cusp taller than lingual cusp, distal edge of buccal cusp longer than mesial edge, buccolingual axis through cusp apices divides occlusal surface asymmetrically (distal part is larger/bulkier than mesial part), root complex deep buccolingually, roots may arc distally toward apices.

> P^1: Buccal and lingual cusps more subequal in size and height; buccal and lingual cusp apices closer together, forming more acute valley between cusps; evidence of double-rootedness more common.

> P^2: Buccal and lingual cusp apices farther apart, forming more obtuse and less "V"-shaped valley between cusps; roots less divided, more tapered toward apex/apices.

Lower: Low buccal cusp and even lower or nonexistent lingual cusp; root single and slender, with marked taper.

> **Side:** Buccal cusp taller than lingual cusp, distal edge of buccal cusp longer than mesial edge, buccolingual axis through buccal cusp apex subdivides crown in larger distal and smaller mesial portions, root may be curved distally toward tip.

P_1: Buccal cusp dominates crown, lingual development typically no larger than a tubercular swelling, lingual surface may bear low central ridge coursing to swelling, root slender and conspicuously narrow compared to crown.

P_2: Lingual cusp more developed but still lower than buccal cusp, root variably stouter.

Molars

More than two cusps, two buccolingually broad or three somewhat more conical roots.

Upper: Three cusps arranged in a triangular configuration, with a somewhat conical root emanating from region of each cusp; a fourth cusp (hypocone) may be present in the distolingual corner of the crown.

Side: Buccal cusps higher than lingual cusps; crown's lingual face more bulbous than buccal face; preprotocrista stouter than postprotocrista (if present); anterior fovea (in front of preprotocrista); hypocone, if present, in distolingual corner of tooth; two roots buccal and the third lingual and somewhat mesial; Carabelli's cusp/pit, if present, on lingual face of protocone (which is lingual and somewhat mesial); roots may curve distally toward their tips.

M^1: Largest with most marked features (if present, hypocone, cristae, conules, Carabelli's cusp/pit); roots distinctly separate, most splayed (one or more roots may protrude beyond the margins of the crown).

M^2: Slightly to somewhat smaller overall and occlusal features of detail less pronounced than on M^1; roots, although possibly distinctly separate for much of their lengths, more closely appressed toward their tips (becoming noticeably narrower than the crown).

M^3: Slightly to somewhat smaller overall and occlusal features less developed than on M^2; hypocone often absent; metacone may be reduced relative to paracone; roots typically shorter and undivided along much of their lengths, with root apices converging to a pointed configuration; in adults, only one (mesial) interstitial wear facet.

Lower: Four cusps arranged in rectangular configuration; a pair of buccolingually broad, somewhat mesiodistally compressed roots, one aligned under protoconid–metaconid pair and the other under entoconid–hypoconid pair; a small fifth cusp (hypoconulid) may be present distally.

Side: Lingual cusps higher than buccal cusps, more expanded at their bases; crown's buccal face more bulbous than lingual face; anterior fovea more pronounced than posterior fovea (if present); hypoconulid (if present) distal or slightly buccal to midline, sandwiched between entoconid and hypoconid; roots may curve distally toward their tips.

M_1: Largest, with most pronounced hypoconulid (if present); roots distinctly separate, most splayed.

M_2: Slightly to somewhat smaller (at least mesiodistally shorter) than M_1, with smaller hypoconulid (if present); foveae smaller; roots closer or even partially undivided.

M_3: Slightly to somewhat smaller (at least mesiodistally shorter) than M_2, with even smaller hypoconulid (if present); foveae even smaller; roots may be shorter and probably partially or even completely undivided (root tips may be separate); in adults, only one (mesial) interstitial wear facet.

Deciduous Teeth

Since permanent incisors, canines, and premolars develop from and then replace deciduous, predecessor teeth, it is not surprising that ontogenetically related teeth will share similar qualities or features. In general, though, the smaller-crowned deciduous teeth (Figures 7–3, 7–4, and 7–7 to 7–9) have more clearly and/or crisply detailed features than their successors. Deciduous tooth enamel tends to be creamier or milkier in color rather than bright white, as in newly erupted, normal permanent teeth (perhaps because they develop rapidly and in utero).

Deciduous tooth roots tend to be much shorter relative to crown height, have relatively larger root canal openings, and typically taper much more drastically toward their apices than permanent tooth roots. In general, deciduous tooth roots often seem too small for their crowns—an incongruous association that is accentuated further by the neck of a deciduous tooth, especially a multicusped, multirooted tooth, being distinctly constricted ("pinched," "waisted"). In single-rooted deciduous incisors and canines, the root is narrow and the crown appears to expand markedly out from the neck. The peculiarities of deciduous tooth roots are in some way related to the facts that (1) the jaws in which deciduous teeth develop and erupt are small and crowded (with developing successional tooth crowns beneath and often between the roots), (2) there is less bone in which deciduous teeth can anchor themselves, and (3) the useful life of a deciduous tooth is much shorter than a permanent tooth's. It is not surprising that deciduous tooth roots tend to be wavy or deflected in ways that make root orientation unreliable for siding teeth.

Deciduous incisor and canine crowns look essentially like miniature versions of their successors but have more detail or characteristic morphology. For example, the lateral flare of the distal edge of **upper deciduous incisors** tends to be more exaggerated. The buccal face is more noticeably convex and the lingual surface is more concave or excavated, with a more pronounced swelling near its base. If there is *shoveling*, it will typically be marked and more often present on not only central but also lateral incisors.

Lower deciduous incisors may be more dramatically straight along their sides than their permanent successors, but they may also veer in the opposite direction, with their distal edges developing a lateral flare. In spite of the latter possibility, lower deciduous incisors are distinguishable from upper deciduous incisors: they are less bulbous buccally and less convex or excavated lingually, and their roots are more disproportionately smaller. Newly erupted lower deciduous incisors are distinctive in that their occlusal edges often bear a greater number of pronounced mamelons; sometimes mamelon development is so marked that the occlusal edge appears partially segmented into lobes. One distinguishes central from lateral and right from left upper as well as lower incisors with the same characteristics used for permanent incisors.

Although markedly smaller, **deciduous canines** tend to exaggerate features seen in their permanent successors: for example, they have more bulbous buccal surfaces; more swelling in general around the base (especially dc^1); and more delineation of lingual surface features (particularly on dc^1) such as lingual tubercles, marginal ridges (margocristids), and central "keels." Nevertheless, deciduous canines are sufficiently similar to permanent canines in shape that the same criteria can be used to identify the tooth as upper, lower, right, or left.

Upper deciduous molars typically develop three roots. Otherwise, they are heteromorphic. The occlusally somewhat triangular **anterior (first) deciduous molar** is much longer mesiodistally along its buccal than mesial side. The crown is dominated by a *paracone*, lingually across from which is a variably smaller *protocone*; these cusps tend to be broadly separated from one another. Blunt *pre-* and *postparacristae* course along the buccal edge of the tooth; each terminates in a stylar-like swelling (that associated with the preparacrista is typically more pointed and style-like). The postparacrista is variably longer than the preparacrista.

A low, blunt, keel-like crest that courses down the face of the paracone and up the opposing face of the protocone often reaches the apices of these cusps. Distinct *distal (posterior)* and *anterior foveae* bound this central "keel" on either side; they are often mesiodistally longer than mere creases and accentuate the crown's raised mesial and distal margins. The posterior fovea is variably larger than the anterior fovea. This size disparity corresponds to the overall pattern of anterior deciduous molars: the distal margin tends to be more swollen or arcuately distended than the mesial margin, which courses more directly to the style-like buccal terminus of the preparacrista. Thus, the bulk of the tooth lies on the distal side of an imaginary line drawn between the apices of paracone and protocone.

Clues to identifying dm^1 include paracone larger than protocone; posterior fovea larger than anterior fovea; distal margin distended and arcuate, mesial margin straighter; buccal terminus of preparacrista more style-like than that of postparacrista; and two of three roots lie on the buccal side of the tooth. Even though it is basically a two-cusped tooth, dm^1 is easily distinguished from its bicuspid successor by, for example, its smaller size, greater number of roots, and slightly more detailed occlusal morphology. Viewed in conjunction with the rest of the molar tooth class, this tooth clearly conforms to the size–shape gradient (e.g., Figure 7–10).

Given its ontogenetic relationship to M^1 and the molar class in general, the **upper posterior** (in humans, **second) deciduous molar** of all heterodont mammals is a distinctly *molariform* tooth, which, although smaller and having a pinched neck and root complex characteristic of deciduous teeth, is often misidentified as a permanent molar.

In addition to general differences in size, neck, and root, a human's dm^2 differs from M^1 in other ways. In particular, the crown is more trapezoidal than square and its features more exaggerated. For example, the trigon and its basin are more pronounced and defined, and the cusps and broadly divergent protocristae more crisply delineated; when present, the paraconule and metaconule are more distinct. The hypocone is usually quite large relative to the overall size of the crown; it is markedly swollen or distended distally, thereby elongating the tooth (particularly along its lingual side) and enlarging the talon basin. The crown's mesial margin (which can easily be confused with the preprotocrista) often swings anteriorly away from the protocone, creating a variably large anterior fovea between it and the preprotocrista. A small style- or conulelike swelling may occur midway along the mesial margin. If Carabelli's cusp and/or pit are present, they will usually appear disproportionately large relative to the size of the crown.

One can **identify an isolated dm^2** by focusing on the hypocone and features of the trigon. Roots also provide clues: since, as in permanent molars, a root develops above each trigon cusp, two will be buccally placed. Other criteria that apply to permanent molars can be used to identify jaw and side of origin.

Similarity between the human **anterior (first)** and **posterior (second) lower deciduous molars** goes no further than the development of two mesiodistally compressed, buccolingually broad roots, one situated mesially and the other distally. Although dm$_1$ is clearly part of a molar class gradient, in crown morphology, it is arguably the most distinctive tooth of the entire dentition. It is double-basined; the anterior moiety (trigonid) is somewhat triangular and smaller than the ovoid or teardrop-shaped posterior part (talonid). In its simplest form, dm$_1$ presents itself as a bottom-heavy figure eight; in its most complex form, it is submolariform.

dm$_1$ is dominated by its protoconid and variably smaller metaconid; the apex of the latter tends to lie slightly distal to the apex of the former. Protoconid and metaconid bases are typically confluent and sometimes so closely melded that only their tips are visible. A distinct paracristid (paraconid shelf) descends from the protoconid's apex and usually courses directly forward slightly before it "kinks" (sometimes quite sharply) and swings back to contact the base of the metaconid (if it does not continue up toward its apex). Thus, dm$_1$ has a very large, enclosed, asymmetrical trigonid basin that is characterized by a long buccal margin and a lingually and distally directed mesial margin.

The dm$_1$ talonid basin extends from behind the melded protoconid and metaconid. It is much larger and elongate, as well as more rounded, symmetrical, and uniformly enclosed than the trigonid basin; its buccolingual breadth or narrowness is variable. In contrast to the trigonid basin, the crest that circumscribes the talonid basin (emanating from the protoconid and eventually terminating on the metaconid) is generally not punctuated with distinct, well-developed cusps. Rather, this continuous margin is often unadorned; at most, it may bear three variably developed cusplike swellings that could, perhaps, be identified as *entoconid, hypoconid*, and *hypoconulid*. In Native North American populations in particular, a distinctive crest courses from the protoconid down into the talonid basin (e.g., Dahlberg, 1949).

The lower **posterior** (in humans, **second**) **deciduous molar** may be mistaken for a permanent molar. Its true identity is revealed by its smaller size, more distinctly constricted neck, more flared sides of crown, short and markedly splayed roots, and exaggeration of various occlusal features. For example, if present, the hypoconulid and grooves that delineate it are more crisply defined than on a permanent molar. A dm$_2$ hypoconulid is larger relative to the other cusps than that on a M$_1$ (which is consistent with a molar-class size–shape gradient) and typically more buccally emplaced. The anterior fovea, too, is often more crisply delineated and relatively larger compared to the overall size of the

crown than on M$_1$. In addition to features used to analyze permanent lower molars, the position of the hypoconulid and easily identified anterior fovea are important **clues** to identifying dm$_2$.

Enamel: Normal and Other Features

Like bone, enamel results from **mineralization**, a process in which **hydroxyapatite** is deposited within a **cellular matrix**. The major differences between bone and enamel formation are that (1) the cells that produce teeth and their components are derivatives of neural crest–derived ectomesenchyme (bone is mesodermal in origin) and (2) enamel does not have a definitive organic precursor (bone does). During the early phase of enamel deposition (*amelogenesis*), the ameloblast moves away from the dentine's surface and its secretory end begins laying down a layer of **enamel matrix** between it and the dentine. Enamel matrix is a combination of organic and inorganic—**enamel crystallite**—material. Subsequent to laying down the enamel matrix, the ameloblast's secretory end changes shape, becoming conical with an extended, pointed end; now identified as **Tomes' process**, it invades the enamel matrix and starts laying down enamel. In addition, an ameloblast is at least functionally analogous to an osteoclast in that it appears to remove material from the enamel matrix. The final phase of mineralization—producing hard enamel— occurs very rapidly. Since rates and lengths of time of enamel deposition are not the same in all mammals, enamel thicknesses can differ (e.g., pigs, humans, and orangutans have thick molar enamel, whereas lemurs, chimpanzees, and gorillas do not).

Mammals differ from reptiles in enamel structure in developing crystallite patterns called **enamel prisms**; reptile enamel is *aprismatic*. But while we can identify enamel prisms and even different enamel prism patterns in different mammals, the underlying process is not fully known. It does seem, though, that enamel prism patterns emerge because the ameloblast's movement away from the dentine layer is restricted by the stellate reticulum, which compresses and changes the parallel flow lines the enamel matrix would otherwise follow. In mammals, enamel crystallites are oriented along the *nonparallel flow lines* of the secreted enamel matrix. Prisms are defined as a consequence of the development of **interprismatic regions**, which result from the resorption of enamel matrix in areas that lie opposite the secretory sides of Tomes' processes.

Scanning electron microscopy made simple the task of studying crystallite or **enamel prism patterns** in mammals. Two enamel prism patterns commonly identified on the basis of a rough interpretation of a cross section of prism shape within an interprismatic space are a *"circular" pattern (pattern 1)* and a *"keyhole" pattern (pattern 2)* (e.g., Boyd, 1971; Figure 7–11).

Sometimes it appears that more than one enamel prism pattern occurs on the same tooth. This illusion is probably due to the degree to which the ameloblasts bend or *decussate* as they move (Figure 7–12). After all, a single ameloblast lays down a continuous tract of enamel from the dentinoenamel juncture to the surface of the crown. An ameloblast's undulating path also creates patterns of cross-striations—so-called **Hunter-Schreger bands**—that are observable under reflected light (Figure 7–13).

Hypocalcification in regions of the crown results from interfering with the process of amelogenesis. The most common defects are **gross enamel hypoplasias** (e.g., Goodman and Armelagos, 1985; Goodman and Rose, 1991; Lukacs, 1989; Skinner and Goodman, 1992), in which the enamel is thinner than surrounding, "normal" regions of enamel; they are most often noted on the buccal surfaces of teeth. Total lack of enamel in a region of a crown is uncommon.

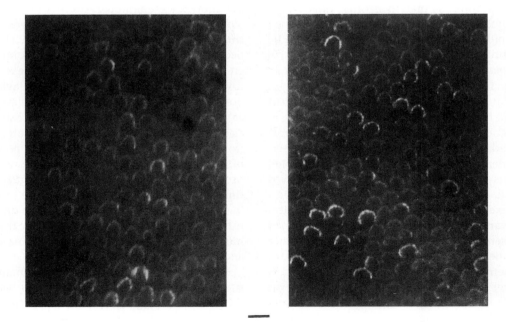

5 μm

Figure 7–11   Confocal images of cross-sectioned enamel prisms (illustrating the "keyhole" pattern) in two fossil (Eocene) primates: (*left*) *Cantius mckennai* and (*right*) *Notharctus* sp. (*courtesy of E. Dumont*).

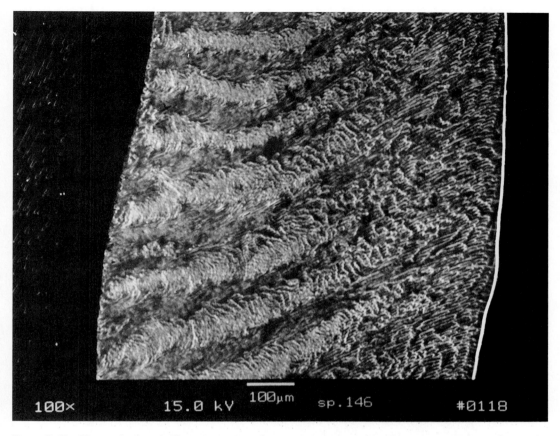

Figure 7–12   Photomicrograph illustrating prism decussation: cross section through a right lower first molar of an Old World monkey (*Cercocebus torquatus*) (*courtesy of E. Dumont*).

Major factors affecting amelogenesis are stresses that interfere with proper growth: nutritional imbalances, vitamin D deficiency, hypoparathyroidism, and serious childhood illnesses (e.g., rheumatic fever). Given the finite time period over which teeth form, stresses affecting amelogenesis will be "recorded" only until puberty. A stress or trauma may produce a **localized** effect (to one or a few, but not necessarily neighboring, teeth; *localized enamel hypoplasia*) or be more **systemic**, affecting the majority of teeth (*systemic enamel hypoplasia*). Teeth may record multiple and/or recurrent insults (although the exact causes cannot be determined).

The most common hypoplasia—**chronological** or **linear hypoplasia**—is seen as horizontal grooves across the enamel surface, most frequently in the middle and cervical thirds of the crown and more often on anterior than posterior permanent teeth. In severe cases, the grooves constrict (Figure 7–14) and even encircle the crown (Figure 7–10, see also Figure 9–18). Although there is not a universally accepted chronology of tooth formation (see Chapter 8), attempts have been made to estimate the age(s) at which an individual suffered the stresses underlying linear enamel hypoplasia (e.g., see Goodman and Rose, 1991; Murray and Murray, 1989; Skinner and

Figure 7–13 Backscattered electron microscopic images of prism cross-striations (Hunter-Schreger bands) (*arrows*): (*top*) a prosimian primate (*Galago alleni*) and (*bottom*) a sac-winged bat (*Taphozous mauritianus*) (*courtesy of E. Dumont*).

Goodman, 1992). Assuming constant rates of crown growth—and that the same tooth takes the same amount of time to form regardless of size differences between individuals—the age of stress insult is calculated by measuring the distance of the hypoplastic groove from the dentinoenamel juncture and matching this measurement to a standardized chart.

A **foramen caecum** is another form of hypoplasia that most frequently occurs on lower molars as a deep, conical pit at the base of the crease between protoconid and hypoconid (Figure 7–14). It may be localized to only one or two teeth or found throughout a molar series, either uni- or bilaterally. The osteologist should take care not to confuse

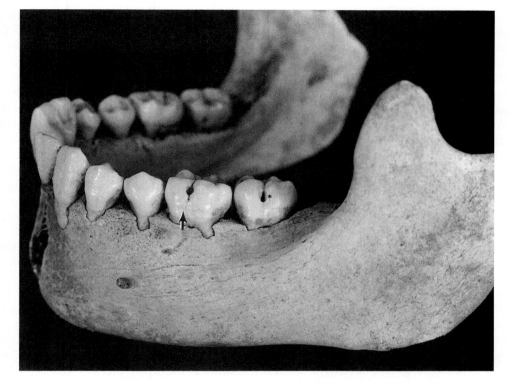

Figure 7–14   Hypodontia (lack of M_3) and examples of hypoplasia: grooves indicative of linear hypoplasia are noted on all teeth (*arrow*) and a foramen cecum (*asterisk*) is present buccally on each lower molar; also note on M_1 (*just distal to the arrow*) an enamel extension coursing down between the buccal roots (prehistoric, Pennsylvania).

a foramen caecum with a carious lesion (see Chapter 10). Broadly shallow hypoplastic pits may also develop on the buccal surfaces of anterior teeth; they are etiologically the same as (systemic) linear enamel hypoplasias.

Supernumerary Teeth and Tooth Agenesis

Although most individuals develop the number of teeth typical for their species, the occasional individual may develop more or fewer. Additional, or **supernumerary**, teeth may appear as either rudimentary or fully formed, recognizable structures. The suggested mechanisms behind supernumerary tooth development are summarized below;

the result is **hyperdontia.** The process of tooth loss is called **tooth agenesis**; the result is **hypodontia** (Figure 7–14).

If teeth are viewed as individually independent, developmentally discrete units, then hyper- and hypodontia must be interpreted as resulting from isolated, usually de novo developmental events. If tooth classes are seen as the products of proliferation of an initial stem progenitor cell mass, a model becomes available that can account for both hyper- and hypodontia.

The simplest example of supernumerary teeth is when teeth are added at the end of the molar tooth class (humans occasionally develop fourth molars). In such cases, the supernumerary tooth is fully formed and complements the size–shape gradient established in the normal molar tooth series. When only part of a supernumerary molar crown forms, the cusps present correspond to the cusps that would calcify first during normal development (i.e., the anterior cusps) (Schwartz, 1984). Although a rudimentary or morphologically incomplete molar is traditionally interpreted as having split off from the tooth germ developing in front of it, there is an alternative explanation: since the molar-class teeth appear to "clone" posteriorly, a supernumerary tooth, or at least a truncated supernumerary tooth, will develop if the developmental potential of the posteriorly proliferating cell mass is maintained and there is sufficient space in the jaw for a tooth to form (Schwartz, 1984). For example, some mammals grow a lifetime of "supernumerary" teeth: for example, the phalangeroid marsupial *Peradorcas* normally develops seven or eight supernumerary molars in each quadrant of each jaw, while the manatee adds new molars posteriorly as worn teeth are shed anteriorly. As such, one can hypothesize that competence for tooth formation is maintained in the posteriorly proliferating cell mass; in *Peradorcas* and the manatee, for example, so-called supernumerary molars complement in size and morphology the molar teeth that preceded them. In the case of the incomplete supernumerary structure, the truncated "tooth" also morphologically complements the molar size–shape gradient.

Dental agenesis or tooth "loss" can be explained in terms of the contraction or inhibition of the potential and competence for tooth formation in the proliferating cell mass of a tooth class. Third molar agenesis—which, for example, occurs with relatively high frequency among Inuits and Aleuts (Davies, 1972; Moorrees, 1957)—would thus result from the truncation of a posteriorly enlarging molar tooth class. In the case of marmosets (a group of New World monkeys distinguished from others by third molar agenesis), X-rays reveal that third molars begin

to develop but are resorbed before reaching the point at which they would be retained as stunted structures or grow into complete teeth (Hershkovitz, 1977). Thus, it would seem that in marmosets molar-class cells are competent to initiate but not to sustain third molar development. Updating earlier suggestions based on notions of cellular competence (Lumsden, 1979; Osborn, 1970, 1973; Schwartz, 1984), the appearance of extra or fewer teeth is related to degrees of regulatory overexpression or suppression imposed upon an established developmental system rather than to separate, unrelated phenomena. Embryological studies of human tooth development are compatible with the suggestion that truncation/inhibition of developmental competence and cellular proliferation, rather than a complete loss of structure, can lead to a reduction in tooth number within a tooth class.

Ooë (e.g., 1965, 1969, 1971) demonstrated the presence in humans of potential successors to both premolars as well as the first and second molars (he did not study the third molar) that are resorbed at a rudimentary stage of development. In all cases, the resorbed dental structures conformed to the "textbook picture" of successional tooth development: each arose lingually from the external dental epithelium of a predecessor tooth germ. Although premolars and molars are not normally shed and replaced, Ooë's work provides an explanation for individuals in whom a "third" dentition does develop (e.g., Figure 7–15). That is, diphyodonty in humans (and, presumably, other mammals) results from truncating/inhibiting tooth development in a polyphyodont condition to maximally two *functional* sets of teeth.

Tooth Eruption and Root Formation

The most frequently used criteria for determining the age of preadult individuals are stages of crown and root formation, eruption of the primary dentition, and replacement

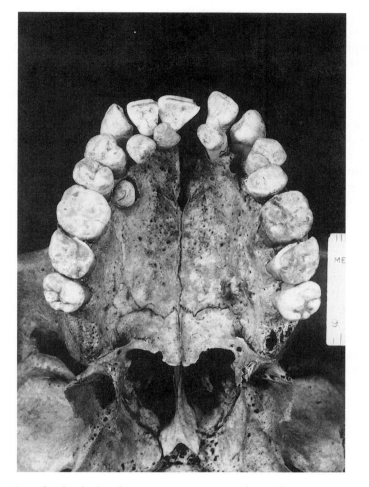

Figure 7–15   Upper jaw of individual with supernumerary antemolar teeth representing a "third" dentition (erupting lingual to the normally present permanent or secondary teeth) (prehistoric, Alaska).

of the primary by the secondary dentition. Usually the data—whether in chart or diagram form—are presented without consideration of the developmental background. Both are provided here.

Although there continue to be numerous studies detailing the eruption sequences of mammalian dentitions, the actual process of tooth eruption is not entirely known, although its potential elements seem to have been isolated (e.g., Ten Cate, 1989; Ten Cate and Osborn, 1976). Until fairly recently,

opinion was that tooth eruption was caused by forces that built up during root formation as the elongating root pushed against a so-called cushion hammock ligament that was supposed to stretch across the alveolus under the root. Although there is no such ligament, root formation remains a possible cause of tooth eruption. There was the theory (still maintained in some quarters) that bone remodeling during growth of the upper and lower jaws is the proximate cause of tooth eruption. It appears instead that maxillary and mandibular bone resorption, deposition, and redeposition in the regions of erupting teeth are consequences, not causes, of tooth eruption. Although the buildup of hydrostatic pressure has been suggested as the force behind tooth eruption, this is not fully supported by experimental studies.

It appears that, although other factors may come into play, the periodontal ligament, which arises coincident with the onset of root formation, is primarily responsible for tooth eruption. It is not exactly clear how it, or some set of its cell types, does this (possibly by contraction); but experimental evidence suggests that this ligament must be present for tooth eruption to occur. In addition, continued tooth movement after eruption (such as compensatory eruption as tooth crowns are worn down) appears to be maintained by the periodontal ligament. Conservatively and in descending order of importance, tooth eruption seems to result from contraction of the periodontal ligament, proliferation of periapical tissues associated with root growth —either individually or collectively—and, perhaps, fluid pressures from vascular tissues.

In most instances, resorption of alveolar bone is sufficient to allow the crown to pass through it. As a tooth erupts, alveolar bone redevelops around it. A successional tooth causes the hard and soft tissue of the tooth it will replace to be resorbed. The cells responsible for the resorption of tooth-related structures are sometimes identified as *odontoclasts* (which are essentially identical to osteoclasts) (Ten Cate, 1989). Although odontoclasts can resorb all dental tissues, including enamel, the deciduous tooth is affected only where contacted by the successional tooth. In general, the periodontal ligament (i.e., support) of the tooth that will be shed is rapidly lost.

The onset of **root development** is associated with the appearance of *Hertwig's root sheath*, which develops as an outgrowth of the cervical loop during the late bell stage of tooth formation. The "edge" of the cervical loop circumscribes the perimeter or *diaphragm*, which, at the onset of root formation, lies at the presumptive neck of the crown. During root development, Hertwig's root sheath grows down between the tooth follicle and the dental papilla, encasing the dental papilla and pulp's neurovascular supply. As the root grows, the diaphragm becomes separated from the region of the tooth neck and becomes the aperture at the end of the elongating root. In a single-rooted tooth, the diaphragm persists as the opening in the root tip (*primary apical foramen*) through which pulpal nerves and arteries course. In a multirooted tooth, the once single diaphragm becomes subdivided, with pulpal innervation and vascularization coursing through two or more *secondary apical foramina*.

Root multiplication begins with the development of projections (*diaphragmatic processes* or *interradicular tongues*) from the edge of the diaphragm that extend toward and converge upon one another in the center of the pulp cavity. Elongating interradicular tongues delineate "bays"; as "bays" differentiate, blood vessels (and possibly nerves) segregate into groups. With the coalescence of interradicular tongues in the center of the pulp cavity, the bays become enclosed and form individual diaphragms (*radicular rings*); each ring corresponds to a presumptive root and each "captures" its own neurovascular bundle. Thus, as foramina elsewhere in the skeleton form *around* nerves and arteries, roots form around the nerves and arteries that serve a tooth. Root length is a function of the rate of interradicular tongue development: when rapid, a root will be short; the longer the process, the longer the root.

From a developmental perspective with evolutionary implications, reduction in root number is a consequence not of root fusion but of a decrease in number of interradicular tongues. In the context of dental anthropological studies that focus on root number, teeth (such as upper premolars and upper and lower molars) with "partially fused roots with separate apices" are examples not of root fusion but of delayed or inhibited interradicular tongue development.

Junction lines form between abutting interradicular tongues. At times, the juncture may be incomplete, creating *pulpoperiodontal canals* that transmit blood vessels and nerves. Pulpoperiodontal canals occur most frequently in deciduous molars. Since the primary plexus of blood vessels (and possibly nerves) segregates into smaller bundles as "bays" form, it is obvious how perturbations or irregularities in interradicular tongue expansion could isolate and "capture" a vascular branch or two.

Extraneous enamel deposits may be laid down on the interradicular tongues where roots arborize; such apparent *hyperamelogenesis* is presumably the result of activated enamel-producing potential of the root sheath. The extraneous deposit may be raised and three-dimensional [taking the form of a small, semi-isolated sphere—an *enamel pearl* (Figure 7–16)] or low, thinly tapering *enamel extension* that is continuous with the enamel of the crown (Figure 7–14). The latter configuration most frequently occurs on the buccal sides of upper and lower permanent molars (with the thin spit of enamel coursing between root clefts); a somewhat truncated enamel extension may develop on upper molars above the cleft between the lingual and mesiobuccal roots as well as between the lingual and distobuccal roots (personal observations). Interradicular enamel can interfere with proper attachment both of the periodontal ligament's interradicular group of fibers and of the oral epithelium. These weaknesses can be aggravated by pulpal inflammation and lead to pocketing.

Preliminary observations on a sample of 500 human crania (Schwartz, unpublished data) indicate that enamel extensions tend to occur on molar teeth that are also high-crowned (i.e., the enamel extends farther down the tooth all around). In addition, in the same individual, fully or virtually single-rooted teeth (which do not develop cervical interradicular tongues and thus cannot form enamel extensions) also appear to be high-crowned.

Other anomalies include *fossacoronoradicular* and *syndesmocoronoradicular teeth*. The former is characterized by a hollow in the junction line between enamel and cementum that invaginates into the crown and may also extend for the entire length of the root; this hollow occurs more frequently buccally than lingually. A syndesmocoronoradicular tooth is affected on its lingual side by such a drastic invagination of the cementoenamel junction into the crown of the tooth that enamel continuity is disrupted; a groove may also extend from the hollow either partially or for the entire length of the root.

Root elongation results from the induction by Hertwig's sheath of the proliferation of odontoblasts, which in turn produce the root's dentine. There are two phases of root development and elongation: **eruptive** and **penetrative**. During the eruptive phase, the root elongates such that its tip remains at a constant level (relative to the alveolar bone) and the crown of the tooth is "pushed" toward the occlusal plane. The eruptive phase continues until the tooth comes into occlusion with its "mate." At this point, the root may be approximately two-thirds developed. The penetrative phase is defined when the root tip begins extending into the basilar bone of the jaw. The two phases leave telltale signs on the root: the portion of the root formed during the eruptive phase is smooth; that formed during the penetrative phase is roughened and corrugated. Sometimes a line or crease around the circumference of the root delineates the two phases. Premature or abnormal contact with another tooth or bony surface can initiate the penetrative phase, causing impacted or imbedded teeth. If there is subsequent loss of periodontal tissue, *ankylosis* or fusion of the root to the surrounding bone can occur.

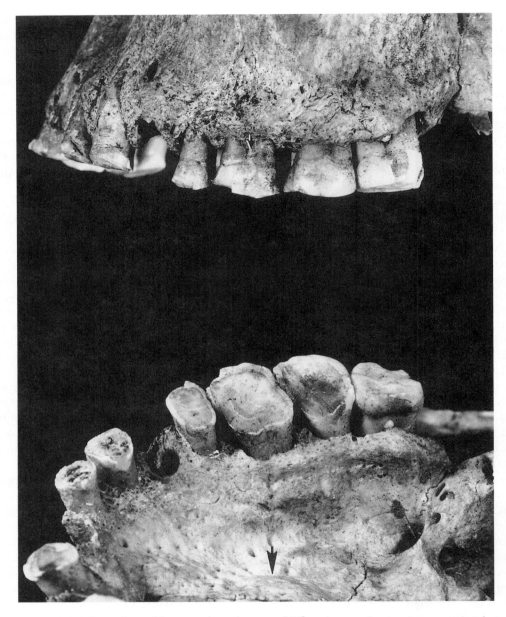

Figure 7–16 (*Top*) Enamel pearl between buccal roots of M³ and enamel extensions coursing between buccal roots of M¹⁻² (*small arrow*); (*bottom*) enamel pearl on lingual root of same M³; also note, for example, severe attrition encroaching on pulp cavities of M¹⁻², varying degrees of pulp cavity exposure on P², the canine, and I², remodeling buccally of the alveolar crest in the region of the premolars and molars [probably due to apical inflammation (see Chapter 10)], and the presence of a palatal torus (*large arrow*) (prehistoric, Alaska).

Kovacs (1971) devised a formula to express the relative length of the root: *length of root × 100/length of tooth = index of relative length of root*. He also suggested that deciduous, rather than permanent, teeth are more securely anchored in the jaws.

Crown Development: An Overview

The work of experimental embryologists such as Glasstone (e.g., 1967), Kollar and Baird (e.g., 1971), and Miller (e.g., 1971) first turned attention to the potential roles in tooth development of two **cell types: ectomesenchyme** (or simply **mesenchyme**) and **epithelium** (defined as **oral epithelium** in the presumptive embryonic jaws). Ectomesenchyme derives from **neural crest cells** that apparently follow cleavage planes and/or pathways established by pioneer nerves (see Lumsden, 1980) as they migrate from the embryonic *neural crest* to invade the presumptive jaws and come into close proximity with the oral epithelium.

What ensued thereafter was a history of debate over which cell type carried the information to induce tooth formation. For a while, ectomesenchyme was favored because in vitro experimentation indicated that tooth germs could be generated from the association of any, not just oral, epithelium and oral ectomesenchyme (e.g., Glasstone, 1967; Kollar and Baird, 1971). Miller's studies (e.g., 1971) suggested that oral epithelium induced tooth formation in the presence of ectomesenchyme, which was later seemingly demonstrated by experiments using very early mouse embryos (Lumsden, 1988). Kollar and Fisher (1980) even generated tooth germs in vitro by associating chick oral epithelium with mouse oral ectomesenchyme.

From experimental evidence, it appears that all the different tissues that eventually make up a tooth are derived from the cell mass that initiates tooth formation. For example, extirpated presumptive tooth-forming ectomesenchyme and oral epithelium, when grown in vitro but especially in vivo, produce teeth in which enamel and dentine are differentiated (e.g., Kollar and Baird, 1971). Ten Cate and Mills (1972) transplanted presumptive tooth-forming regions of fetal mice into different areas (e.g., skull, back) of host animals, with the result that teeth with differentiated crowns and roots formed, "erupted" through the epidermis of the hosts, and developed their own periodontal ligaments and surrounding alveolar bone. These latter results are particularly relevant to the observation that, when a tooth is lost from the jaw, only the alveolar bone in the affected region is resorbed, not the basilar bone of the mandible or the bone of the maxilla proper.

The onset of **tooth development** (Figure 7–17) is heralded microscopically by a condensation of ectomesenchyme in the region in which a tooth will form, as well as by a thickening of the overlying oral epithelium. This occurs in humans at approximately 6–7 fetal weeks. Subsequent events are subject to alternative interpretations.

One suggestion is that as the oral epithelium invades the neighboring ectomesenchyme, it continues to thicken longitudinally in each jaw quadrant, resulting in a **primary epithelial band**, ultimately bifurcating into buccal and lingual extensions. The latter is the **dental lamina**, which will become associated with the developing tooth germ. Eventually, the dental lamina expands into the **cap** and then **bell** shapes associated with later phases of tooth development. The buccal division—**vestibular lamina**—will continue to invaginate into the presumptive jaw, creating the **vestibule of the mouth** and delineating the lip.

The alternative, and perhaps more viable, interpretation of early tooth development is that there is a "tug-of-war" between the proliferating ectomesenchyme and oral epithelium, with the latter "trying" to contain the former. The epithelium expands and, as it passes through the cap and bell stages, eventually encapsulates most of the ectomesenchymal mass. In this model, there is no

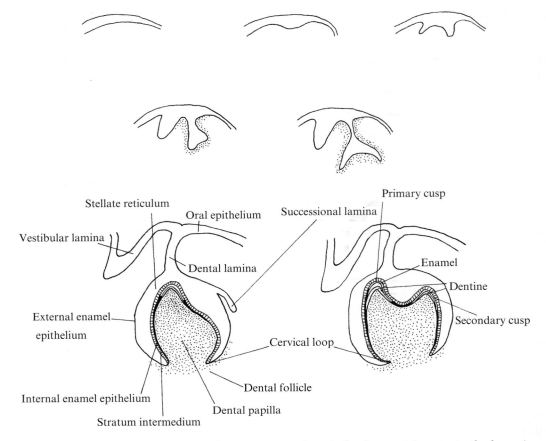

Figure 7–17 Schematic representation of major aspects of tooth development (as seen in the lower jaw, lingual is to the right): (*from top left to bottom right*) the oral epithelium thickening and eventually producing two projections, a vestibular lamina (*on the left*) and dental lamina (*on the right*); the dental lamina eventually expands into cap and then bell stages as it "captures" ectomesenchyme (*stippled*); (*bottom row*) (*left*) example of a single-cusped tooth, (*right*) example of a multicusped tooth (see text for further elaboration).

primary epithelial band and the vestibular and dental laminae develop as independent structures.

As the epithelial "cap" enlarges, its originally polygon-shaped cells, which maintained contact along their sides, transform into star-shaped cells, which contact each other only at their "points." The **intercellular spaces** that separate these star-shaped cells are swollen by water that is attracted by certain macromolecules (*hydrophilic acidic mucopolysaccharides*). This transformation results in a cellular layer—**stellate reticulum** ("starlike network")—whose functions may include protecting the developing tooth germ and maintaining an environment around the dental organ that provides the space necessary in which teeth can form properly. The stellate reticulum's outer layer is bounded by the **external (outer) dental (enamel) epithelium**, which is a continuation of the oral epithelium. The external dental epithelium is confluent with an **internal (inner) dental (enamel) epithelium** that is cytologically distinguished by its low, cuboidally shaped columnar cells. Eventually, another layer of cells—**stratum intermedium**—differentiates and interdigitates with cells of the internal dental epithelium. At the bell stage, the internal and external dental epithelial juncture creates the **rim** or **cervical loop** of the **enamel organ**, which in turn surrounds the **dental papilla** (Figure 7–17). The cervical loop constrains tooth germ expansion. The entire tooth germ then becomes surrounded by the **dental follicle**, which is derived from ectomesenchymal cells that had been displaced from the dental papilla. Nerves and arteries pass through the "base" of the dental papilla and pervade the dental follicle. When a tooth erupts, its dental follicle gives rise to the periodontal ligament.

A tooth that develops in association with the dental lamina—that is, is connected directly to the oral epithelium via the dental lamina—is a **primary tooth**. In humans and all other mammals, "permanent molars" are developmentally primary teeth because they arise sequentially from the posteriorly elongating lamina that earlier gave rise to the "deciduous molars" (Ooë, 1979). One can appreciate this by representing it in a *dental formula for the human primary dentition* (compare with Figures 7–3 and 7–4):

$$\frac{2.1.5}{2.1.5}$$

A **successional tooth** arises *lingually* from the external dental epithelium of the tooth it will eventually replace (Ooë, 1965; cf. Osborn, 1971; Tonge, 1976) (Figure 7–17); it is *not* connected directly to the oral epithelium. In diphyodont animals such as humans, a successional tooth is also a **secondary tooth**. The lingual extension from the external dental epithelium of a primary tooth that gives rise to a secondary tooth is the **successional lamina** (Figure 7–17) [thus, a tertiary tooth (see Figure 7–15), e.g., would arise from the successional lamina of a secondary tooth]. In polyphyodont animals, such as sharks, each tooth of each generation develops via a successional lamina that is a lingual outgrowth

of the external dental epithelium of the tooth it will replace.

Since erupted secondary teeth in diphyodont mammals are normally not shed, they can also be referred to as *permanent teeth*. In humans, these teeth are the successional incisors, canines, and premolars (but permanent molars are developmentally primary teeth). Obviously, not all permanent teeth have the same developmental history. Thus, while there might be historical reasons for referring to the teeth of an adult individual as "permanent," they may not be serially homologous structures. The *dental formula for the human secondary dentition* (compare with Figures 7–3 and 7–4) is as follows:

$$\frac{2.1.2}{2.1.2}$$

Crown shape is initiated during the bell stage (Figure 7–17). Grossly, it results from a flexing or folding of the internal dental epithelium that results from the antagonistic process of surface area increase—due to continuing cell division and proliferation—and restriction of outward growth by the cervical loop. The flexure in the internal dental epithelium occurs at a quiescent, nondividing region, which is pushed into the stellate reticulum and creates the tip or apex of a **primary cusp**; in humans and apes, for example, the lower molar protoconid is the first cusp to develop. The region of the primary cusp is also called the **enamel knot**, which is a controlling signaling center for subsequent aspects of tooth development (see review by Tucker and Sharpe, 2004). It is also at this stage that epithelial cells closest to the mesenchyme differentiate into enamel-producing cells (**ameloblasts**) and adjacent mesenchymal cells differentiate into dentine-producing cells (**odontoblasts**), the latter of which begin depositing dentine in the region of the primary cusp/enamel knot (Ten Cate and Osborn, 1976). Ameloblasts deposit enamel on top of dentine, first at the apex of a newly forming cusp and then down the presumptive cusp's sides, following on the heels of dentine deposition; their border is the **dentinoenamel juncture. Cusp height** is achieved as the internal dental epithelium expands into the dental papilla and elongates the sides of the developing cusp.

Additional or **secondary** and **tertiary** cusps arise as primary cusps do: cell division stops in a region of the internal dental epithelium [a **secondary** or **tertiary enamel knot** or developmental controlling center (Tucker and Sharpe, 2004)], which then flexes and pushes into the stellate reticulum; odontoblasts and ameloblasts differentiate and deposit, respectively, dentine and enamel. The degree to which cusp separation is achieved is a function of the extent to which the internal dental epithelium continues to proliferate and enlarge cusp size. Tooth shape is finalized as dentine fills in the basins between cusps. Enamel deposition may reflect details of dentinal surface topography, obscure these details, or add features to a crown.

Molecular Regulation of Crown Development: A Brief Overview

During the transformation of oral epithelium into dental lamina, the former cells signal the mesenchyme, which initiates cell condensation. Signals from the mesenchyme instruct the epithelium to invaginate into it and begin tooth bud formation. However, prior to dental lamina and tooth bud formation is an initiation period that involves signaling from the epithelium that establishes tooth type and position in the developing jaws.

Most of the information on the molecular regulation of mammalian tooth development comes primarily from experiments with mouse dentition (e.g., Tucker and Sharpe, 2004). But even though mice, like other rodents, are dentally very specialized mammals—only developing primary lower teeth (a pair of continually growing, buccolingually compressed teeth at the front of the jaw and right and left M_1 to M_3) and perhaps only one successional

premolar in an otherwise similarly configured upper dentition—some generalities appear applicable to humans.

The most important signaling molecules are **fibroblast growth factor** (FGF) and **bone morphogenetic protein** (BMP; both mentioned in Chapter 1 with regard to bone development), whose interaction is critical for establishing the spatial patterning of various developing tissues, not only those associated with teeth. Early in facial development, *Fgf-8* and *Bmp-4* are expressed in complementary domains in the epithelium of developing mandibular and maxillary primordia; they induce and/or maintain mesenchymal expression of various growth and transcription factors. FGF-8 induces and/or maintains, for example, the growth factors *activin β* and *Fgf-3* and the transcription factors *Alx-3, Barx-1, Dlx-1, Dlx-2, Pax-9, Msx-1, Lhx-6*, and *Lhx-7. Lhx-6* and *-7* define the anteroposterior axis of the mandible prior to appearance of tooth primordia, which also implies a role of FGF-8 signaling in inducing odontogenetic potential in the mesenchyme of the presumptive jaws. BMP-4 induces and/or maintains expression of *Msx-1* and *Dlx-2* in the mesenchyme; when synthesized in the oral epithelium, BMP-4 antagonizes the Fgf-8 inductive signal, restricting Fgf-8 expression to the presumptive odontogenetic (tooth-producing) mesenchyme.

FGF-8 and BMP-4 can establish spatially restricted domains of homeobox-containing genes in the mesenchyme. For example, BMP-4 expression in the presumptive "incisor" (anterior tooth) region of mice induces the mesenchymal expression of *Msx-1* and *Alx-3* in that region but also inhibits FGF-8-induced expression of *Barx-1* (whose expression is consequently restricted to the mesenchyme of the presumptive molar region). As seen in mice, the spatial information necessary to determine tooth type appears to be due to overlapping domains of gene expression: *Msx-1, Msx-2*, and *Alx-3* in the presumptive "incisor" field and *Barx-1* and *Dlx-2* in the presumptive molar field. Interestingly, experimental studies in mice demonstrate that lack of maxillary molar tooth development results from mutating *Dlx-1* and *Dlx-2*, which are coexpressed distally in the maxilla and mandible. The protein Noggin-induced inhibition of BMP-4 signaling results in the loss of mesenchymal expression of exclusively anterior-tooth region genes and in the ectopic expression of *Barx-1* in this region, resulting in the development of molariform teeth in the "incisor" region. These experiments have been taken as indicating that overlapping domains of *Msx* and *Dlx* genes might "code for" the canine and premolars of other mammals (McCollum and Sharpe, 2002), while others have been seen as indicating that premolar morphology may result from the expression of *Lhx-6* and *-7* and *Dlx-2* but not of *Barx-1* (see Tucker and Sharpe, 2004).

While it is clear that a tooth's morphology depends ultimately on different pathways of expression and/or interaction of various transcription factors, cell-signaling proteins, and genes—which may tempt one to generate a globally applicable theory of tooth formation—we must not forget that studies on mice deal only with primary teeth. Humans, like other mammals, develop secondary teeth, of which the premolars are some. Although the contents of cells are similar, it is their differential expression and/or recruitment that leads to structural difference. Consequently, while the outline of the molecular regulation of tooth development may be generally known, we might suspect that there should be differences—at least with regard to notions of overlapping domains of gene expression—in the development of primary versus secondary teeth.

Coincident with the formation of the dental lamina and of the tooth bud is the activation of members of the **Wnt** family, **epidermal growth factor (EGF), Jagged,** and **hepatocyte growth factor (HGF)**; interaction between **sonic hedgehog (Shh)** and *Wnt7b* defines the region of oral versus dental epithelium. In fact, it appears that the Shh protein regulates proliferation of the cells that produce the tooth bud and that its expression is restricted to future sites of cell proliferation and tooth bud formation. In dental epithelium, overexpression of Wnt7b leads to suppression of *Shh* expression, with subsequent truncation of tooth development. Wnt10 targets *Lef-1* (a transcription factor gene), which is expressed in the dental epithelium and mesenchyme throughout tooth development; in mice without functional Lef-1 protein, tooth development is arrested at the bud stage (which implies that Lef-1 signaling acts on the dental mesenchyme during the bud stage of tooth morphogenesis).

The initiation phase of tooth development ends with the establishment of the tooth bud, when *Bmp-4* expression and the potential to direct tooth development switches from the dental epithelium to the dental mesenchyme. FGF-8 may also be involved in this transition. With this shift, the inner surface of the epithelial bud (from the internal dental epithelium) surrounds the condensing mesenchymal cells of the dental papilla, producing the bud and then cap stages. At this point, BMP-4 signaling induces epithelial cells at the tooth bud's tip to express p21 (an inhibitor of cell proliferation), which causes them to stop dividing; this produces the tightly packed epithelial cells identified as the primary cusp or enamel knot. Cells of the primary enamel knot express the transcription factor Msx2 as well as numerous signaling factors, such as members of the WNT family, FGF-4, and BMP-2, -4, and –7. The expression domains of FGF-9 and Shh are now restricted to cells of the enamel knot, which, within 24 hours in mice, undergo apoptosis (cell death) and are removed from the cap stage of the tooth. Secondary and tertiary enamel knots, which very quickly appear in areas where other cusps and structures will form, express *Fgf-4*. It appears that primary enamel knots organize the development of tooth shape by regulating the spatial and temporal activation of secondary and tertiary enamel knots, which in turn regulate the growth of the individual cusps that underlie tooth shape. At least from studies on mice, cusp and crest position and shape—fundamentals of tooth form and shape—appear to be correlated to some extent with, and affected by, the cell-signaling protein **ectodysplasin** (**Eda**) (Kangas et al., 2004). In addition, overexpression of Eda in mice leads to development of a stunted supernumerary tooth, while loss of Eda function leads to tooth agenesis (Tucker and Sharpe, 2004). Mutations involving the transcription factor Msx-1 have also been implicated in different configurations of tooth agenesis in humans: M3 plus P2 and of P1 and P2 (in association with clefting). [For a general overview of the molecular regulation of development see Carroll et al. (2005) and Wolpert et al. (2002).]

Chapter 8

Aging

Overview

The life history of an individual is often subdivided into phases on the basis of various developmental and morphological changes (e.g., Krogman, 1962; Krogman and İşcan, 1986; Stewart, 1979). For example, Saunders (2000 and earlier references therein) recognizes subadult and adult age categories. Since tooth-measurement data are virtually useless and unfused epiphyses often not recovered archaeologically or, if they are, not easily correctly reassociated with diaphyses, Saunders focuses on stages of tooth formation and lengths of diaphyses for determining the ages of subadults. She recommends that adult ages are best determined on the basis of iliac auricular and pubic symphyseal morphology, aspects of intact cranial morphology, and measurements of postcrania.

Although one reviewer of the 1995 edition of this text commented that Western osteologists do not typically cite Acsádi and Nemeskéri (1970), their three phases are useful, even though any subdivision of a lifetime of growth- and age-related change, which is a continuous process, will be somewhat artificial. Indeed, patterns of ontogenetic and postontogenetic transformation of any given skeletal unit—teeth, skull, postcranium—are not invariant (e.g., sometimes a tooth erupts before the tooth it "should" erupt after) and the timing of skeletal unit transformation often overlaps rather than forms a neat sequence of events. But if we think of phases as generally represented by classes of organizing criteria, they can have some utility. The inclusion of different approaches also serves an important learning experience. As seen from the various approaches summarized below, there is more of a subjective element to determining age than one (including the public) would expect (see review by Jackes, 2000)—a fact that legal and forensic demands on producing definitive conclusions in reports and courtroom testimony and limiting one's analysis to one or two popular estimating techniques tend to submerge. Yet, in spite of the fact that Acsádi and Nemeskéri's (1970) approach may not commonly be used by Western osteologists, in one of the few studies based on a skeletal sample (Spitalfields) of indisputably documented age (as well as sex and life history), Molleson (1995; also see Molleson and Cox, 1993) found that the mean adult age at death was most consistent with Acsádi and Nemeskéri's (1970) complex method.

Since an osteologist's goal should be to use as many criteria as possible to estimate age at death (and not to embrace an inflexible age-stage program), we turn to Acsádi and Nemeskéri's three general phases, to which I have added other potentially useful criteria (prenatal development and ossification of bony elements and teeth are dealt with in the appropriate morphological sections on individual bone and teeth and summarized in tabular form):

1. **Childhood:** Growth and eruption of deciduous and permanent teeth; appearance of centers of ossification; growth of cranial bones—for example, the petrosal (petromastoid region), temporal, and elements of the occipital.
2. **Juvenile age:** Union of epiphyses with diaphyses, unification of the os coxa, closure of the spheno-occipital synchondrosis.
3. **Adulthood:** Closure of cranial sutures; changes in the surface of the pubic symphysis, auricular surface, and sternal end of the rib; structural changes in the spongy substance of the humeral and femoral proximal epiphyses; patterns and degrees of tooth wear.

As in sexing skeletal remains (see Chapter 9), variability, not consistency, is the rule in determining age at death. Even under normal and developmentally uneventful conditions, the rates at which various components of one individual develop and take shape, erupt or fuse, become incorporated into a larger entity, or change thereafter with increasing age can vary considerably between females and males, individuals of the same sex, and individuals of different populations (Jackes, 2000; Saunders, 2000; Schwartz, 1995). Furthermore, and again paralleling the sexing of skeletal remains, generalizations concerning estimation of age at death that dominate textbooks and review articles on human skeletal analysis are based on region- or site-specific, rather than globally representative, samples (Jackes, 2000; Saunders, 2000; Schwartz, 1995). Consequently, one must use caution when applying aging (or any other) criteria and standards to one's study group, especially because the data from which those criteria derive may not be from a relevant sample. Finally, one must not forget that individual life histories complicate the task of determining age at death. Hormonal imbalances, socioeconomic factors (e.g., diet and some disease or other assaults to the body's systems), as well as the relative onset and pace of aging may interfere with "normal" patterns of tooth and bone formation. The assessment of age at death is an approximation, which is usually most accurate for younger ages because there is more developmental activity throughout the skeleton in young individuals. As arenas of skeletal activity decline and idiosyncrasies of individual life histories impress themselves upon the skeleton, inaccuracy in determining age increases.

Of importance, however, before turning to age-estimating criteria of childhood is the identification of an individual's birth, or at least the departure of the fetus from the womb (through either the birth canal or cesarian section). In this regard, a histological approach that can be used to corroborate visual assessment of tooth crown and skeletal development is the detection of the neonatal (natal) line in thin-sectioned tooth crowns (Eli et al., 1989; Levine et al., 1979; Smith and Avishai, 2004; Whittaker and Richards, 1978). This distinct, microscopically visible "line" delineates the fetal from the postnatal phases of enamel deposition and records a period of up to a few days of decreased mineralization as well as of alteration of enamel prism orientation. Since the development of the neonatal line results from the trauma of birth, its presence is independent of a neonate's size or lunar age. The absence of a neonatal line, however, need not be taken as representing a fetus or a stillbirth for the reason that the newborn may die prior to its formation.

Table 8–1   Average Fetal Age Based on Average Diaphyseal Length (mm)

Age (fw)	Humerus	Ulna	Radius	Femur	Tibia	Fibula
12	8.8	7.2	6.7	8.5	6.0	6.0
14	12.4	11.2	10.1	12.4	10.2	9.9
16	19.5	19.0	17.2	20.7	17.4	16.7
18	25.8	23.9	21.5	26.4	23.4	22.6
20	31.8	29.4	26.2	32.6	28.5	27.8
22	34.5	31.6	28.9	35.7	32.6	31.1
24	37.6	35.1	31.6	40.3	35.8	34.3
26	39.9	37.1	33.4	41.9	38.0	36.5
28	44.2	40.2	35.6	47.1	42.0	40.0
30	45.8	42.8	38.1	48.7	43.9	42.8
32	50.4	46.7	40.8	55.5	48.6	46.8
34	53.1	49.1	43.3	59.8	52.7	50.5
36	55.5	51.0	45.7	62.5	54.7	51.6
38	61.3	55.9	48.8	69.0	60.1	57.6
40	64.9	59.3	51.8	74.4	65.2	62.0

Source: Fasekas and Kósa (1978).
Abbreviation: fw, fetal week.

Fetal age can be estimated using measurements of cranial and post cranial elements (see Fazekas and Kósa, 1978). Diaphyseal lengths of long bones are provided in Table 8–1. Tooth formation data can be found later in Tables 8–4 and 8–5.

Childhood

Tooth Formation Sequences: A Brief Overview

Postnatal aspects of tooth growth and eruption can be broken down into general phases:

1. The toothless phase of infancy (0–7 months).
2. The teething phase, that is, eruption of deciduous teeth (7 months to 2 years).
3. The (use) phase of the deciduous teeth (2–6 years).
4. The phase of eruption of the permanent teeth and replacement of deciduous teeth, that is, the phase of the mixed dentition (6–12 years).
5. The (use) phase of the permanent teeth (12+ years).

"Teething" and "mixed dentition" are most helpful because they are associated with easily identifiable changes in relative states of crown and root formation and positions of teeth within the jaw and/or above the alveolar margin. Although the fifth and the latter part of the fourth phase (particularly with regard to M3) are relevant to determining juvenile and adult ages, respectively, they are listed here for completeness.

Visual assessment of tooth activity is basically limited to evaluation of the emergence of the crown above the alveolar margin. This is particularly true of secondary teeth, of which, at best, cusp tips are barely visible through preeruption perforations in the alveolar margin (foramina through which gubernacula of successional teeth course) that lie lingual to the erupted primary teeth. Primary tooth (including permanent molar) crown development can be assessed with varied success directly through the incomplete alveolar margin. Radiography is the most effective approach (especially with regard to cost and time) by which to analyze relative states of crown and root formation as well as of erupting teeth. Often, the type of X-ray unit available (e.g., dental, "oven," hospital) and the limitations on settings of kilovolts, milliamps, and length of exposure time will affect the degree with which one can control the crispness (high contrast) or softness (greater spectrum of grays) of the radiograph. Also, the type of X-ray film used and the recommended developing chemicals can influence the contrast. Depending on bone and tooth density, high kilovolt and milliamp settings with short exposure times often produce crisper radiographs than longer exposure times at lower kilovolt and milliamp settings. Radiographic units capable of fluoroscopy provide instantaneous but generally fuzzy images. Computed tomographic (CT) scanning, which can capture digital data that can be used for three-dimensional reconstruction, is useful only to the degree that the scanned slices (usually 2 mm) do not exceed the level of definition necessary to perform the analysis. In addition, access to CT-scanning equipment is still often limited to the good will of medical and industrial facilities and availability of CT equipment (usually late at night or early morning). A newer technology, micro-CT scanning, has the potential to capture finer levels of morphological detail that can be used in three-dimensional reconstructions.

As with any attempt to use parts of the skeleton to assess age (or sex), the osteologist is limited by the published samplings of sexes, age groups, and populations. Although there is a long history of studies on correlating tooth formation/eruption with an individual's age, many of them are not comparable; and as Smith (1991) pointed out in her valuable review, there is still much work to be done. Indeed, Saunders' (2000) summary of the literature on differences in actual and relative timing between various populations makes clear the absence (as in virtually every aspect of human skeletal analysis) of critical information: for example, in the same area of the mid-southern United States, tooth formation

was more advanced in African-Americans than European-Americans, as are black compared to white South Africans and Latinos, African-Americans, and European-Americans compared to French Canadians. Saunders (2000) as well as Smith (1991) also point out that many studies on tooth eruption (emergence) were based on penetration of the gingiva (not the alveolar margin) and that the timing of eruption rather than tooth formation may be more easily disrupted (e.g., by hormonal, nutritional, and even social factors).

Nevertheless, it is also fair to say that various osteologists have their favorite dental-aging chart or table. For example, some reviewers of this revision noted the absence in the first edition of Ubelaker's (1999) oft-reprinted diagrammatic representation of stages of tooth formation and eruption. I now include this diagram (Figure 8–1), but reiterate that the part illustrating permanent tooth eruption is based on Native North Americans and various other "nonwhite" groups, while the information on earlier phases is derived primarily from studies on white North American groups. Tables 8–2 to 8–4 offer a variety of potentially useful kinds of data on the timing of dental growth and eruption from which student and professional alike may extract information relevant to their particular needs. Although not providing information on stages of tooth formation between the onset of mineralization and completion of crown and root, Ten Cate (1989; Table 8–5) includes prenatal tooth development and distinguishes between timing in females and males. Even in light of the caveats and pitfalls mentioned above about tooth-formation and eruption-sequence data, it is still considered a better approximation of chronological age than other aspects of skeletal maturity (e.g., Demirjian, 1980; Moorrees et al., 1963a; Saunders, 2000).

As a key to deciphering relative tooth/root formation, Figure 8–2 presents Winkler et al.'s (1991) elaboration on Demirjian's (1980) eight phases. Since any attempt to delineate discrete stages in a developmental continuum

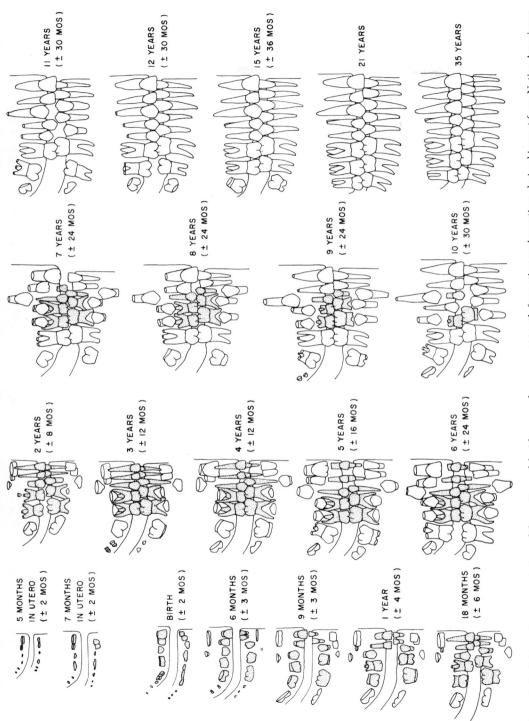

Figure 8–1   Schematic representation of the pattern of deciduous and permanent tooth formation and eruption, derived in part from Native Americans (copyright © D. H. Ubelaker; reprinted with the author's permission).

Table 8–2   Chronology of Tooth Eruption

	Females		Males	
Tooth	Upper	Lower	Upper	Lower
Deciduous dentition				
di1	9 m 0 d ± 2 m	7 m 9 d ± 2 m	8 m 14 d ± 2 m	7 m 6 d ± 2 m
di2	10 m 23 d ± 2 m	11 m 23 d ± 2 m	10 m 23 d ± 2 m	11 m 9 d ± 3 m
dc	18 m 5 d ± 2 m	13 m 6 d ± 3 m	17 m 18 d ± 3 m	18 m 0 d ± 3 m
dm1	14 m 23 d ± 2 m	14 m 27 d ± 2 m	14 m 24 d ± 2 m	16 m 3 d ± 2 m
dm2	24 m 24 d ± 6 m	24 m 18 d ± 3 m	24 m 7 d ± 6 m	23 m 28 d ± 3 m
Permanent dentition				
I1	7 y 0 m 0 d ± 0 y 9 m	6 y 1 m 23 d ± 0 y 8 m	7 y 2 m 13 d ± 0 y 11 m	6 y 3 m 23 d ± 0 y 8 m
I2	7 y 11 m 14 d ± 0 y 11 m	7 y 1 m 8 d ± 0 y 9 m	8 y 4 m 12 d ± 0 y 0 m	7 y 4 m 20 d ± 0 y 9 m
C	10 y 6 m 1 d ± 1 y 1 m	9 y 8 m 10 d ± 1 y 2 m	11 y 1 m 23 d ± 1 y 2 m	10 y 8 m 28 d ± 1 y 2 m
P1	10 y 1 m 8 d ± 1 y 3 m	10 y 0 m 3 d ± 1 y 4 m	10 y 7 m 6 d ± 1 y 4 m	11 y 6 m 8 d ± 1 y 3 m
P2	10 y 8 m 7 d ± 1 y 5 m	10 y 1 m 27 d ± 1 y 4 m	11 y 4 m 12 d ± 1 y 4 m	11 y 6 m 8 d ± 1 y 3 m
M1	6 y 3 m 15 d ± 0 y 10 m	6 y 2 m 22 d ± 0 y 11 m	6 y 4 m 1 d ± 0 y 10 m	6 y 2 m 26 d ± 0 y 9 m
M2	11 y 11 m 26 d ± 1 y 2 m	11 y 6 m 10 d ± 1 y 4 m	12 y 6 m 1 d ± 1 y 3 m	12 y 0 m 7 d ± 1 y 3 m

SOURCE: Modified from Moyers (1959).
Abbreviations: d, day; m, month; y, year.

Table 8–3   Chronology of Tooth Eruption

	Deciduous		Permanent	
Tooth	Average Age (Months)	Normal Range (Months)	Tooth	Normal Range (Years)
Upper dentition				
di¹	7	8–11	I¹	7–8
di²	9	8–11	I²	8–10
dc	18	16–24	C	11–12
dm¹	14	9–21	P¹	10–11
dm²	24	20–36	P²	10–12
			M¹	6–7
			M²	12–13
			M³	17–30
Lower dentition				
di₁	6	4–8	I₁	6–7
di₂	8	7–12	I₂	7–10
dc	18	16–25	C	9–10
dm₁	12	9–21	P₁	10–12
dm₂	22	20–36	P₂	11–12
			M₁	6–7
			M₂	11–13
			M₃	16–30

SOURCE: Modified from Vallois (1960).

is arbitrary, one must extrapolate between phases in order to evaluate relative states of formation/eruption more accurately.

Tooth Formation and Eruption Sequences: A Discussion

The common order of deciduous tooth eruption is $di_1 \rightarrow di^1 \rightarrow di^2 \rightarrow di_2 \rightarrow dm1 \rightarrow dc \rightarrow dm2$, although variations do occur. There are no apparent sexual differences in the order, but the timing of deciduous tooth eruption tends to be earlier in males than females. It is the reverse with the permanent teeth: they tend to erupt earlier in females than males. Olivier (1960) provided more detailed information on deciduous and permanent teeth than the timing of eruption of individual teeth. This offers the possibility of more precisely reconstructing the age of a pre- or postnatal individual, even if all dental elements are not present (e.g., having been lost postmortem during secondary burial or excavation).

Table 8–4 Chronology of Tooth Formation and Eruption

		Deciduous Dentition				
Tooth	Appearance of Tooth Germ (Fetal Months)	Percent Crown Formation at Birth	Crown Fully Formed (Months)	Root Fully Formed (Years)	Reabsorption of Root (Years)	Exfoliation (Years)
di1	4–4.5	Upper: 83 Lower: 60	Upper: 1.5 Lower: 2.5	1.5 1.5–2	4–5	6–7
di2	4.5	Upper: 66 Lower: 60	Upper: 2.5 Lower: 3	1.5–2	4–5	7–8
dc	5	33	9	3.25	6–7	10–12
dm1	5	Cusps continuous	5.5–6	2.5	4–5	9–11
dm2	6	Cusps still distinct	10–11	3	4–5	10–12

	Permanent Dentition		
Tooth	Appearance of Tooth Germ	Crown Fully Formed (Years)	Root Fully Formed (Years)
M1	At birth	2.5–3	9–10
I1	3–5 months	4–5	9–10
I2	Upper 10–12 months Lower 3–4 months	4–5	10–11
C	4–5 months	6–7	12–15
P1	1.5–2 years	5–6	12–13
P2	2–2.5 years	6–7	12–14

SOURCE: Modified from Olivier (1960).

Moorrees et al. (1963a,b) discriminated deciduous tooth formation/age stages more finely than Olivier (1960) but only for lower dcs and dms, their successors, and the permanent molars; they also collected data on the timing of deciduous tooth root resorption. Although only sampling modern white children from Ohio, Moorrees et al. found differences between girls and boys and that most variability occurred during root formation.

Demirjian (1980), whose study was limited to the permanent dentition, argued that one could accurately assess an individual's age by radiographically analyzing only one side of the upper and lower dentitions because growth and eruption sequences were essentially the same for upper and lower teeth. He advocated using mandibular teeth, which are easier to radiograph. In addition, since the development of right and left lower teeth is supposed to be highly correlated, Dermirjian suggested

that one could study only one side of the mandibular dentition (he chose the left).

Demirjian (1980) assessed relative states of tooth formation by dividing the process for each tooth into eight stages. He constructed tables for girls and boys of maturity scores correlated with stages of tooth formation that can be used comparatively between individuals without assuming accurate dental ages. Precise dental ages can be reconstructed by applying the sum total of the maturity scores to the (50th centile or median) age curves obtained in the study. He correctly pointed out the difficulty in assessing absolute values of crown and root growth, which can vary considerably between individuals, and recommended assessing the relationship between crown and root size when possible. Evaluating relative states of tooth formation also increases the age range within which the analysis can take place. If age determination were based solely

Table 8–5   Chronology of Tooth Formation and Eruption

Tooth	Mineralization Onset	Crown at Birth	Crown Complete	Crown Complete: F	Crown Complete: M	Emergence
di^1	14 weeks iu	83%	1.5 months			7.5 months
di$_1$	18 weeks iu	60%	2.5 months			6 months
di^2	16 weeks iu	66%	2.5 months			9 months
di$_2$	18 weeks iu	60%	3 months			7 months
Upper dc	17 weeks iu	33%	9 months			18 months
Lower dc	20 weeks iu	33%	9 months			16 months
dm^1	12.5–15.5 weeks iu	Cusps united	6 months			14 months
dm$_1$	12.5–15.5 weeks iu	Cusps united	5.5 months			12 months
dm^2	12.5–19 weeks iu	Cusps separate	11 months			24 months
dm$_2$	12.5–18 weeks iu	Cusps separate	10 months			20 months
M^1	At birth			2.6 years	2.7 years	
M$_1$	At birth			2.6 years	2.7 years	
M^2	2.5–3 years			6.3 years	6.7 years	
M$_2$	2.5–3 years			6.3 years	6.7 years	
M^3	7–9 years			12.7 years	13.3 years	
M$_3$	8–10 years			12.8 years	13.3 years	
I^1	3–4 months			3.3 years	3.7 years	
I$_1$	3–4 months			3.3 years	3.6 years	
I^2	10–12 months			3.8 years	4.0 years	
I$_2$	3–4 months			3.7 years	4.0 years	
Upper C	4–5 months			4.1 years	4.8 years	
Lower C	4–5 months			4.1 years	4.9 years	
P^1	1.5–1.75 years			5.1 years	5.8 years	
P$_1$	1.75–2 years			5.0 years	5.6 years	
P^2	2.2–2.5 years			5.9 years	6.3 years	
P$_2$	2.2–2.5 years			5.9 years	6.3 years	

SOURCE: Modified from Ten Cate (1989).
Abbreviations: iu, in utero; F, female; M, male.

on tooth emergence, the analysis would be restricted, at least in boys, to the periods of approximately 6–7 years (for the incisors and M$_1$) and approximately 10.5–11.5 years (for the canine, premolars, and M$_2$). Variability in timing of M$_3$ crown and root formation makes this tooth an unreliable indicator of age.

Although time-saving in terms of generally assessing maturational age, using one quadrant of one jaw, or even only one jaw, to represent the entire picture of an individual's history of dental growth and eruption likely results in loss of information—with regard, for example, to intra- as well as interpopulational differences when standards of known age are lacking for comparison. Furthermore, as more studies on variation in patterns of dental growth and eruption emerge (see, e.g., Winkler et al., 1991, and references therein), it is becoming increasingly obvious that differences in timing between upper and lower teeth are real, as was pointed out by Kronfeld (1954), Moyers (1959), Olivier (1960), Schranz (1959), and Vallois (1960).

Approaching the topic of age determination from a different perspective, Spalding et al.

Emergence: F	Emergence: M	Root Complete	Root Complete: F	Root Complete: M	Eruption Sequence
		1.5 years			3
		1.5 years			1
		2 years			4
		1.5 years			2
		3.25 years			8
		3.25 years			7
		2.5 years			6
		2.25 years			5
		3 years			10
		3 years			9
7.2 years	7.8 years		9.2 years	10.1 years	13 (3)
7.2 years	7.8 years		9.2 years	10.0 years	12 (2)
11.8 years	12.4 years		13.6 years	14.6 years	23 (13)
11.8 years	12.5 years		13.8 years	14.8 years	24 (14)
17.8 years	17.4 years		18.8 years	18.2 years	26 (16)
17.7 years	17.4 years		18.3 years	18.5 years	25 (15)
7.4 years	8.3 years		9.3 years	10.6 years	15 (5)
7.3 years	7.3 years		8.1 years	9.2 years	11 (1)
8.1 years	9.1 years		9.7 years	11.1 years	16 (6)
7.3 years	8.1 years		8.8 years	9.9 years	14 (4)
9.4 years	11.0 years		11.9 years	13.7 years	18 (8)
9.2 years	10.9 years		11.4 years	13.5 years	17 (7)
9.7 years	11.1 years		11.8 years	13.5 years	19 (9)
9.9 years	11.2 years		11.9 years	13.3 years	20 (10)
10.6 years	11.6 years		12.6 years	13.8 years	21 (11)
10.6 years	11.9 years		12.8 years	14.0 years	22 (12)

(2005) considered determining an individual's date of birth from levels of the carbon 14 isotope (^{14}C) in tooth enamel. This preliminary study was provoked by the fact that the atmospheric level of ^{14}C, which plants incorporate in the form of carbon dioxide and which plant-consuming animals subsequently incorporate into their bones and teeth, increased during the period 1955–1963 as a result of atomic bomb testing. ^{14}C levels began to decline slowly and exponentially thereafter, which means that this method could be useful well into the future.

Since enamel does not change after it is formed (in contrast to bone, which is constantly remodeling until an individual dies —which is why ^{14}C analyses are typically employed to estimate time of death), the level of enamel-borne ^{14}C is potentially useful for determining an individual's age of birth. Using 12 years for the final formation of M3 crowns, Spalding et al. (2005) concluded that absence of increased levels of ^{14}C would indicate a date of birth prior to 1943. In a sample of 22 individuals of known age who were born during or after 1955, Spalding

Molar   Premolar   Canine   Incisor

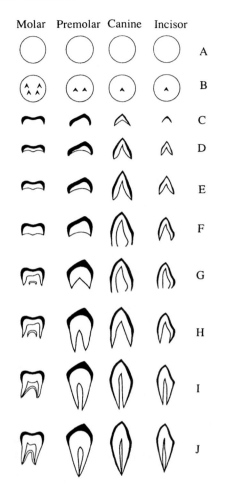

Figure 8–2 Schematic representation of relative "stages" of tooth crown and root formation: *A*, Tooth crypt present; *B*, initial calcification; *C*, crown one-quarter formed; *D*, crown one-half formed; *E*, crown three-quarters formed; *F*, crown fully formed; *G*, onset of root formation (whether single- or multirooted); *H*, apical foramen/foramina broad, walls of root canals funnel-shaped, root outline more definitive, with root length equal to or slightly greater than crown height; *I*, apical foramen/foramina still somewhat open, walls of root canals parallel; *J*, apical foramen/foramina "closed" (after Winkler et al., 1991).

et al. achieved an age estimate with an average absolute error of only 1.6 ± 1.3 years. In order to increase the accuracy of estimating time of birth, more than one tooth per individual should be analyzed to determine whether the ^{14}C levels correlate with the post-1955 rise or post-1963 decline in the ^{14}C curve.

Cranium

Various cranial regions or elements are often recovered in archaeological and forensic situations and present themselves as potential indicators of the age of young individuals. The petrosal, squamosal, and elements of the sphenoid as well as occipital are often sufficiently preserved in post- and prenatal individuals, even after being exposed to extreme heat and then centuries of internment in waterlogged conditions, to be useful (e.g., see Schwartz, 1989).

Details of petrosal development are presented in Chapter 3 (from Schwartz, 1995; see also Scheuer and Black, 2000). Important events include the following:

1. During sixth fetal month: superior semicircular canal ossifies, carotid canal visible as fissure superiorly, jugular notch is distinct.
2. About eighth fetal month: carotid fissure enlarges, ossifies inferiorly into carotid canal; tympanic ring begins to fuse at least with squamosal; ossification center for mastoid region arises.
3. At birth: petrosal incompletely ossified anteriorly, internal acoustic meatus and subarcuate fossa approximately same size, mastoid region small but identifiable, horizontally oriented tympanic ring attached to petrosal at both cornua.
4. First year: tympanohyal portion of styloid process fuses to temporal; petrous and squamous portions unite, perhaps as early as first postnatal month; foramen of Huschke defined; tympanic ring begins to extend as bony tube.
5. By fifth year: foramen of Huschke ossified, tympanic ring vertically oriented, mastoid process clearly discernible, vaginal process emerges around base of styloid process.

As summarized earlier (from Schwartz, 1995), the elements that form the adult sphenoid follow a general pattern of coalescence:

1. At birth: lesser wings united to body and greater wings to pterygoid plates, sinuses tiny.
2. By end of first year: all elements united; jugum formed by union of lesser wings; there is some variation during the first year in the sequence in which the elements become one (Scheuer and Black, 2000).

With regard to the occipital, earlier stages of development are described in Chapter 3 (from Schwartz, 1995; see also Scheuer and Black, 2000). Relevant events are summarized here.

1. Fifth fetal month: posterior end of basilar part begins to assume "Y" shape at anterior end of foramen magnum; anterior end of basilar part (contribution to spheno-occipital synchondrosis) broadens and flattens; occipital and nuchal planes coalesce centrally, leaving a central and a pair of lateral (mendosal) fissures.
2. Fifth postnatal month: width of basilar part surpasses maximum length.
2. After seventh fetal month: lateral parts become longer than basilar part (previously of equal lengths).
3. First year: jugular process forms posterior to jugular foramen.
4. First to third years: hooked extension grows from inferior (condylar) limb of lateral part to join the intracranial (jugular) limb, forming condylar canal.
5. Third to fourth years: mendosal fissures closed.
6. Fourth year: lateral parts fuse with squama.
7. About sixth year: occipital becomes totally unified.

Weaver (1979) divided the growth of the *petrosal* (petromastoid region) and *tympanic plate* of the temporal bone into six stages, the first four of which he felt could be associated with ages. The sample he used, derived from the Grasshopper Pueblo skeletal series, consisted of 179 temporal bones from individuals for whom dental age could be estimated. As such, Weaver used one set of estimates (dental age) to establish another set of estimates

(tympanic plate age). In and of itself, this should cause the casual user to be cautious of using temporal data alone, whether or not one is applying these criteria to skeletal remains from non-Southwest Native North Americans (but see The Complex Method of Determining Adult Age, later).

The stages and ages Weaver (1979, pp. 264–266) delineated are as follows:

1. Fetal or newborn: petromastoid portion present but tympanic ring undeveloped.
2. Newborn to 0.5 years: "U"-shaped, incomplete tympanic ring partially coalesced with lateral margin of petromastoid.
3. 1.0–2.5 years: tympanic ring well coalesced inferiorly with petromastoid, which markedly extends laterally beyond the ring anteriorly and posteriorly (i.e., forming a "U"-shaped and incomplete tympanic plate).
4. 1.0–2.5 years: tympanic plate more fully realized laterally and anteroposteriorly through the closure of the "U" into an "O," which leaves a patency (presumptive foramen of Huschke) in the middle of the "O," and its lateral edge remains medially jagged and indented ("U"-shaped).
5. No age determined: lateral edge of the tympanic plate more laterally extended, its margin more smoothly defined, and the patency more completely ossified.
6. No age determined: tympanic plate now forms well-defined external acoustic meatus; its floor is typically completely ossified.

Although the illustration of and the morphological criteria for this sequence are oft repeated in print, some emendations are in order. Particularly important is the fact that the *tympanic ring* is an ossified structure prior to birth. But because the tympanic ring in the neonate is secured by soft connective tissue to the basicranium and typically does not fuse with the basicranium until the end of the first postnatal year, this element is often missing not only in prepared specimens but also especially in specimens that have endured burial and subsequent excavation. Nevertheless, the groovelike impression left by the tympanic ring along the lateral margin of the expanding bulla (i.e., the "petromastoid" portion) is a clue to this bone's pre- and neonatal existence. It is upon fusion of the tympanic ring to the lateral margin of the bulla that, in archaeological and other potentially destructive conditions, one more frequently finds the tympanic ring in place. It is also with fusion of the ring to the bulla that the ring begins its ossification laterally as a bony tube; this process of ossification and extension is most aggressive in the regions along the ring's anterior and posterior cornua.

Other aspects of petrosal (petromastoid) and temporal bone development should be taken into consideration as well. For example, in the neonate, the *arcuate eminence* is not low and rounded as in the adult but is a prominent elevation created by the underlying and somewhat laterally aligned superior semicircular canal. In the neonate, the anterior face of the eminence descends more or less vertically from the elevation of the semicircular canal; at this time, the posterior face of the eminence is more gently sloping. Also in contrast to the adult, the neonatal *subarcuate fossa* is quite large; that is, it is a vast opening subtended by bone surrounding the *superior semicircular canal*. The *internal acoustic meatus* is more anteriorly displaced away from the subarcuate fossa in the neonate than in the adult, and the entire *petrosal bone* is unfused to the *squamous portion of the temporal bone*. This fusion does not usually occur until later on, during the first postnatal year.

The components of petrosal and temporal bone development that thus present themselves as potentially useful in assessing young ages are as follows: (1) changes in the shape and relative size of the arcuate eminence; (2) changes in the shape, relative position, and relative size of the subarcuate fossa and of the internal acoustic meatus; (3) fusion of the petrosal to the squamosal portion and the cornua of the tympanic ring with the basicranium (by the end of the first year); (4) fusion of the tympanic ring to the lateral edge of the bulla and its growth laterally as well as inferiorly, with the latter correlated with the diminution of a presumptive *foramen of Huschke* (by the end of the fifth year). Although the mastoid process becomes an extension of the petrosal region, once the latter has coalesced with the squamous portion of the temporal (to form the presumptive petromastoid portion), the development of any elevation that could be identified as a mastoid process does not usually occur until toward the end of the second year. It thus would seem that, if dental ages from a given sample could be correlated with morphological changes in the petrosal/temporal bone, additional criteria for assessing age in the absence of preserved dental remains are available.

Redfield (1970) developed a scheme of *occipital bone* formation based on samples he analyzed as well as on his review of the literature. Collectively, these sources demonstrated considerable variability in time of fusion of (1) the mendosal fissures (the remnants of the different developmental origins of the occipital and nuchal planes) and (2) the lateral and basilar parts with one another as well as with the rest of the occipital bone. Perhaps the most valuable piece of data from Redfield's effort is the indication that all occipital elements fuse into a single "unit" by the age of 7 years. Prior to the age of 7 years, according to Fazekas and Kósa (1978), the mendosal fissures usually close by the third year. However, this closure may be delayed until the fourth year, when the lateral parts typically fuse with the squamous portion of the occipital.

Postcranium

Important changes occur in the *os coxa*. In the neonate, the pubis, ischium, and ilium converge within the presumptive *acetabulum* and are separated from one another by the thick *triradiate cartilage*. By the sixth postnatal month, the acetabulum is transformed into a shallow cuplike structure, into which the three bones extend farther. Prior to ossification in the acetabulum, the pubis and ischium unite around the obturator foramen; union is marked by the sixth year and essentially complete by the ninth. Between 9 and 14 years, ossification proceeds across the triradiate cartilage, first between pubis and ilium, then between ilium and ischium, and finally between pubis and ischium. Ages up to 12 years can be estimated using diaphyseal lengths of long bones (Tables 8–6 and 8–7).

Juvenile Age

Overview

Juvenile age may be taken as beginning at about 15 years, well *after* the *eruption of M2* (12 years). Sometimes the completed eruption of M3 is used as a marker of this period, but this is of limited value since this tooth may be completely erupted in individuals as young as 17 and as old as 30 years. A somewhat more reliable criterion is closure of the

Table 8–6   Average Postnatal Age Based on Diaphyseal Length (mm) for Females

Age(yrs)	Humerus	Ulna	Radius	Femur	Tibia	Fibula
0.125	71.8 ± 3.6	65.3 ± 3.1	57.8 ± 2.8	87.2 ± 4.3	70.3 ± 4.6	66.8 ± 4.4
0.25	80.2 ± 3.8	71.2 ± 3.1	63.4 ± 2.8	100.8 ± 3.6	80.8 ± 4.6	77.1 ± 4.1
0.5	86.8 ± 4.6	75.7 ± 3.8	67.6 ± 3.4	111.1 ± 4.6	88.9 ± 5.3	84.9 ± 5.2
1	103.6 ± 4.8	89 ± 4.0	78.9 ± 3.4	134.6 ± 4.9	108.5 ± 4.8	105 ± 5.1
1.5	117 ± 5.1	98.9 ± 4.4	87.5 ± 4.0	153.9 ± 6.4	124 ± 5.6	121.3 ± 5.9
2	127.7 ± 5.8	107.1 ± 4.8	95 ± 4.5	170.8 ± 7.1	138.2 ± 6.5	136 ± 6.8
2.5	136.9 ± 6.1	113.8 ± 5.2	101.4 ± 5.0	185.2 ± 7.7	150.1 ± 7.0	147.9 ± 7.1
3	145.3 ± 6.7	120.6 ± 5.4	107.7 ± 5.2	198.4 ± 8.7	161.1 ± 8.2	159.4 ± 7.9
3.5	153.4 ± 7.1	127.2 ± 5.7	113.8 ± 5.5	211.1 ± 10.0	171.2 ± 8.7	169.6 ± 8.3
4	160.9 ± 7.7	133.1 ± 5.8	119.2 ± 5.7	223.2 ± 10.1	180.8 ± 9.5	179.5 ± 9.1
4.5	169.1 ± 8.3	139.3 ± 6.6	125.2 ± 6.6	235.5 ± 11.4	190.9 ± 10.5	189.4 ± 10.2
5	176.3 ± 8.7	144.6 ± 7.1	130.2 ± 6.9	247 ± 11.5	199.9 ± 11.4	198.6 ± 11.1
5.5	182.6 ± 9.0	149.1 ± 7.2	134.6 ± 7.2	257 ± 12.2	207.9 ± 12.5	206.5 ± 11.7
6	190 ± 9.6	154.9 ± 7.4	140 ± 7.4	268.9 ± 13.5	217.4 ± 12.6	216 ± 12.2
6.5	196.7 ± 9.7	159.9 ± 7.9	144.7 ± 7.8	279 ± 13.8	226.3 ± 13.6	224.3 ± 13.4
7	202.6 ± 10.0	164.8 ± 8.3	149.3 ± 8.0	288.8 ± 13.6	234.1 ± 14.1	232.1 ± 13.4
7.5	209.3 ± 10.5	170.1 ± 8.5	154.3 ± 8.4	299.8 ± 15.2	243.2 ± 15.0	240.8 ± 14.5
8	216.3 ± 10.4	174.9 ± 8.7	158.9 ± 8.7	309.8 ± 15.6	251.7 ± 15.6	248.8 ± 14.8
8.5	221.3 ± 11.2	179.1 ± 8.8	162.8 ± 8.8	318.9 ± 15.8	259.1 ± 15.6	256.1 ± 15.2
9	228 ± 11.8	184.3 ± 9.52	167.6 ± 9.3	328.7 ± 16.8	265.5 ± 17.1	263.7 ± 16.3
9.5	234.2 ± 12.9	189.7 ± 10.4	172.2 ± 10.2	338.8 ± 18.6	276.6 ± 18.7	272.2 ± 17.6
10	239.8 ± 13.2	194.4 ± 10.6	176.8 ± 10.4	347.9 ± 19.1	284.3 ± 19.3	279.4 ± 18.3
10.5	245.9 ± 14.6	200 ± 12.4	181.8 – 11.8	356.5 ± 21.4	292.4 ± 21.4	287.2 ± 20.4
11	251.9 ± 14.7	204.7 ± 12.0	186 ± 11.7	367 ± 22.4	300.8 ± 21.2	294.4 ± 19.8
11.5	259.1 ± 15.3	211.3 ± 13.1	192 ± 12.1	378 ± 23.4	310.5 ± 21.4	303.8 ± 20.7
12	265.6 ± 15.6	216.4 ± 13.3	196.9 ± 12.7	387.6 ± 22.9	318.2 ± 21.7	311.1 ± 20.8

SOURCE: Maresh (1970).
Abbreviation: yrs, years.

Table 8–7   Average Postnatal Age Based on Diaphyseal Length (mm) for Males

Age(yrs)	Humerus	Ulna	Radius	Femur	Tibia	Fibula
0.125	72.4 ± 4.5	67 ± 3.5	59.7 ± 3.3	86 ± 5.4	70.8 ± 5.4	68.1 ± 5.3
0.25	80.6 ± 4.8	73.8 ± 3.4	66 ± 3.3	100.7 ± 4.8	81.9 ± 5.3	78.6 ± 4.9
0.5	88.4 ± 5.0	79.1 ± 3.7	70.8 ± 3.5	112.2 ± 5.0	91 ± 5.2	87.2 ± 4.8
1	105.5 ± 5.2	92.6 ± 4.4	82.6 ± 4.0	136.6 ± 5.8	110.3 ± 5.2	107.1 ± 5.5
1.5	118.8 ± 5.4	102.3 ± 4.6	91.4 ± 4.4	155.4 ± 6.8	126.1 ± 6.0	123.9 ± 6.2
2	130 ± 5.5	109.7 ± 4.9	98.6 ± 4.7	172.4 ± 7.3	140.1 ± 6.5	138.1 ± 6.7
2.5	139 ± 5.9	116.6 ± 5.2	105.2 ± 4.8	187.2 ± 7.8	152.5 ± 6.8	150.7 ± 7.1
3	147.5 ± 6.7	123.4 ± 5.6	111.6 ± 5.3	200.3 ± 8.5	163.5 ± 7.7	162.1 ± 7.7
3.5	155 ± 7.8	129.1 ± 6.4	116.9 ± 6.2	212.1 ± 11.4	172.8 ± 9.8	171.6 ± 9.6
4	162.7 ± 6.9	135.6 ± 5.6	123.1 ± 5.6	224.1 ± 9.9	182.8 ± 9.0	181.8 ± 8.7
4.5	169.8 ± 7.4	141 ± 5.6	128.2 ± 5.6	235.7 ± 10.5	191.8 ± 9.2	190.8 ± 8.8
5	177.4 ± 8.2	147 ± 6.1	133.8 ± 6.1	247.5 ± 11.1	201.4 ± 9.9	200.4 ± 9.6
5.5	184.6 ± 8.1	152.6 ± 6.7	138.9 ± 6.4	258.2 ± 11.7	210.3 ± 10.7	209 ± 10.2
6	190.9 ± 7.6	157.5 ± 6.2	143.8 ± 5.9	269.7 ± 12.0	218.9 ± 10.0	217.5 ± 9.6
6.5	197.3 ± 8.1	162.2 ± 6.8	148.3 ± 6.4	280.3 ± 12.6	227.8 ± 11.6	226 ± 10.5
7	203.6 ± 8.7	167.3 ± 7.0	153 ± 6.7	291.1 ± 13.3	236.2 ± 11.8	234.2 ± 11.3
7.5	210.4 ± 8.9	172.2 ± 7.4	157.9 ± 6.9	301.2 ± 13.5	244.2 ± 12.4	242.1 ± 11.8
8	217.3 ± 9.8	177.3 ± 7.4	162.9 ± 7.1	312.1 ± 14.6	253.3 ± 12.9	251 ± 12.4
8.5	222.5 ± 9.2	181.6 ± 7.1	166.8 ± 6.6	321 ± 14.6	260.6 ± 12.3	257.7 ± 11.8
9	228.7 ± 9.6	186.4 ± 7.9	171.3 ± 7.4	330.4 ± 14.6	268.7 ± 13.4	265.6 ± 13.0
9.5	235.1 ± 10.7	191.7 ± 8.3	176.1 ± 7.4	340 ± 15.8	276.9 ± 14.4	273.8 ± 13.8
10	241 ± 10.3	196.2 ± 8.5	180.5 ± 7.9	349.3 ± 15.7	284.9 ± 14.2	281.3 ± 13.9
10.5	245.8 ± 11.0	200.4 ± 8.8	184.4 ± 8.4	357.4 ± 16.2	292 ± 15.1	287.8 ± 14.6
11	251.7 ± 10.7	205.1 ± 9.2	188.7 ± 8.5	367 ± 16.5	298.8 ± 15.0	294.9 ± 14.6
11.5	257.4 ± 11.9	209.8 ± 9.9	193 ± 9.2	375.8 ± 18.1	306.8 ± 16.5	301.7 ± 16.0
12	263 ± 12.8	214.5 ± 10.2	197.4 ± 9.6	386.1 ± 19.0	315.9 ± 17.0	310.1 ± 16.4

SOURCE: Maresh (1970).
Abbreviation: yrs, years.

spheno-occipital synchondrosis, which begins at approximately 17 years and may be complete between 22 and 25 years (see Figure 9–28). Krogman and İşcan (1986) found that the spheno-occipital synchondrosis is typically closed at 23 years.

Epiphyseal Union

The most accurate source of information for determining ages within the juvenile period is the sequence of *fusion of epiphyses with long bones* and the *unification of the three bones of*

the os coxa and its epiphyses into one bone. Vallois (1960) modified Martin's scale, as it was originally used to record cranial suture closure, for use in assessing stages of epiphyseal fusion.

1. Stage 0: Open, no fusion, metaphyseal region between epiphysis and diaphysis is cartilaginous.
2. Stage 1: Ossification between epiphysis and diaphysis encompasses approximately one-fourth of the circumference (as a total of individual areas of union).

Table 8–8   Chronology of Ossification and Fusion of Some Epiphyses in a European Sample

Bone	Epiphysis	Appearance of Ossification Center (Years)	Union Completed (Years)
Clavicle	Sternal	16–20	21–25
Humerus	Proximal	1	18–22
	Distal	1–2	14–15
Radius	Proximal	4–7	14–18
	Distal	1–2	21–23
Ulna	Proximal (olecranon)	10–12	15–17
	Distal	4–6	18–20
Femur	Distal	1	17–20
	Greater trochanter	—	17–20
	Lesser trochanter	—	16–20
Patella	—	3–5	—
Tibia	Proximal	—	17–20
	Distal	2	16–19
Fibula	Proximal	3–5	17–20
	Distal	2	16–19

SOURCE: Modified from Vallois (1960).

3. Stage 2: Ossification encompasses approximately one-half of the circumference.
4. Stage 3: Ossification encompasses approximately three-fourths of the circumference.
5. Stage 4: Metaphyseal line is present only in traces around the circumference.

Vallois' (1960) data on recent Europeans (analyzed collectively, not by sex) are presented in Table 8–8. The lower age for each element corresponds to stage 1 and the higher to stage 4. Scrutiny of the reported sequence of fusion reveals, for example, that the distal humeral epiphysis is the first to completely unite with a diaphysis, while the last three to fuse are the humeral head, distal radial epiphysis, and then sternal end of the clavicle. Stewart's (1934) data on Inuits and other Native North Americans as well as Johnston's (1961) on an archaeological collection from the Indian Knoll site indicate the reverse sequence of fusion for the humeral head and distal radial epiphysis. Although they analyzed a large sample (N = 375), McKern and Stewart (1957) based their oft-cited study solely on American male soldiers of the Korean War; in

addition, they analyzed only one end of each long bone.

As a general rule of thumb, complete union of epiphysis and diaphysis seems usually to occur 1–2 years earlier in females than males (Krogman, 1962; Stewart, 1979). A continuing and unfortunate aspect of extant epiphyseal union studies, however, is that a global sampling, with distinctions consistently made between males and females, is still lacking. Thus, until this lacuna is filled, the osteologist is forced to apply age estimates of epiphyseal union (which themselves may have been derived from the application first of other sets of age estimates) from one population to another, perhaps totally inappropriate, population.

According to various authors (e.g., Flecker, 1942; Johnston, 1961; Todd, 1930), *ossification of the pubis, ischium, and ilium within the acetabulum* may be fairly well along at least as early as, if not even slightly earlier than, similar states of union of the first long bone epiphyses. In these studies, the apparent age at complete union within the acetabulum in males is 14 years; in females, complete union can occur between 11.5 and 13 years and may even be retarded until 18 years. According to Scheuer and Black (2000), acetabular coalescence occurs earlier in females (commencing at c. 11 and completing at c. 15 years) than in males (commencing at c. 14 and completing at c. 17 years). Ossification in the *iliac crest, ischial tuberosity, anterior inferior iliac spine,* and *pubic symphysis* broadly coincides with the onset of puberty. Coalescence of these epiphyses with the *os coxa* may begin at approximately 16–17 years and may be completed between as early as 23 and as late as 25 years, slightly after the union of the *head of the humerus* and more or less in synchrony with the *spheno-occipital synchondrosis,* the *distal end of the radius* (in some populations), and the *sternal end of the clavicle.* The onset of puberty in different populations—whether from the distant past, recent past, or present—is not, of course, a constant since it can be affected by diet and other socioeconomically based factors within and across generations of the same population. Thus, although the relationship of relative times of ossification and union of bony elements may be more consistent across human groups, one should exercise caution in assigning definitive ages, especially if there is any concern about developmental timing differing from "the expected."

Stewart (1979) suggested that epiphyseal union commonly proceeds from joints of the elbow to those of the hip, ankle, knee, wrist, and, finally, shoulder. For purposes of aging, an attempt at summarizing ossification center appearance/epiphyseal union data for elements of each bone is presented in Table 8–9. Additional information on development growth for each bone is provided in preceding chapters.

Adulthood

Pubic Symphysis

In 1920, Todd introduced a method for estimating adult age based on changes of the pubic symphysis that proceed from a break-down of its initially horizontally *wavy* or *ridged pubic symphyseal surface* characteristic of *young adults* to the *pitted* or *granular surface* of *old individuals*. Concurrent with the symphysis' transformation to granularity is the development of a *lipped rim* along its *dorsal margin* (forming a dorsal "plateau") and a beveling of its *ventral border* that eventually becomes a *rampartlike* feature via bony extensions from the symphysis' superior and/or inferior extremity(ies). Todd's sample consisted entirely of males ranging in age from 18 to 50+ years, a span of years he divided into 10 age-related phases. Todd's (1920, see pp. 301–314) method involved an overall evaluation of the symphyseal region. His phases and correlated ages are presented in Figures 8–3 and 8–4 and Table 8–10.

In an attempt to refine Todd's method, McKern and Stewart (1957) subdivided the pubic symphysis into *three components: symphyseal rim* and *dorsal* and *ventral demifacets*. In reality, the component "symphyseal rim" is an overall assessment of symphyseal change, which includes details about rim formation and deterioration and about surface transformation. One should not expect to find evidence of a symphyseal rim in very young or very old adults because it forms and then breaks down with increasing age. Dorsal and ventral demifacets are delineated by a longitudinal elevation or disruption of the horizontally oriented ridge pattern (of varying definition and straightness) that courses more or less down the center of the symphyseal face. In McKern and Stewart's method, features of the breakdown of the horizontal ridge pattern (which they referred to as *billowing*) and of the development of a dorsal plateau are evaluated together, as are aspects of the breakdown of the horizontal ridging and the development of a ventral rampart.

In determining age using McKern and Stewart's method, each symphyseal component is analyzed separately and assigned a stage from 0 to 5. The assigned stages are summed, and the total is converted to an age

Table 8–9   Chronology of Postcranial Bone Formation

Bone Element	Ossification Begins	Fusion Begins	Fusion Complete
First cervical vertebra			
Posterior arch	7–10 f/wk		3–4 yr
Lateral mass	7–10 f/wk		
Anterior arch	≤1 yr		5–9 yr
Second cervical vertebra			
Body	4 f/mo		
Arches	7–8 f/wk		3–6 yr
Dens (base)	4–6 f/mo		4–6 yr
Dens (tip)	2–3 yr		12 yr
Epiphyseal plate	17 yr		
True vertebrae			
Arches	7–9 f/wk (beginning in cervicals)		≥1 yr (beginning in lumbars)
Transverse process	Puberty–16 yr		≤20 yr
Spinous process	Puberty–16 yr		≤20 yr
Body	8–10 f/wk (beginning in lower thoracics–upper lumbars)		3–6 yr (beginning in cervicals and/or thoracics)
Epiphyseal plate	16–17 yr		20–25 yr
Sacrum			
Vertebrae	6–9 f/wk (beginning superiorly)		
Arches	6–10 f/mo		7–15 yr
Arch/body			6–9 yr (beginning caudally)
Lateral part (costal process)	6 f/mo–≥birth		Puberty
Epiphyseal plates	Puberty–16 yr		Puberty–16 yr
Auricular surface	18–20 yr		
Intervertebral disc	18–30 yr (beginning caudally)		
Coccyx			25–30 yr (beginning caudally)
First segment	≤1 yr		
Second segment	4–10 yr		
Third segment	10–15 yr		
Fourth segment	14–20 yr		
Ribs			
V–VII	7–9 f/wk		
Other ribs	≥9 f/wk (rapidly)		
Head	(14)16–20 yr	≥17 yr	20–25 yr
Tubercle	(14)16–20 yr	≥17 yr	20–25 yr
Sternum			≤Puberty (beginning inferiorly)
Manubrial sternebra	3–6 f/mo		≥25 yr
2nd–4th sternebrae	3–7 f/mo		
5th sternebra	≤1 yr		
6th (xiphoid) sternebra	5–18 yr		
Scapula			
Body	<Birth		
Spine	<Birth		
Glenoid cavity	c. 10 yr	10 yr	Puberty
Head of coracoid process	1–2 yr	10 yr	Puberty
Base of coracoid process	Puberty		20 yr
Lateral center of acromion	16 yr	10–14 yr	22–23 yr
Medial center of acromion	15 yr	10–14 yr	22–23 yr
Vertebral border	Puberty	Puberty	20–23 yr
Medial angle	16–18 yr	19–20 yr	22–23 yr
Inferior angle	16–18 yr	19–20 yr	22–23 yr

Table 8–9 (*continued*)

Bone Element	Ossification Begins	Fusion Begins	Fusion Complete
Clavicle			
Diaphysis	6–8 f/wk		
Acromial end	c. 20 yr		c. 20 yr
Sternal end	16–20 yr	21–22 yr	25–30 yr
Humerus			
Diaphysis	2 f/mo		
Head	c. Birth	20 yr	25 yr
Greater tubercle	7 mo(f)/1 yr(m)–3 yr	20 yr	25 yr
Lesser tubercle	7 mo(f)/1 yr(m)–5 yr	20 yr	25 yr
Medial epicondyle	5–7 yr	12 yr	19–20 yr
Lateral epicondyle	12–13 yr	16–17 yr	18 yr
Trochlea	10 yr	16–17 yr	18 yr
Capitulum	5 mo(f)/7 mo(m)–2 yr	16–17 yr	18 yr
Radius			
Diaphysis	8 f/wk		
Proximal epiphysis	5 yr	Puberty	15–18 yr
Distal epiphysis	1–2 yr	17–20 yr	20–23 yr
Radial tuberosity (if present)	14–15 yr		
Ulna			
Diaphysis	8 f/wk		
Olecranon	7–14 yr	≥16 yr	≤23 yr
Distal epiphysis	5–7 yr	≥21 yr	≤25 yr
Scaphoid (body)	5–7 yr		
Lunate (body)	4–5 yr		
Triquetrum (body)	2–3 yr		
Pisiform (body)	9–11 yr		
Trapezium (body)	5–7 yr		
Trapezoid (body)	5–7 yr		
Capitate (body)	Birth–1 yr		
Hamate (body)	Birth–1 yr		
Metacarpals (I–V)			
Diaphysis (I)	9 f/wk		
Proximal epiphysis (I) (94%)	2–3 yr		18–20 yr
Distal epiphysis (I) (6%)	2–3 yr		18–20 yr
Diaphysis (II–V)	9 f/wk		
Distal epiphysis (II–V)	2–3 yr		18–20 yr
Proximal epiphysis (II) (rare)			18–20 yr
Proximal manual phalanges (I–V)			
Diaphysis (I–V)	9 f/wk		
Proximal epiphysis (I–V)	1–3 yr	14 yr	18–25 yr
Middle manual phalanges (I–V)			
Diaphysis (I–V)	11–17 f/wk		
Proximal epiphysis (I–V)	2–4 yr	14 yr	18–25 yr
Distal manual phalanges (I–V)			
Diaphysis (I–V)	7–8 f/wk		
Proximal epiphysis (I–V)	2–4 yr	14 yr	18–25 yr
Os coxa (innominate)			
Ilium	2–3 f/mo		
Ischium	3–5 f/mo		
Pubis	4–6 f/mo		

Table 8–9   (*continued*)

Bone Element	Ossification Begins	Fusion Begins	Fusion Complete
Acetabulum		9–12 yr	13 yr (f), 14–18 yr (m)
Ischiopubic ramus		≤6 yr	8 yr
Ischioiliac ramus			17 yr
Ischial tuberosity	13 yr (f), 15 yr (m)	16–17 yr	23–25 yr
Pubic symphysis	Puberty	16–17 yr	23–25 yr
Anterior inferior iliac spine	Puberty	16–17 yr	23–25 yr
Iliac crest	12 yr (f), 13 yr (m)	16–17 yr	23–25 yr
Femur			
Diaphysis	7 f/wk		
Lesser trochanter	(9)13–14 yr	15 yr	18–20 yr
Greater trochanter	(1–)4 yr	≥15 yr	18–20 yr
Head	≤1 yr	≥15 yr	18–20 yr
Distal epiphysis	7 f/mo–birth	≥15 yr	(18)20–23 yr
Patella	2–6 yr		
Tibia			
Diaphysis	8 f/wk		
Distal epiphysis	2 yr	16 yr	18–20 yr
Proximal epiphysis	9 f/mo	16 yr	20–23 yr
Tuberosity	≤39f/wk–birth	8–12 yr (f) 10–13 yr (m)	19 yr
Fibula			
Diaphysis	8 f/wk		
Proximal epiphysis	3–4 yr	16 yr	23–25 yr
Distal epiphysis	9 mo (f), 1 yr (m)–2 yr	16 yr	20 yr
Talus			
Body	6 f/mo		
Posterior process (os trigonum if unfused)	8–11 yr		
Calcaneus			
Body	12 f/wk–7 f/mo		
Calcaneal epiphysis	(4)7–10 yr	12 yr	(16)20–22 yr
Cuboid (body)	9 f/mo–birth		
Lateral cuneiform (body)	1 yr		
Medial cuneiform (body)	2–4 yr		
Intermediate cunciform (body)	3–5 yr		
Navicular (body)	3–5 yr		
Metatarsals (I-V)			
Diaphysis (I–V)	8–10 f/wk		
Proximal epiphysis (I)	3–8 yr	12 yr	18–22 yr
Distal epiphysis (II–V)	3–8 yr	12 yr	18–22 yr
Proximal pedal phalanges (I–V)			
Diaphysis (I–V)	9–10 wk		
Proximal epiphysis (I–V)	2–10 yr	14 yr	17–21 yr
Middle pedal phalanges (I–V)			
Diaphysis (I–V)	4 mo		
Proximal epiphysis (I–V)	2–10 yr	14 yr	17–21 yr
Distal pedal phalanges (I–V)			
Diaphysis (I–V)	9–10 wk		
Proximal epiphysis (I–V)	2–10 yr	14 yr	17–21 yr

Abbreviations: f/wk, fetal week; f/mo, fetal month; wk, week; yr, year; (f), female; (m), male.

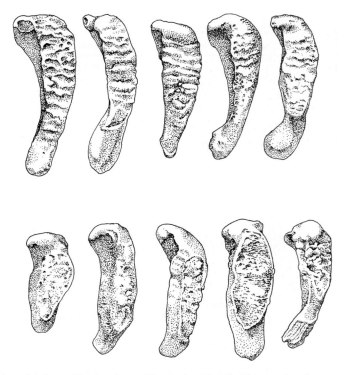

Figure 8–3   Examples of right pubic symphyses illustrating Todd's 10 age-related stages of change: (*top row, left to right*) stages 1–5, (*bottom row, left to right*) stages 6–10; see Tables 8–10 for description and associated ages of each stage (adapted from Todd, 1920).

or age range by comparison with a table calculated from their sample of individuals of known age. The *dorsal demifacet* is defined as *component I*, the *ventral demifacet* is *component II*, and the *symphyseal rim* is *component III*. As change in the dorsal demifacet often precedes noticeable change in the ventral demifacet, which begins to change before the symphyseal rim, scoring typically proceeds from components I to II to III.

Sixteen years later, Gilbert and McKern (1973) attempted to provide more accurate criteria by which one could assess the age of females using pubic symphyseal morphology. For, after applying Todd's method to a sample of females of known age, they concluded that the sequence of transformation of overall symphyseal morphology, as well as of a symphysis' components, was sufficiently different

in females that Todd's age estimates were off by ±10 years. Consequently, Gilbert and McKern devised a set of criteria for analyzing the female symphyseal region that followed the three-component system of McKern and Stewart (1957), in which each component was broken down into six stages (0–5) of age-related morphological change. The calculation of average age was also done the same way. The total score obtained was converted to age by comparison with a table of female-specific age brackets.

Questions concerning the accuracy and applicability of these landmark studies continue to arise. The most obvious is that age ranges can be exceedingly broad (as seen in the tables presented above). More specifically, there was the problem of replicability in applying McKern and Stewart's and particularly

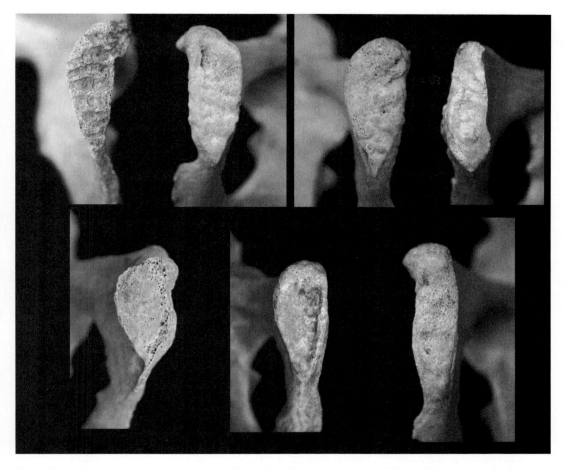

Figure 8–4   Pubic symphyses ranked according to Todd's stages: (*top row, left to right*) stage 2, stage 3, stages 3–4, stage 6; (*bottom row, left to right*) stage 8, stage 9, stage 10 (University of Pittsburgh teaching collection).

Gilbert and McKern's standards to other large samples derived from individuals of known age at death. Meindl et al. (1985) tested these criteria against a sample of African-American and white American females and males of known age at death and concluded that Todd's method could be applied with confidence to individuals of either sex and of either group. They also suggested that Todd's 10 phases could be compressed into five stages (Table 8–11). [Angel et al. (1986) had proposed that Todd's 10 phases could be collapsed into five and also that the most accurate component of McKern and Stewart's

analysis was component III (overall symphyseal change, which, of course, is the focus of Todd's system), which could be collapsed into fewer age categories.] It would thus appear that a reorganized version of Todd's criteria for determining the age at death of adults on the basis of changes in the pubic symphysis remains among the best currently available [at least up to the age of 35 when, as Meindl and Russell (1998) subsequently suggested, pubic symphyseal morphology may begin to lose its value in assessing age]. It would be interesting to see if the applicability of these criteria would hold up against samples from

Table 8–10   Todd's 10 Stages of Pubic Symphyseal Age-Related Change

1. *18–19 years (first postadolescent phase)*: Clearly defined horizontal ridges and furrows, margin and ventral beveling as well as ossific nodules lacking.

2. *20–21 years (second postadolescent phase)*: Dorsal furrows filling in with finely textured bone and dorsal margin becoming defined, occasional development of ossific nodules, hints of ventral beveling.

3. *22–24 years (third postadolescent phase)*: Destruction and filling-in of horizontal ridge/furrow pattern, dorsal margin more clearly delineated, onset of ventral beveling and rarefaction (porous beveled strip of bone along ventral border) obvious, ossific nodules present.

4. *25–26 years [24–26 years as emended by Brooks (1955)] (fourth phase)*: Continued destruction of ridge/furrow pattern, dorsal margin complete, ventral beveling more pronounced (increased rarefaction), lower extremity of symphyseal surface becoming delineated.

5. *27–30 (26–27) years (fifth phase)*: Lower extremity of symphyseal surface more clearly defined and upper extremity becoming delineated, continued ventral beveling.

6. *30–35 (27–34) years (sixth phase)*: Symphyseal surface granular, upper and lower extremities even more clearly delimited with partial (central) or complete development of ventral rampart.

7. *35–39 (34–38) years (seventh phase)*: Rarefaction of symphyseal surface and ventral border decreasingly active, mineralization into attendant tendons and ligaments (hyperostotic activity).

8. *39–44 (38–42) years (eighth phase)*: Symphyseal surface and ventral border smooth; extremities clearly defined, adding to definition of ovoid perimeter of modified symphyseal surface.

9. *45–50 (42–51) years (ninth phase)*: Ovoid perimeter modified into thin rim (similar to rim around glenoid fossa of scapula) (dorsal and ventral margins thus lipped).

10. *50 + (51 +) years (tenth phase)*: Erosion with possible osteophytic outgrowth of symphyseal surface; breakdown of ventral, lipped margin.

Source: From Todd (1920).

Table 8–11   Summary of the Revision by Meindl et al. of Todd's Stages of Pubic Symphyseal Age-Related Changes

1. *Todd's stages 1–5 = preepiphyseal phase*: Modal phase, 20–29 years, crisply defined billowing without ventral rampart formation (age range in sample 18–37 years); 20–25, crisply defined billowing without ventral bevel and defined lower extremity; 26–29, reduced billowing with little ventral rampart formation or definition of lower extremity (overall 18–25, marked billowing; 24–37, reduction in billowing; 24–37, ventral rampart formation; ≥25, appearance of distinct lower extremity; ≤29, indistinct lower extremity; 21–30, ossific nodules without ventral rampart formation, clearly defined horizontal ridges and furrows; margin and ventral beveling as well as ossific nodules lacking).

2. *Todd's stage 6 = active epiphyseal phase*: Modal phase 30–35 years, active onset and completion of ventral rampart formation.

3. *Todd's stage 7 = immediate postepiphyseal phase*: Modal phase, 36–40, symphyseal surface and ventral margin become fine-grained and dense.

4. *Todd's stage 8 = predegenerative (maturing) phase*: 40–44, quiescent period with no change in symphyseal surface.

5. *Todd's stages 9–10 = degenerative phase*: 45–50, formation of thin rim around perimeter of symphyseal surface, with lipping of dorsal margin and mineralization in ligamentous/tendonous attachment sites (also in females, further erosion of symphyseal surface, possibly due to postmenopausal osteoporosis); the broader the symphyseal surface (e.g., in males), the less the change that may occur with age (and thus less correct information forthcoming about age above 40 years).

Source: Meindl et al. (1985).

Table 8–12   Brooks and Suchey's Phases of Pubic Symphyseal Age-Related Change (with Approximate Mean Age and Range)

1. *Female (19.4, 15–24), male (18.5, 15–23)*: Well-billowed surface with ridges and furrows onto pubic tubercle; well-marked horizontal ridging; upper and lower extremities not defined, but nodules may be present in upper region; ventral beveling may have begun.

2. *Female (25, 19–40), male (23.4, 19–34)*: Some ridges/furrows may persist, upper and lower extremities becoming defined even if nodules not present, formation of ventral rampart may begin from either extremity.

3. *Female (30.7, 21–53), male (28.7, 21–46)*: Surface may retain some ridges/furrows or be smooth, nodules (perhaps coalescing) mark upper extremity and may occur on ventral border, dorsal plateau completely and lower extremity and ventral rampart almost fully delineated.

4. *Female (38.2, 26–70), male (35.2, 23–57)*: Ridges/furrows may persist in fine-grained, typically now oval surface, which may be completely rimmed, excluding pubic tubercle; dorsal border may show lipping, and syndesmophytes may occur near surface's inferior portion.

5. *Female (48.1, 25–83), male (45.6, 27–66)*: Rimming with some dorsal lipping and syndesmophytes ventrally surrounds slightly depressed surface, rim erosion may have begun superoventrally.

6. *Female (60, 42–87), male (61.2, 34–86)*: Depressed, sometimes irregularly shaped surface may be pitted/porotic; rim erosion and ventral syndesmosis (marked) continues; pubic tubercle may appear further isolated.

Source: Brooks and Suchey (1990).

other regions. If so, one could then apply them with a certain degree of confidence to skeletal material derived from archaeological or even more obscure contexts.

Based on a sample from Los Angeles and not without the problem of broad age ranges [e.g., individuals grouped into the 40–50 age range using the Todd method would be grouped into a 35–65 age range (Jackes, 2000)], a popular approach to using changes in the pubic symphyseal surface to estimate age is Suchey's (Brooks and Suchey, 1990; Suchey and Katz, 1986), which distinguishes between males and females (Table 8–12). While the same general descriptions can be applied to symphyses of either sex, female estimated age is typically older than male estimated age with the exception of the last phase (6), in which mean male age is slightly greater than mean female age. The disparity between mean female and mean male age increases from phase 2 to phase 5 (see Buikstra and Ubelaker, 1994, p. 37). While Brooks and Suchey (1990) provide illustrations of symphyseal surfaces to accompany the descriptions of phases, those presented here in Figure 8–3 should suffice as immediate sources of reference for changes in morphology; but images of the Diane France casts of the Brooks and Suchey reference specimens for males and females are also available on the accompanying CD. As with all such approaches, phases and their description and illustration are more generally "rules of thumb" than fixed criteria. That having been said, various sets of examples of symphyseal surface change, for female and male, available through France Casting (www.francecasts.com) can be used as study guides for any of the approaches summarized here.

Recalling the caveat expressed earlier about subjectivity in estimating age, one must note that Jackes (2000, p. 433), who compared the various methods for determining age from pubic symphyseal morphology, concluded "that there is no certainty of adult age based on pubes." Nevertheless, as Lovejoy (personal communication, August 2005) reminds us, as a late epiphysis, the pubic symphysis' "ventral rampart" is a reliable indicator of age at least until 35 years (Meindl and Russell, 1998), if not even as late as 40. Given the central place pubic symphyseal morphology has held in osteological analyses, it would be an important learning experience for the student to apply the various methods and evaluate the different results.

Auricular Region (Ilium)

In the same volume of the journal in which Meindl et al. (1985) critiqued the use of pubic symphyseal morphology for estimating age, Lovejoy et al. (1985b) introduced a new method for assessing age based on changes in the *auricular* and contiguous *area of the posterior portion of the ilium.* Although Lovejoy et al. (p. 15) admitted that age changes in the auricular surface "are somewhat more difficult to interpret than those used in pubic symphyseal aging," they rightly suggested that "the rewards are well worth the effort" in light of the fact that intact auricular regions are more frequently preserved than pubic symphyses in archaeological and often in forensic contexts.

Features of interest of the auricular region are the *apex* as well as the *superior* and *inferior demifacets* of the auricular surface, the *preauricular sulcus*, and the *retroauricular area.* The often somewhat peaked or bluntly pointed *apex* is the most anterior extension of the auricular surface (i.e., the posterior terminus of the arcuate line). If you mentally project a continuation of the arcuate line across the auricular surface, you would delineate, above this imaginary line, the *superior demiface* and, below it, the *inferior demiface.* The *preauricular sulcus*, if present, will be located lateral to the anterior margin of the auricular surface, facing into the space of the greater sciatic notch. The *retroauricular area* is the expanse of roughened bone that lies between the auricular surface and iliac crest.

Changes in the auricular surface—that is, the ilial contribution to the sacroiliac joint —accumulate with age and must not be confused with changes in the same general region that may result from degenerative osteoarthritis and/or the onset of osteophytic growth. True age-related change of the subchondral bone of the ilial auricular surface results from an increase over time in the amount of fibroid cartilage that covers the surface, regardless of any osteoarthritic or osteophytic change.

Lovejoy et al. (1985b, p. 18) delineated aspects of auricular surface change with these terms: *porosity* (with *microporosity* referring to fine, barely visible perforations and *macroporosity* to large, oval, irregular perforations ranging in diameter from 1 to 10 mm); *grain* (e.g., with "markedly grainy" looking like fine sandpaper); *billowing* (adopted from pubic symphysis analyses to refer to the *transverse ridges* across the auricular surface, which can range from being topographically large or finely grained to barely visible); and *density* (e.g., "dense" meaning that the subchondral bone is compact, smooth, and lacking "grain").

Figure 8–5 presents a sampling of auricular surfaces at various ages for males and females; in practice, the assessment of age is independent of an individual's sex. (As with the illustrations of the pubic symphysis, examples of the auricular surface are only illustrative and not meant to serve as "type specimens" for each age bracket; color images of Lovejoy et al.'s specimens are provided in the accompanying CD.) Lovejoy et al. (1985b, p. 26) recommend that one should rely most heavily on aspects of only the auricular surface itself (not including surrounding features). Common sense dictates which end of the age bracket the assessment should favor; for example, "[i]n the case of a coarsely grained surface, but one that still retains some billowing, the former indicator is paramount, but the latter should be used to reduce the age estimate slightly (within the mode)" (Lovejoy et al., p. 27). One may refine age assessment slightly up or down in accordance with characteristics of the apex and retroauricular area. Finally, and although these criteria are in general applicable to individuals of either sex, one should be aware of the degree to which the preauricular sulcus is excavated (being typically more so in females) because a well-developed preauricular sulcus can skew age estimates. The age ranges (in years) and

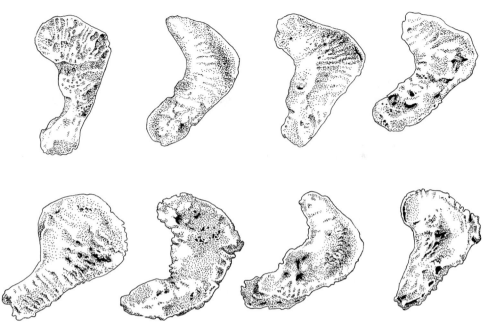

Figure 8–5   Examples of auricular surfaces of left ilia illustrating age-related stages of change: (*top row, from left to right*) stages 1–4, (*bottom row, left to right*) stages 5–8; see text for description and associated ages of each stage (adapted from Lovejoy et al., 1985b).

associated changes Lovejoy et al. determined are as follows:

1. 20–24: "Youthful appearance"; that is, well-defined, broad transverse billows covering most of the auricular surface, which is finely granular with no porosity; absence of retroauricular and apical activity.

2. 25–29: Still rather "youthful" looking (i.e., distinct transverse pattern) but with some loss of billowing and concomitant increase in *striae* (i.e., striations) as well as of coarseness in grain; absence of retroauricular and apical activity.

3. 30–34: More so in the inferior than the superior demiface is noticeable replacement of billowing by striae and continued loss of the transverse pattern, resulting in a smoother appearance; coarseness of surface graininess also increases, with possible development of patches of microporosity;

activity is more likely to occur in the retroauricular region than at the apex.

4. 35–39: Breakdown of transverse patterning, almost complete obliteration of striae and especially of billowing, and a fairly uniform increase in the coarseness of granularity across the entire surface; minimal activity still in the retroauricular region and even less at the apex.

5. 40–44: Billowing absent, remaining striae obscure; continued breakdown of transverse patterning; patches of denser texture, with possible microporosity, appear within the coarsely granular surface; activity at the apex is more frequent but more evident in the retroauricular region.

6. 45–49: Absence of billowing, striae, and transverse patterning; the surface and its margins distinctly irregular due to the almost complete transformation of granularity and microporosity to a denser texture;

activity at the apex slight to moderate but still more prevalent in the retroauricular region.

7. 50–59: More extensive irregularity of the surface and its margins, with transformation from granular to dense texture and occasional development of macroporosity; lipping of inferior margin of auricular facet common, often extending below body of ilium; activity in retroauricular region, and sometimes even in apical region, moderate to marked.

8. 60+: Extensive topographic and marginal irregularity, with marginal lipping and macroporosity; overall deterioration of subchondral bone prevalent; activity at the apex is often marked but remains even more marked in now largely osteophytic retroauricular region.

Jackes (2000) evaluated two studies that sought to test the Lovejoy et al. auricular surface method of estimating age. Although these two—which were essentially statistically identical in sample size as well as in sex and age representation—achieved similar distributions of auricular surface stages, Jackes thought that "stage" and "real" ages were not associated. From this, she (p. 431) concluded that "[t]he estimated ages are not characteristic of the target sample, they are in some way reflecting the built-in biases of the method." In other words, she claimed, the method, not the sample, controlled the estimates. Nevertheless, Jackes did offer a positive suggestion for using auricular surface morphology in estimating age: the sexes should be evaluated separately. It may, however, be the case that these criticisms of the Lovejoy et al. method arose from improper use or incomplete understanding of the morphological criteria summarized above (Lovejoy, personal communication, August 2005; again, see the images provided by Lovejoy in the CD).

Most recently, Igarashi et al. (2005) presented another approach to using changes in auricular surface morphology that also appears to be more accurate for older individuals. From scrutiny of auricular surface features of 700 Japanese of known age, they developed variables for multiple regression analysis that require the investigator to determine only either "presence" or "absence" of each character, rather than to assess relative degrees of morphological change.

Igarashi et al. (2005) delineated four features they defined as relating to "relief" and five relating to the "texture" of the auricular surface, as well as four aspects of hypertrophied (hyperostotic, reactive) structure around the auricular surface. One obvious difference between (at least some of the features defined by) Igarashi et al. and Lovejoy et al. is that the former often delineate topographic features by their surficial depth (i.e., troughs, especially in their "relief" features), whereas the latter emphasize surficial height (i.e., peaks). "Texture" features are distinguished on the basis of whether the penetration of pores remains in the cortical bone or deepens into the spongy bone—if the latter, whether pores are not profuse and separated widely by fields of cortical bone or more numerous and separated by thin septa of cortical bone (much thinner than a pore).

Relief features are as follows:

1. Wide groove (WG): a surface with wide transverse grooves with wide, flat bottoms; cf. Lovejoy et al.'s "transverse patterning."
2. Striation (ST): narrow/fine, "V"-shaped (inferiorly tapering) transverse grooves; cf. Lovejoy et al.'s "striae."
3. Roughness (RO): uneven surface lacking regular structure, that is, wide or "V"-shaped grooves, or a general "bumpiness."
4. Flatness (FL): smooth surface lacking topographic relief.

Texture features are as follows:

1. Smoothness (SM): smooth surface lacking topographic relief (appears descriptively equivalent to FL), confined to cortical bone.

2. Fine granularity (FG): surface with very shallow depressions (so shallow that they do not cast well-defined shadows when light is shined on them), confined to cortical bone.
3. Coarse granularity (CG): surface with depressions sufficiently deep to cast well-defined shadows when light is shined on them (compare with the texture of the fabric crepe de chine), still confined to cortical bone.
4. Sparse porosity (SP): surface with pores that just penetrate into the spongy bone; there is more compact than porous bone.
5. Dense porosity (DP): surface with more porous than cortical bone; pores penetrate more deeply into the spongy bone.

Hypertrophied (hyperostotic, reactive) features surrounding the auricular region are as follows:

1. Dull rim (DR): the auricular margin has been remodeled such that it is no longer crisply delineated as a rim (assessment apparently restricted to the rim of the superior demifacet of the auricular surface).
2. Lipping (LP): the auricular margin has been further remodeled such that it is distended beyond the table of bone on which the surface sits (compare with vertebral lipping; assessment apparently restricted to the inferior extremity of the inferior demifacet of the auricular surface).
3. Tuberosity (TB): a node(s) or spine(s) that develops along the margin that subtends the (region of a) preauricular sulcus and/or the posterior extremity of the superior demifacet.
4. Bony bridge (BB): often found connecting the sacrum and the ilium along the anterior margin of the sacroiliac joint.

Igarashi et al. (2005, p. 337) suggest "that the mode of chronological change in the auricular surface differs between males and females," whose auricular surface morphology is the more variable. With regard to texture and hypertrophied features, FG, DP, and TB appear to be the most useful for aging males, with FG being typical of younger and DP and TB typical of older individuals. FG and DP are most useful for aging females, again with FG being typical of younger and DP typical of older individuals. With regard to relief features, WG and FL are useful in aging males, whereas RO is better for aging females.

Although they developed a multiple regression equation for aging males separately, females separately, and individuals regardless of sex that uses all variables, Igarashi et al. (2005) found that the margin of error in estimating age did not increase significantly using a subset of features that were selected using multiple stepwise regression. The parameter estimates for all features (the full model, FM) and for the subset of features (the reduced model, RM) for aging males, females, and the combined-sex sample are listed in Table 8–13.

The full (multiple regression) equation is as follows:

$$Age = \Sigma \text{ WG, ST, RO, FL, SM, FG, CG, SP,}$$
$$\text{DP, DR, LP, TB, BB, constant}$$

Igarashi et al. (2005, pp. 330–331) provide an example of how to use this equation: "For example, the age of a male with an auricular surface on which WG, ST, FL, SM, and BB are absent and RO, FG, CG, SP, DP, DR, LP, and TB are present is assessed as follows":

$$Age = 0.0 - 7.74154 - 8.47209 + 1.44147$$
$$+ 4.04147 + 7.48609 + 2.99929$$
$$+ 4.24496 + 9.39888 + 0.0$$
$$+ 45.43998 = 58.83851$$

For the reduced models, one sums the specified features for male, female, and combined sexes.

In order to include this study in this edition, one of my graduate students (M. E. Kovacik) and I tried to select specimens from our teaching collection that would illustrate Igarashi et al.'s features. Some of them—especially BB, LP, WG, ST, and RO—were

Table 8–13   Parameter Estimates for Estimating Age

Feature	Male FM	Male RM	Female FM	Female RM	Both FM	Both RM
Wide groove (WG)						
Absent	0.0	0.0	0.0		0.0	
Present	−2.98539	−3.14461	0.15729		−1.19373	
Striation (ST)						
Absent	0.0	0.0	0.0	0.0	0.0	0.0
Present	−4.80846	−4.83121	−3.00191	−3.69695	−5.16653	−5.14979
Roughness (RO)						
Absent	0.0		0.0	0.0	0.0	0.0
Present	−7.74154		−41.27099	−27.40760	−13.99149	−13.42038
Flatness (FL)						
Absent	0.0	0.0	0.0		0.0	0.0
Present	9.17629	9.41175	−6.38202		4.10925	4.44508
Smoothness (SM)						
Absent	0.0		0.0	0.0	0.0	0.0
Present	−2.06441		−33.02554	−32.03711	−7.99967	−8.90605
Fine granularity (FG)						
Absent	0.0	0.0	0.0	0.0	0.0	0.0
Present	−8.47209	−8.57549	−8.73319	−9.44228	−9.54778	−9.84780
Coarse granularity (CG)						
Absent	0.0		0.0	0.0	0.0	
Present	1.44147		9.05453	8.69938	2.26084	
Sparse porosity (SP)						
Absent	0.0	0.0	0.0		0.0	0.0
Present	4.04147	4.28213	6.70445		5.98487	5.93993
Dense porosity (DP)						
Absent	0.0	0.0	0.0	0.0	0.0	0.0
Present	7.48609	7.64942	10.00568	10.32028	9.40567	9.62396
Dull rim (DR)						
Absent	0.0	0.0	0.0	0.0	0.0	0.0
Present	2.99929	3.22778	6.03079	5.74976	5.18270	5.47507
Lipping (LP)						
Absent	0.0	0.0	0.0		0.0	0.0
Present	4.24496	4.32498	−2.38901		4.56494	4.49070
Tuberosity (TB)						
Absent	0.0	0.0	0.0		0.0	0.0
Present	9.3988	9.45809	1.423425		3.31935	3.39652
Bony bridge (BB)						
Absent	0.0				0.0	
Present	6.40646				2.64062	
Constant	45.43998	38.65828	87.34695	79.44228	55.74522	56.53739

SOURCE: Igarashi et al. (2005).

Abbreviations: FM, full model; RM, reduced model.

readily identified when the relevant specimen was uncovered (Figures 8–6 and 8–7). Dense porosity was presumably identified only upon microscopic scrutiny, when pores that penetrated well beneath the thin cortical bone could be discerned (Figure 8–6). We could not with confidence distinguish between sparse porosity and aspects of granularity or between smoothness and flatness other than by assuming from Igarashi et al.'s published images

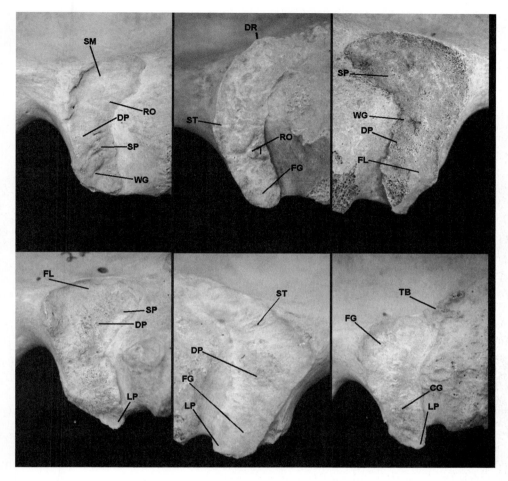

Figure 8–6   Ilial auricular regions illustrating some of Igarashi et al.'s (2005) features: (*top left*) wide groove (*WG*), sparse porosity (*SP*), dense porosity (*DP*), smooth (*SM*), and roughened (*RO*); (*top middle*) dull rim (*DR*), fine grain (*FG*), striation (*ST*), and RO; (*top right*) SP, DP, flattened (*FL*), and WG; (*bottom left*) SP, DP, lipping (*LP*), and FL; (*bottom middle*) ST, DP, some LP, and FG; (*bottom right*) tubercle (*TB*), LP, FG, and coarse grain (*CG*) (University of Pittsburgh teaching collection).

that flatness would be found superiorly near a crisply delineated rim and that smoothness could be found on a flat or contoured surface between other features (as in the illustration, between wide grooves) (Figure 8–6). We found only one specimen that approached the condition of having a dull rim (Figure 8–6), but presumably more obvious cases would not be difficult to mistake.

Perhaps these difficulties were specific to the specimens at hand. Nevertheless, such

ambiguities in identification (which may be alleviated by a brief course or apprenticeship in the specifics of each feature) may signal caution when using this method for estimating age. Flatness and sparse porosity are not applicable to the "female reduced model," but smoothness and fine and coarse granularity are. Smoothness and degrees of granularity are not included in the "male reduced model," but sparse porosity and flatness are. Clearly, there is much morphology in the auricular

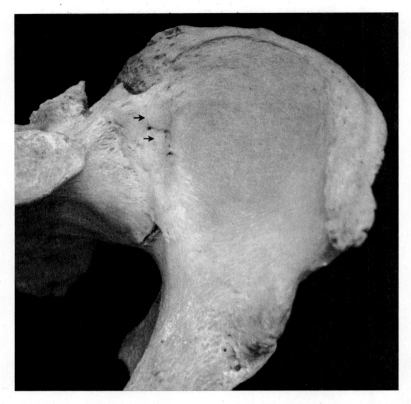

Figure 8–7   Bridging (*BB*) of sacrum and ilium (*arrows*) (University of Pittsburgh teaching collection) (after Igarashi et al., 2005).

region of the ilium and a need to systematize its analysis in a way that can easily be practiced by the novice as well as professional.

Rib (Sternal End)

The age-estimating potential of the rib's sternal end was first pointed out by McKern and Stewart (1957) and reemphasized by Kerley (1970). Subsequent studies (İşcan et al., 1984a,b, 1985, 1987; İşcan and Loth, 1986, 1989; Loth and İşcan, 1987), which achieved similar results using right and left ribs, demonstrated differences in the timing of change between white American females and males and perhaps between samples of African-Americans and white Americans. In the latter case, and in apparent correlation

with differences in timing of pubertal onset, age changes in the rib's sternal end begin earlier in females (beginning at approximately 14 years) than males (approximately 17 years). İşcan et al. (1984a,b) also suggested that ossification in a rib's sternal end is more extensive at an earlier age in African-Americans than in white Americans, even though in general the bones, including ribs, of African-Americans do not become thinner and more fragile with age to the extent that they do in white Americans. Although corroborating İşcan et al.'s results in aging individuals using the fourth rib's sternal end, Russell et al. (1993, p. 53) found that African-Americans showed not an acceleration in rib end changes but "a nonsignificant trend for the rib changes to be delayed compared to" white Americans. Thus, if this approach is to be applied and until a global sampling of human populations becomes available, only the criteria for assessing age of white American females and males should be used with any confidence. While originally studying the right fourth rib, İşcan et al. claimed to achieve accurate age assessments using the third and fifth ribs with criteria established for the fourth rib.

Although Dudar et al. (1993) found that sternal rib end morphology correlated well with known age, other studies have been less enthusiastic. In the known age and sex sample from Spitalfields, Molleson and Cox (1993) found for the second rib (which had the highest survival rate) that analysis by sex was somewhat useful: there was an age correlation of .772 for males but only .345 for females. Jackes (2000), however, concluded that age estimated from sternal rib end morphology did not correlate at all with other age markers.

The utility of sternal rib end analysis depends on preservation of this region, which, as it becomes more fragile over time as buried bone decalcifies, can affect accuracy in distinguishing age phases (Kemkes-Grottenthaler, 1996; Schwartz, 1995). Careful excavation of burials can, however, lead to recovery of these regions. Furthermore, in forensic cases in which rodent or small-carnivore gnawing has not altered sternal rib end morphology, analysis is possible since (at least in my experience with cases in western Pennsylvania) the time period involved in the partial or even complete skeletalization of a victim is usually shorter than the time necessary to decalcify and break down ribs. Bearing in mind the limitations but also the osteologist's obligation to pursue as many avenues of investigation as possible, the criteria are summarized here.

A rib's sternal end can be thought of as a billowy surface that, with age, becomes cuplike as it increasingly embraces the costal cartilage that connects it with the sternum (see Figure 8–8). Changes in this "cup" can affect its depth, floor, walls, and rim or margin.

Specific criteria are (1) *smoothness versus sharpness* of the rim, (2) *scalloping* along the *rim*, (3) *projections* from the *rim*, (4) the formation of a *pit* in or *indentation* of the *floor*, (5) the *shape* of the *pit* or *indentation* ("V" or "U"), and (6) thickness and *solidness versus brittleness* and fragility of the *walls* of the pit.

Age-related changes of the sternal end of at least ribs III, IV, and V of white American females and males are summarized below. Age ranges are listed in parentheses for females first, then for males. Of note is that females reach phases 0–4 at earlier ages than males; males pass through phases 5–8 sooner than females. In general, it also appears that the description of a "V"-shaped pit applies more frequently to females than males, although individuals of either sex could be described as having a "U"-shaped pit. İşcan et al. portray males as having a narrow "U"-shaped pit and females as having a "V"-shaped pit (the reader should bear this distinction in mind in reading the descriptions). More specific differences between females and males are noted as necessary within the discussion of each phase.

1. Phase 0 (13 years, 16 years): Sternal surface unexcavated, billowy, or ridged and seemingly "wrapped" in an extra layer of bone; rim continuous and smooth; bone generally solid and smooth.
2. Phase 1 (14–15, 17–19): Onset of change in regularity or evenness of rim as well as in sternal surface, with some indentation possible.
3. Phase 2 (16–19, 20–23): Development of "V"- (or narrow "U"-) shaped pit in sternal end, with the "V" being created between what are now thick anterior and posterior "walls"; rim still smooth and rounded with scalloping possible; floor may retain some billowing.
4. Phase 3 (20–24, 24–28): "V"-shaped indentation wider and becoming "U"-shaped as walls thin (still fairly thick in males); the floor of the indentation or pit begins to assume an inwardly arced configuration

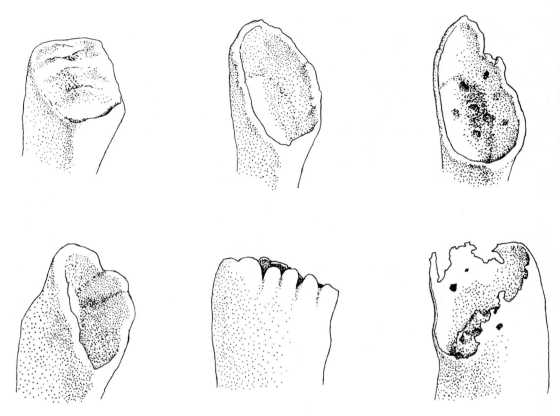

Figure 8–8   Examples of age-related features of the sternal end of rib (typical of ribs III–V): (*top left*) showing features of phase 0, (*bottom left*) phases 2–3, (*top middle*) phase 4, (*bottom middle*) phase 5, (*top right*) phases 6–7, and (*bottom right*) phase 8; see text for descriptions and associated ages (after İşcan and Loth, 1986).

along its superoinferior (long) axis; rim still smooth and rounded but scalloping prevalent (perhaps more irregular in males).

5. Phase 4 (24–32, 26–32): Depth of pit, broad "V" or "U" shape, and superoinferior arc further accentuated; walls thinner and edges possibly flared; rim still smooth but scalloping deteriorating in definition; bone begins to show change in solidness.

6. Phase 5 (33–46, 33–42): Broad "V" or "U" shape further accentuated as walls continue to thin; rim now more sharp than smooth and rounded, with loss of scalloping; floor of pit at least partially covered with dense layer; bone continues to lose solidness and become brittle (perhaps less so in males).

7. Phase 6 (43–58, 43–55): Progressive deepening of pit and widening and flaring of "V" or "U" as walls thin; rim increasingly irregular and sharp and possibly adorned with sharp, pointed projections; floor of pit may show signs of porosity (especially in males); bone overall continues to deteriorate.

8. Phase 7 (59–71, 54–64): Pit typically flared and "U"-shaped, with markedly thin walls and often bearing irregular bony projections internally; pit decreasing in depth in females but remaining deep in males; rim irregular, sharp, and bearing bony projections, particularly at its superior and inferior margins; there is a prevailing

deterioration of bone solidness and an increase in porosity, including in floor of pit.

9. Phase 8 (70+, 65+): Bone deterioration extremely marked, in general as well as in the floor of the pit; bony projections prevalent in floor of "U"-shaped pit (obliterating central arc) as well as on markedly thinned, irregular rim, especially at its superior and inferior margins; the pits of males that lack bony projections are very deep; perforations ("windows") may occur in the walls.

Other Postcranial Elements

Age-related changes in the *vertebral column* have received less attention than those in other parts of the skeleton. The most useful aspect of age-related vertebral change is the development of bony spicules or spurs (*osteophytes*, the process is *osteophytosis*) around the perimeter of the superior and inferior margins of lumbar vertebrae (creating *lipping*, see Chapter 10). In fully expressed cases of osteophytosis, osteophytes form a projecting rim or lip around the margin that goes from right to left pedicles.

Stewart (1958) devised a scale of 0 to +++ to reflect a gradient of "lipping absent" to "maximum lipping." The degree of osteophyte development is assessed separately for superior and inferior margins, and then the two scores are averaged; this could yield, at times, total average scores of "something and one-half" (e.g., ++$^1/_2$). Stewart's sample of known age and sex was heavily white- and male-biased (368 Korean War dead and 87 males but only 17 females from the Terry Collection). He suggested cautiously that at most one could say that, in the case of white Americans, individuals below 30 years typically yield average scores of less than ++, with individuals above 40 yielding average scores higher than ++.

Weisl's (1954) study of the articular (auricular) surface of the sacrum [the sister facet of the auricular surface of the ilium (Lovejoy et al., 1985b)] provided only broad, general clues to determining an individual's age. That is, topographic elevations of the sacral auricular surface are confined in young individuals to the superior and inferior portions and increase in size and number with age.

Skull

The historically oldest approach to determining adult age relies on degrees of cranial suture closure. This approach has shortcomings, but if a skull is all that one has to work with, it is probably more consistently reliable than using relative degrees of tooth wear.

As Meindl and Lovejoy (1985) point out, the use of cranial suture closure as an estimate of an individual's age—which Todd and Lyon (1924, 1925a–c) believed they had perfected —fell out of favor during the 1950s largely because it became increasingly apparent that this approach could not provide easy and accurate results. For instance, Singer (1953, p. 59) concluded from his attempt to replicate Todd and Lyon's accuracy of age determination that "with techniques available at present, an assessment regarding the precise age at death of any individual, gauged only on the degree of closure of the vault sutures of the skull, is a hazardous and unreliable procedure." It has, however, become overwhelmingly apparent that no one method of assessing an individual's age can be considered eminently and consistently "the most reliable" and that as many regions and features as possible should be analyzed in order to increase the level of accuracy (e.g., see Acsádi and Nemeskéri, 1970). As such, analysis of cranial suture closure is again being used in the determination of age.

Todd and Lyon (1924, 1925a–c) considered endocranial closure more reliable in the assessment of earlier adult years because it tends to precede ectocranial closure, the timing of which seemed to be excessively variable. Endocranial suture closure still tends to be favored over ectocranial closure, but

Table 8–14   Age at Death (in Years) Based on Suture Closure

Mean Closure Stage	Mean Age	Mean Deviation	Range
0.4–1.5	28.6	13.08	15–40 (juvenile–young adult)
1.6–2.5	43.7	14.46	30–60 (young–middle adult)
2.6–2.9	49.1	16.40	35–65 (young–middle adult)
3.0–3.9	60.0	13.23	45–75 (middle–old adult)
4.0	65.4	14.05	50–80 (middle–old adult)

SOURCE: Based on Acsádi and Nemeskéri (1970).

Table 8–15   Age Ranks and Associated Age Ranges (in Years)

Age Rank	Age Range
1	15–19
2	20–24
3	25–29
4	30–34
5	35–39
6	40–44
7	45–49
8	50–54
9	55–59
10	60–64
11	65–69
12	70–74
13	75–79
14	80–84
15	85–89

SOURCE: Based on Acsádi and Nemeskéri (1970).

the criteria now used are appreciably more sensitive than Todd and Lyon's (e.g., see Acsádi and Nemeskéri, 1970; Krogman and İşcan, 1986).

Acsádi and Nemeskéri (1970, pp. 115–121) determined age on the basis of the average assessment of different components of the coronal, sagittal, and lambdoid sutures endocranially. In order to use their technique, the coronal suture on either side of bregma is divided into three segments (C_1, C_2, C_3, proceeding laterally from bregma), the sagittal suture into four segments (S_1, S_2, S_3, S_4, proceeding posteriorly from bregma), and the lambdoid suture on either side of lambda into three segments (L_1, L_2, L_3, proceeding laterally from lambda). Each segment corresponds to a naturally occurring (and easily identified) portion of a suture that is distinguished by its own particular configuration and degree and tightness of interdigitation. Martin's scale is used for scoring closure: 0, open; 1, incipient closure; 2, closure in progress; 3, advanced closure; and 4, obliterated. Each sutural segment is scored for degree of closure. All segments are averaged to achieve a score of closure for each suture. These average scores are then averaged, which yields "the mean of closure stage." Age is estimated either by referring to Table 8–14 or by using the following regression equation:

$$y = 1.1627 + 0.4212x - 0.0171x^2$$

where y is the mean of closure stage and x is the age rank [from which the associated age can then be determined (Table 8–15)].

From a sample of 236 crania of known age and sex from the Hamann-Todd Collection, Meindl and Lovejoy (1985) developed a method for assessing age using degrees of ectocranial closure of cranial vault and lateral anterior sutures. They (p. 58) chose to concentrate on ectocranial suture closure because such synostotic events are "far more closely associated with extreme age (for which new forensic standards are most needed)." After investigating the accuracy in age determination of 17 ectocranial sites (each defined as a 1 cm circle whose center is the specific landmark or point), Meindl and Lovejoy

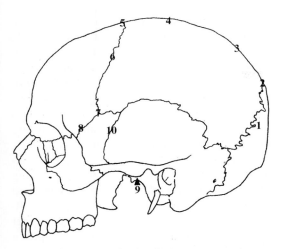

Figure 8–9 Ectocranial landmarks of the lateral and vault systems used by Meindl and Lovejoy (1985) in estimating age at death: *1*, midlambdoid; *2*, lambda; *3*, obelion; *4*, anterior sagittal; *5*, bregma; *6*, midcoronal; *7*, pterion; *8*, sphenofrontal; *9*, inferior sphenotemporal; and *10*, superior sphenotemporal; see text for descriptions.

(pp. 59–60) concluded that the following sites were viable (Figure 8–9):

1. Midlambdoid: Centered at the midpoint of each half of the lambdoid suture (i.e., in the pars intermedia of the lambdoid suture).
2. Lambda: Centered at lambda (in the pars lambdica of the sagittal and lambdoid sutures).
3. Obelion: Centered at obelion (in the pars obelica of the sagittal suture).
4. Anterior sagittal: Centered at a "point on the sagittal suture at the juncture of the anterior one-third and posterior two-thirds of its length (usually near the juncture of the 'pars bregmatic' and 'pars verticis' of the sagittal suture)."
5. Bregmatica: Centered at bregma (in the pars bregmatica of the coronal and sagittal sutures).
6. Midcoronal: Centered at the midpoint of each half of the coronal suture (in the pars complicata of the coronal suture).

7. Pterion: Centered at pterion ("the region of the upper portion of the greater wing of the sphenoid, usually the point at which the parietosphenoid suture meets the frontal bone") (see the Glossary for alternative definitions of "pterion").
8. Sphenofrontal: Centered at the midpoint of the sphenofrontal suture (alternatively identified as the region of pterion, see the Glossary).
9. Inferior sphenotemporal: Centered at a "point of the sphenotemporal suture lying at its intersection with a line connecting both articular tubercles of the temporomandibular joint."
10. Superior sphenotemporal: centered at a "point on the sphenotemporal suture lying 2 cm below its juncture with the parietal bone."

Meindl and Lovejoy (1985) identified ectocranial sites 1–7 as belonging to the "vault system" and sites 8–10, when taken in conjunction with sites 6 and 7, as the "lateral system." A scale of 0–3 is used to score each site within each system: 0, open, no evidence of incipient synostosis; 1, minimal to moderate (50%) closure or synostosis across the site; 2, significant (>50%) but not complete closure across the site; and 3, complete closure. A composite score of suture closure is calculated by summing the individual scores for each site within each system. Thus, a separate composite score is obtained for the vault system and another for the lateral system. Composite scores (based on seven sites) for the vault system can range 0–21; composite scores (based on five sites) for the lateral anterior system can range 0–15. A composite score is translated into an approximate age by comparison with the appropriate table (Table 8–16 for the lateral-anterior system and Table 8–17 for the vault system). Because age, sex, *and* population (i.e., African-American and white American) of each individual in their sample were known, Meindl and Lovejoy tested the possibility of error of age prediction due to sex

Table 8–16 Age at Death (in Years) Based on Ectocranial Lateral–Anterior Suture Closure

Composite Score	Mean Age	Mean Deviation	Range
0 (open)			–50
1	32.0	6.7	19–48
2	36.2	4.8	25–49
3, 4, 5	41.1	8.3	23–68
6	43.4	8.5	23–63
7, 8	45.5	7.4	32–65
9, 10	51.9	10.2	33–76
11, 12, 13, 14	56.2	6.3	34–68
15 (closed)			

SOURCE: Based on Meindl and Lovejoy (1985).

Table 8–17 Age at Death (in Years) Based on Ectocranial Vault Suture Closure

Composite Score	Mean Age	Mean Deviation	Range
0 (open)			–49
1, 2	30.5	7.4	18–45
3, 4, 5, 6	34.7	6.4	22–48
7, 8, 9, 10, 11	39.4	7.2	24–60
12, 13, 14, 15	45.2	10.3	24–75
16, 17, 18	48.8	8.3	30–71
19, 20	51.5	9.8	23–76
21 (closed)			40–

SOURCE: Based on Meindl and Lovejoy (1985).

and population and found that neither variable contributed "any measurable bias" (p. 64). Follow-up studies on other populations will hopefully yield similarly accurate results.

Age-Related Changes in Teeth

Because a tooth's morphology is produced before it erupts into the jaw, one cannot study tooth alteration with age in the same manner as alteration of other skeletal elements. However, because teeth erupt in the jaws at different and known times, their occlusal and interstitial surfaces will be subject to differing degrees of attrition. As such and assuming a constant rate of tooth wear throughout life, the degree to which teeth are worn down should reflect the individual's age at death.

But theory and reality do not always coincide. Different diets, extraneous inclusions in food substances (e.g., grit from utensils used for food preparation and/or as found naturally in food), tooth loss/survivorship (e.g., through periodontal disease or carious infection), the use of teeth as tools (e.g., preparing hides, stripping bark, softening sinews and reeds, flaking edges of stone tools), the modification of teeth (e.g., through anterior upper and/or lower tooth ablation or gripping objects such as pipe stems, nails, pins), the introduction of dentistry, and differences between the sexes within any of these categories can result in subtle to extreme differences between and within populations. This, of course, obviates any possibility of developing globally or temporally useful age-related scales of tooth wear patterns. In some instances, however, one might be able to delineate a pattern of tooth wear that persists from one population to the next over time. For instance, Brothwell (1972) suggested that age-related rates and patterns of upper and lower molar wear were relatively constant among British populations from the Neolithic through the Middle Ages.

As Brothwell (1972) also pointed out, the larger the study population and the more ages represented in it, the greater the likelihood of reconstructing an age–wear chronology for that population. At the very least, one might be able to establish a chronological seriation (cf. Lovejoy et al., 1985a) that reflects *relative* ages within the population.

A chronology of dental wear that is associated with (at least estimates of) *absolute* ages can be approximated by linking defined stages of tooth wear with age-related changes known for other parts of the skeleton. If the sample under study includes individuals in whom teeth are still erupting or growing in the jaws and these individuals can be seriated by dental age, one can construct a tooth-wear chronology by noting how much attrition accumulates during the time between the eruption of one tooth and another and so on.

Studies of age-related changes in tooth wear have concentrated on the "adult" dentition (e.g., Lovejoy, 1985; Molnar, 1971), sometimes only the permanent molars (e.g., Brothwell, 1972; Miles, 1963). It would seem, though, that analysis of all teeth—or as many as are preserved—would provide a more complete and more interesting picture of wear. Furthermore, although children might not be as well-represented in samples as adults, individuals of younger age groups should be included in the overall chronology of tooth wear whenever possible—not only for the purposes of establishing an additional criterion for assessing individuals of young age but also for investigating potential differences in rates of attrition which might, for example, reflect shifts in diet or tooth-use activities. Regardless, when attempting to construct a chronology of tooth wear, one must recognize that occlusal surfaces begin to show attrition when they are sufficiently erupted to contact substances being chewed, which occurs before the tooth's occlusal surface becomes level with the occlusal surfaces of neighboring teeth. This point gains further relevance in light of the common complaint about studies of tooth growth and eruption patterns—that is, the incomparability of "alveolar eruption" with "gingival eruption," the latter being the criterion used on living individuals or cadavers. Careful identification of incipient tooth wear—through polarized light or scanning electron microscopy—can provide evidence of an incompletely erupted tooth having emerged occlusally through the gum.

Gustafson (1950) introduced the first rigorous and systematic approach to determining "dental" age that was not based on development or eruption. He advocated investigating changes associated with attrition, periodontosis (i.e., recession of the gums leading to exposure of the root), the formation of secondary dentine in the pulp cavity, the deposition of cementum on the root (cemental annulation), atrophy of the root or roots, and thinning of the root with a concomitant increase in its translucency. Each category is evaluated on a scale of 0 (absent or incipient) to 3 (most marked), and the scores are summed. This total is x in the equation

$$y = 11.43 + 4.56x$$

where y represents age; the standard deviation is ±3.6 years.

There are problems in applying these criteria to skeletal material. The most obvious is that alveolar resorption, which can easily be evaluated on skeletal material, does not necessarily mirror the extent to which gum recession had proceeded; as with correlating alveolar with gingival eruption, data relevant to translating alveolar into gingival resorption are still lacking (Saunders, 2000; Schwartz, 1995). More generally, Miles (1963) found that application of Gustafson's method (38% accuracy ±3 years) did not significantly improve upon his own more intuitive estimates of individuals' ages (34% accuracy ±3 years), while Condon et al. (1986) concluded that cemental annulation was most accurate for individuals younger than 35 years.

Miles (1963) did suggest that the criterion of "root translucency" was a potentially viable feature—that is, there is a general tendency for the root, beginning with the apex, to become thinner and, in longitudinal section, more translucent with increasing age. In a pilot study of upper central incisors of individuals of known age ($N = 118$), Miles plotted length in millimeters of the translucent part of the root (x) against individual age (y) and found that the data could be characterized roughly ($p = .05$, correlation coefficient = .73) by the regression equation

$$y = 21.857 + 4.6169x$$

When he applied this equation to another sample, Miles achieved a 32% (±3 years) rate of accuracy in assessing age. This is hardly an impressive result, but Miles (p. 167) suggested that, if this body of data was increased (and I would include the analysis of additional teeth), the technique could be applied "by those with virtually no previous experience of either root translucency or age determination." The implications would seem to be particularly relevant to forensic investigations in which skeletal remains are typically very recent in origin, leeching and possible redeposition of minerals are less likely to obscure results, and evaluation could be done efficiently.

Molnar (1971) presented a very complete but also very complex approach toward evaluating attritional effects on the "adult" dentition, which he organized into the groups "incisor + canine," "premolar," and "molar." He subdivided the category of tooth wear into eight components, which can be described roughly as the development of wear facets, exposure of dentine (minimal, maximum), presence of secondary dentine (moderate to extensive), loss of crown enamel (at least one side) and development of extensive secondary dentine, and involvement of the roots in occlusion. He also proposed categories of direction of tooth wear and of occlusal-surface form. Unfortunately, Molnar did not correlate stages or states of tooth wear with age, and an attempt to apply his approach to another sample failed (Lunt, 1978).

Lovejoy (1985) was able to seriate a sequence of stages or patterns of tooth wear on the Libben population and found that they were strongly correlated with age (Lovejoy et al., 1985a). A major reason for Lovejoy's success was the presence in his sample of individuals ($N = 132$) who were young enough (6–18 years) to allow him to define functional rates of wear on the anterior teeth (incisors + canines), premolars, and molars. Overall, "dental wear was found to be sufficiently regular to allow the designation of modal wear groups (with attendant age estimations)" (Lovejoy, 1985, p. 54). Upper and lower dentitions were analyzed separately and modal age-related wear patterns derived for each jaw (nine for the maxilla and 10 for the mandible). When he analyzed the sexes separately, Lovejoy found that females had a slightly higher rate of dental attrition. However, one should be prepared to find such differences in other samples. For example, Campbell (1939) delineated sex-related differences in tooth wear among Australian aborigines, with females, who are the major food collectors, having higher rates of attrition because they test the different types of food they collect.

Clearly, if the necessary elements are preserved, study of age-related changes in patterns of dental attrition can be a fruitful component of the analysis of a skeletal population, for not only will another age criterion be available but also details of tooth use among age groups and between sexes can be reconstructed.

The Complex Method of Determining Adult Age (After Acsádi and Nemeskéri, 1970)

Although the history of the study of age-related skeletal changes reveals an overwhelming desire to isolate one consistently accurate set of criteria for any given phase of life, in reality, the use of only one approach is often riddled with unreliability. This, of course, frustrates forensic and medicolegal cases and undermines paleodemographic reconstruction (Jackes, 2000; Schwartz, 1995). In order to overcome this problem, Acsádi and Nemeskéri (1970, p. 122 et seq.) introduced "the complex method of determining the age of adults," which is based on the combined age-related characteristics of different skeletal regions. After evaluating the relative accuracy of the various analyses and methods reported in the literature, they concluded that the four most reliable indicators of age-related change come from analysis of endocranial sutures, the proximal humerus, the proximal femur, and the pubic symphysis (a modification of Todd's stages). Changes in the proximal humerus and femur are primarily internal and involve a decrease in trabeculation and pervasiveness of spongy bone that is correlated with an expansion upward of the medullary cavity and a general increase in cavitation. Externally, the proximal epiphyses atrophy, with concomitant alteration of topographic features, and the cortical bone thins. Acsádi and Nemeskéri identified six phases of age-related change for the proximal region of each bone, which must be determined either radiographically (or fluoroscopically) or by longitudinal sectioning.

The Proximal Humerus: Phases of Age-Related Change

Examples of each phase are illustrated in Figure 8–10. A description of each phase and the ages associated with it (mean age ± standard deviation and actual range in years) follows:

1. 41.1 ± 6.60 (18–68): The apex of the medullary cavity terminates well below the surgical neck; trabeculae are primarily arranged in a radial pattern.
2. 52.3 ± 2.51 (24–76): The apex of the medullary cavity extends to or above the level of the surgical neck and is subtended by a more fragile trabecular system that approximates the appearance of a pointed arch (e.g., ogival).
3. 59.8 ± 3.59 (37–86): The apex of the medullary cavity may reach the epiphyseal (metaphyseal) line; the trabecular system subtending the medullary cavity is distinctly shaped like a pointed arch, while the trabeculae along the metaphyseal line are more vacuous, with thicker walls.
4. 56.0 ± 1.84 (19–79): The apex of the medullary cavity extends to or above the level of the metaphyseal line; trabeculae subtending the medullary cavity are breaking down, as is the spongy bone within the greater tuberosity.
5. 61.0 ± 2.05 (40–84): The apex of the medullary cavity extends variably above the metaphyseal line and is subtended

only by fragments of trabeculae; distinct large spaces (2–5 mm) develop within the greater tuberosity.
6. 61.1 ± 3.39 (38–84): Deterioration within the greater tuberosity forms a large cavity (c. 5 mm) and may reach the cortex; the apex of the medullary cavity often merges with the cavity formed within the greater tuberosity; trabeculae within the epiphysis are thin-walled, and spongy bone is sparsely distributed; externally, features of the proximal humerus are atrophied and the cortex is thin and transparent.

The Proximal Femur: Phases of Age-Related Change

Examples of each phase are illustrated in Figure 8–11. A description of each phase and the ages associated with it (mean age ± standard deviation and actual range in years, when available) is listed below. Age-related changes in the proximal femur begin earlier than in the proximal humerus (mean ages are, respectively, 31.4 and 41.1 years). Change in the proximal femur is slow until the third phase, when the rate of deterioration parallels that in the proximal humerus.

1. 31.4 (18–52): The apex of the medullary cavity lies well below the level of the lesser trochanter; trabeculae are densely packed, obscuring individual trabecular detail.
2. 44.0 ± 2.6 (19–61): The apex of the medullary cavity extends to or above the inferior portion of the lesser trochanter; details of trabecular morphology and patterning are more clearly defined, with most loss of density occurring in the medial region of the femoral neck.
3. 52.6 ± 1.86 (23–72): The apex of the medullary cavity extends to the top of the lesser trochanter; trabeculae continue to deteriorate and become thinner, especially in the medial region of the neck, as does the spongy bone within the greater trochanter.

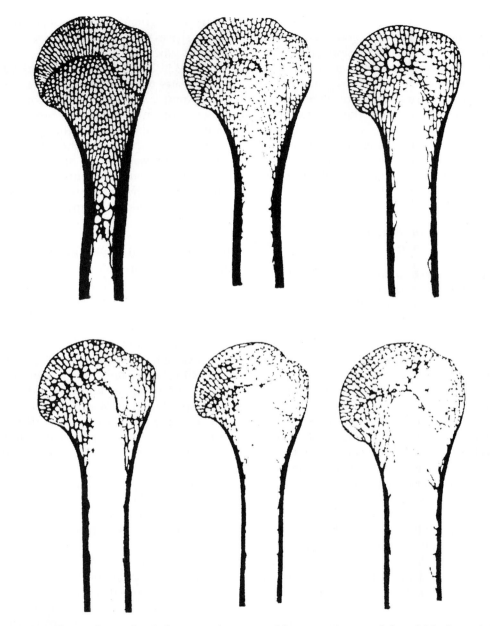

Figure 8–10   Phases of age-related change in the proximal humerus: (*top row, left to right*) phases 1–3 and (*bottom row, left to right*) phases 4–6; see text for description and associated ages (after Acsádi and Nemeskéri, 1970).

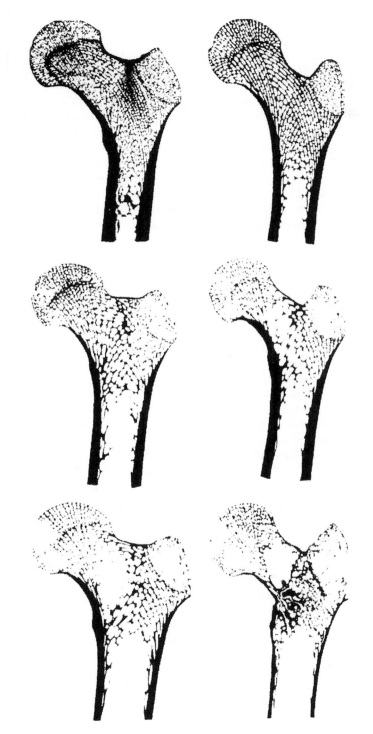

Figure 8–11   Phases of age-related change in the proximal femur: (*top row, left and right*) phases 1 and 2, (*middle row, left and right*) phases 3 and 4, and (*bottom row, left and right*) phases 5 and 6; see text for description and associated ages (after Acsádi and Nemeskéri, 1970).

4. 56.0 ± 2.32 (32–86): The apex of the medullary cavity extends above the top of the lesser trochanter; continued deterioration of trabeculae and spongy bone is pervasive within the greater trochanter, along the metaphyseal line, and in the region just internal to the fovea of the femoral head and creates a large cavity (5–10 mm) in the medial portion of the neck.

5. 63.3 ± 2.17 (38–84): The apex of the medullary cavity extends well above the top of the lesser trochanter; the walls of the medullary cavity bear only vestiges of trabeculae; continued deterioration of trabeculae and spongy bone enlarges the cavities within the greater trochanter (3–5 mm), beneath the fovea, along the metaphyseal line, and within the neck.

6. 67.8 ± 3.64 (25–85): The medullary cavity, which is subtended at best by fragments of trabeculae, may merge with the large cavity in the neck (typically 10 mm); the other cavities are also larger, with the cavity within the greater trochanter often being 5 mm; externally, the features of the proximal femur have atrophied and the layer of cortical bone is thin and transparent.

The Pubic Symphysis: Phases of Age-Related Change

Although Acsádi and Nemeskéri (1970) present their own set of illustrations of the pubic symphysis—which is often reprinted along with all of the others in other texts—this will not be done here, especially since any set of illustrations, like any list of phase/stage descriptions, constitutes a compilation of modal summaries that should be used only as guides and not as universally applicable examples. The reader can refer to the illustrations of the pubic symphysis provided above (Figures 8–3 and 8–4) but is advised to gain experience through study of actual specimens. A description of each phase and the ages (mean age ± standard deviation and actual range in years, when available) associated with it is given below.

1. 26.3 (18–45): The symphyseal surface is convex and transversely billowy (i.e., with horizontal ridges and furrows); the transition from the symphyseal margins (lower and upper extremities) into the inferior and superior pubic rami is smoothly arcuate.

2. 46.5 ± 1.76 (23–69): Deterioration of the symphyseal surface begins to break down ridges and furrows; the dorsal and ventral margins and upper and lower extremities are becoming rimmed.

3. 51.1 ± 1.62 (25–76): The symphyseal surface is primarily granular, with only vestiges of billowing visible; dorsal and ventral margins are almost completely rimmed and upper and lower extremities bear distinct edges.

4. 58.1 ± 2.16 (24–81): The symphyseal surface is flat and its texture granular; dorsal and ventral margins bear sharp rims, and the lower extremity terminates in a distinct ridge.

5. 68.5 ± 2.53 (41–86): The symphyseal surface is porous and appears to have collapsed inward, its "sunkenness" being accentuated by an essentially continuous, raised rim that is continuous with the inferiorly extending, ridgelike lower extremity.

Calculation of Age

Upon comparing these four sources of age-related data, Acsádi and Nemeskéri (1970) found that endocranial suture closure begins earlier and advances rather rapidly at earlier ages than changes in the humerus, femur, and pubic symphysis. Table 8–18 lists the mean ages and the lower and upper limits of the age ranges for each skeletal region.

Acsádi and Nemeskéri (1970) recommend assessing the pubic symphysis first. If the age thus determined is less than 50 years (i.e., phases I and II), they suggest using the age values listed for the lower limit of the age range of each skeletal component; if age estimated from the pubic symphysis is approximately 50 years (i.e., phase III), the mean age

Table 8–18 Ranges (in Years) Predicted by Four Age Indicators

	Lower Limit				Mean				Upper Limit			
Phase	Suture	Symphysis	Femur	Humerus	Suture	Symphysis	Femur	Humerus	Suture	Symphysis	Femur	Humerus
I	23	23	23	23	30	32	33	41	39	40	43	57
II	35	37	35	41	44	44	44	51	52	49	53	61
III	45	46	44	48	53	52	52	57	60	58	59	65
IV	53	54	50	52	60	60	58	59	66	68	56	67
V	58	61	54	54	63	67	63	61	72	75	71	69
VI	—	—	58	55	—	—	67	62	—	—	76	70

SOURCE: From Acsádi and Nemeskéri (1970).

values for each skeletal component should be used; and if pubic symphyseal morphology indicates an age of 50+ years (i.e., phases IV and V), one should use the age values listed for the upper limit of the age range of each skeletal component. The age values obtained for each skeletal component are then averaged in order to eliminate bias from any one source. Thus, for individuals who, on the basis of pubic symphyseal morphology, appear to be younger than 50 years, the lower limits of the age ranges will be averaged (i.e., the sum of the four lower age limits ÷ 4); for individuals approximately 50 years of age, the means of the age ranges will be averaged; and for individuals older than 50 years, the means of the upper limits of the age ranges will be averaged. As Acsádi and Nemeskéri (p. 131) point out, this approach yields the "'most probable' age at death of an individual" and reflects an 80%–85% degree of accuracy with a ±2.5–year margin of error. The average age determined for any individual should be represented with this margin of error.

In an example of their method, Acsádi and Nemeskéri (1970) presented an analysis of the skeleton of a young individual in whom endocranial suture closure was evaluated as being phase 2 while the proximal humerus, pubic symphysis, and proximal femur were phase 1. The lower limits for the age ranges of these phases are 35, 23, 23, and 23, respectively. The total of these figures summed is 104. As 104 ÷ 4 = 26.0, the estimated age is 26.0 ± 2.5 years. The actual age of this individual was 23 years.

The Multifactorial Method of Determining Summary Age (After Lovejoy et al., 1985a)

Lovejoy et al. (1985a) developed a multiple-skeletal-component system for estimating a "summary" age at death of individuals from a sample of 512 individuals from the Hamann-Todd collection for whom reliable documentation on age, etc., was available. Of the potential skeletal sources from which age at death can be calculated, they delineated five as the most appropriate indicators of age for this sample: pubic symphyseal and auricular surfaces, internal structure of the proximal femur, degrees of dental wear, and ectocranial suture closure. These age indicators were used according to the modifications Lovejoy and colleagues suggested in accompanying articles in the same issue of the *American Journal of Physical Anthropology*. Upon applying this method to the Libben site skeletal population, Lovejoy et al. found that they could also include age-related data on the radius, which raised to six the number of age indicators that could be used in assessing this population.

Practically, the number of age indicators that are applicable is determined on a case-by-case basis, depending on specific taphonomic histories and populational norms. For example, assessment of degrees of dental wear may yield valuable comparative information for some populations (e.g., because teeth are retained longer), whereas for others, such data may not be attainable (e.g., because teeth are lost early on due to periodontal disease).

In order to achieve objectivity in data collection, each skeletal age indicator should be analyzed independently (e.g., masking the pubic symphysis when assessing age on the basis of ilial auricular morphology and vice versa). In order to minimize observer error, Lovejoy et al. also advocate seriation of the skeletal material to be analyzed—that is, arrange the material serially in an internally consistent sequence of morphological and inferred age-related change for the population.

Calculation of an individual's *summary age* begins with the determination of age based on each skeletal age indicator (e.g., pubic symphysis, cranial suture closure). The data can be entered into a spreadsheet. All age indicators for all individuals are then used to generate an *intercorrelation matrix* for *principal components analysis*. The *first component factor* score represents *true chronological age*. The weight of an indicator is the *correlation coefficient* (i.e., the correlation of that indicator with the first component factor determined for that age indicator). An individual's *summary age* is calculated by (1) multiplying each age indicator by its weight (i.e., its correlation coefficient), (2) summing them all, and (3) dividing this total by the sum of the weights (i.e., the sum of the correlation coefficients). Summary age can be refined thusly (Lovejoy et al., 1985a, p. 10): for individuals with summary ages of 45–55, recalculate with the lowest age indicator excluded; for individuals with a summary age of 55+, recalculate with the two lowest age indicators excluded.

Of particular note with regard to Lovejoy et al.'s analysis of the Hamann-Todd collection (when summary ages were compared to recorded ages) is that their sample was more diverse geographically, ethnically, and socioeconomically than one would expect for an archaeologically derived skeletal sample. Lovejoy et al. (1985a, p. 12) suggest that "[a]ge determination in archaeological populations is generally more accurate than in modern anatomical collection samples because both environmental and genetic variables are more uniform in the former" and that "[w]hen such age determinations are made by composite methods . . . they may provide more accurate mortality profiles than those derived from living 'primitive' populations."

In their test of this aging method using a skeletal collection of known ages at death, Bedford et al. (1993) found that "[m]ultifactorial age estimates correlated better with real age than did those from any single indicator used" (p. 287) and that "[t]he method produces estimated age distributions which are statistically indistinguishable from those of real age" (p. 297).

Histological Age-Related Changes in Bone

Two of the attributes that Acsádi and Nemeskéri (1970) evaluated in their "complex method of determining adult age" involved age-related changes in trabecular and cortical bone in the humerus and femur. Such changes are also characteristic of other skeletal elements so far studied. However, the only studies that have attempted to correlate structural change in bones with age are those summarized above for the proximal humerus and femur. Other aspects of bone resorption and remodeling will be discussed here as they relate to age estimation.

Specimen Preparation

Internal structural change in the proximal humerus and femur can be studied not only radiographically but also histologically (after preparing a cross section of bone for microscopy). The latter approach focuses on features of osteon populations, which with age increase in number of whole and fragmentary secondary osteons in cortical bone (see Chapter 1). Although cranial bones (Frost, 1987a), ribs, vertebrae, the iliac crest (Stout and Teitelbaum, 1976; Stout and Paine, 1992), and the clavicle (Stout and Paine, 1992; Stout et al., 1996) are amenable to this analysis [and the clavicle and sixth rib together can yield reasonable results (Robling and Stout, 2000)] and since long bones of the leg have been the most extensively analyzed, discussion here focuses on their preparation [following procedures suggested by Blumberg and Kerley (1966), Putschar (1966), Stout (1989), Stout and Teitelbaum (1976), and Ubelaker (1999), who illustrate the procedure], with the caveat that osteon size may differ between populations (Pfeiffer et al., 1995), which, however, may not be an issue in non-stress-bearing bones such as the rib (Pfeiffer, 1998).

Typically, a midshaft cross-sectional slice of c. 1 cm is removed using a fine-toothed saw. Friable material should first be embedded in synthetic resin after being dehydrated in a series of 8-hour alcohol baths (70%, 95%, and then two in absolute alcohol), after which a thin section can be cut, ground, and polished to the required thickness. Unfortunately, a specimen thusly prepared may retain foreign particles that derive either from soil penetration prior to excavation or from the preservative applied during excavation.

An unembedded bone section that is sufficiently solidified to be mounted on the specimen stage of a thin-section saw can be reduced to c. 5 mm in thickness, which is thin enough for one to attempt to remove foreign particles and/or preservatives. Because the polymeric plastic preservatives used in the field are typically dissolved and the resultant solution diluted to the necessary consistency in the organic solvent toluene, a series of toluene baths, with ultrasonic agitation to speed up the process, should redissolve the preservative (small ultrasonic cleaners can be purchased through major biological and chemical supply companies or at the jewelry counters of large—and often discount—department stores). Ubelaker (1999) recommends cleaning thin sections for approximately 50 minutes in a solution of Decal (a commercially available product), with the beaker of Decal suspended in a water bath in the well of an ultrasonic cleaner. The Decal solution is removed by immersing the cleaned specimen in the water bath in the well of the ultrasonic cleaner, after which it is dried overnight. The specimen is then embedded in synthetic resin in preparation for the next phase; placing the container (with specimen and resin) in a vacuum chamber will accelerate the process.

Next, the specimen is ground and polished on both sides to reduce it to a thickness of approximately 75° (50°–90°). Electric grinding and polishing wheels are commercially available from biological supply companies, but both tasks can be done manually. W. von Koenigswald (personal communication, 1980) has used a series of graded sandpapers to grind and then polish embedded teeth for scanning electron microscopic analysis of enamel prism patterns. Kits for polishing plastic-embedded specimens are available from biological supply companies.

The finished thin section is mounted on a slide and protected with a coverslip. The slide should be labeled with information not only about the specimen's origin (e.g., skeletal catalogue or field number) but also about its orientation (i.e., which side is anterior, posterior, medial, or lateral).

Cortical thickness can also be determined via bone-core analysis (Thompson, 1978; also Laughlin et al., 1979), in which a hollow bone-core drill bit is used to extract a 0.4 cm diameter core. This technique leaves the target bone largely intact and unreduced in size, which is particularly important if the specimen is unique or fragmentary. The resultant small piece of bone is cleaned, embedded, and thin-sectioned. As only four loci [at 12, 3, 6 (or linea aspera), and 9 o'clock] of a femoral cross section are analyzed for estimating age [i.e., on the basis of secondary osteons, osteon fragments, non-haversian canals, and circumferential lamellar bone (Kerley, 1965)] and the recommended field size for a locus is less than 2.0 mm (Kerley and Ubelaker, 1978), study of bone cores from these loci can yield results with minimal destruction of bone.

Gross Histological Changes in Cortical Bone

Changes in the cortex of the femur, tibia, and fibula in areas of resorption have been correlated with age (Kerley, 1965, 1970): different cortical areas undergo resorption at different ages. Early in life, when bone size (especially girth) is increasing, resorption is most active in the endosteal region or medullary cavity. As growth ceases and osteoclastic and osteoblastic activity approximate equilibrium, resorption occurs throughout the cortex and, eventually, in the periosteal portion. After a period of inactivity and with advancing age, endosteal resorption increases. The age brackets thus delineated, although quite broad, can be useful in sorting bones into age groups (e.g., as with bones from ossuaries, multi-individual graves, or unintentional mixing of skeletal elements during long-term use and reuse of burial plots).

For analysis, the cortex can be subdivided roughly into inner and outer thirds. From birth through 3 years, resorption is most active in the inner third of the cortex. From the fourth and into the tenth year, intense resorption is distributed fairly uniformly throughout the cortex. During the next 7 years, resorption is primarily restricted to the outer third, near the periosteal surface in particular. After 17 years, osteoblastic and osteoclastic activity typically approach equilibrium. Intense resorption is not noted until later in life. In females, there tends to be an increase in bone remodeling coincident with menopause (Parfitt, 1979), although marked resorption in the inner third of the cortex may begin as early as 40 or as late as 60 years. Although there is some variation in timing in males, intense resorptive activity

in the inner third of the cortex often begins at 40. In both sexes, this phase of osteoclastic activity produces cortical thinning because osteoblastic activity is diminished.

Detailed Histological Changes in Cortical Bone

In attempting to refine age estimates on the basis of histologically recognizable changes in cortical bone, Kerley (1965) studied midshaft femoral, tibial, and fibular cross sections of 126 individuals; individuals in the sample ranged in age from birth to 95 years (median age 35) and represented female and male African-Americans, Asian-Americans, and white Americans. Only pathology-free sections were analyzed—which, regardless of the approach to assessing fields of osteons, should be the specimens of preference (Robling and Stout, 2000). In order to record accurately the cellular life history of the bone under study, Kerley focused on the outer third of the cortex, which, being least affected by intense resorption, presents a more uniform texture throughout an individual's life. Four loci on each section (corresponding to the anterior, posterior, medial, and lateral "borders" of the bone, or 12, 3, 6, and 9 o'clock) were studied microscopically.

Initially, Kerley used a circular microscopic field 1.25 mm in diameter (formed by 10× ocular wide-field lenses in conjunction with a 10× objective lens). Subsequently, he and Ubelaker (1978) found that intermicroscope error could lead to underestimating age, which they suggested could be corrected by using a field of 1.62 mm in diameter (for which they offered a modified set of regression equations for calculating age). The cortical components analyzed are the separate totals across all four fields of the number of (1) whole osteons, (2) fragments of osteons, and (3) non-haversian canals (see Figure 1–10). In addition, they calculated the average percent of circumferential lamellar bone of all four fields.

In Kerley's (also Kerley and Ubelaker's) approach, one counts the number of whole

Table 8–19   Predicting (Regression) Equations for Estimating Age at Death

Bone Elements	Predicting Equation	Mean Square Residual
Femur		
Whole osteons	$Y = 2.278 + 0.187X + 0.00226X^2$	9.19
Osteon fragments	$Y = 5.241 + 0.509 + 0.017X^2 - 0.00015X^3$	6.98
Lamellar bone	$Y = 75.017 - 1.79X + 0.0114X^2$	12.52
Non-haversian canals	$Y = 58.390 - 3.184X + 0.0628X^2 - 0.00036X^3$	12.12
Tibia		
Whole osteons	$Y = 13.4218 + 0.660X$	10.53
Osteon fragments	$Y = -26.997 + 2.501X - 0.014X^3$	8.42
Lamellar bone	$Y = 80.934 - 2.281X + 0.019X^2$	14.28
Non-haversian canals	$Y = 67.872 - 9.0870X + 0.440X^2 - 0.0062X^3$	10.19
Fibula		
Whole osteons	$Y = -23.59 + 0.74511X$	8.33
Osteon fragments	$Y = -9.89 + 1.064X$	3.66
Lamellar bone	$Y = 124.09 - 10.92X + 0.3723X^2 - 0.00412X^3$	10.74
Non-haversian canals	$Y = 62.33 - 9.776X + 0.5502X - 0.00704X^3$	14.62

SOURCE: Modified from Kerley and Ubelaker (1978).

osteons, osteon fragments, and/or non-haversian canals present in all four loci of a section. The total count is regressed, using the appropriate equation. The average percent of circumferential lamellar bone is estimated. Given problems in intermicroscope calibration, Kerley and Ubelaker (1978, p. 546) recommend the following corrections for adjusting the data to a field diameter of 1.62 mm:

The area (πr^2) of the 100× field of the microscope used must be calculated using a stage micrometer and that field size divided into 2.06 mm^2 (area of a field diameter of 1.62 mm) to determine the relationship between the original field size and the field size of the individual microscope being used. All counts of osteons, fragments or nonhaversian canals should then be multiplied by that factor. The estimate of percentage of lamellar bone is not affected.

In his original paper, Kerley (1965) presented graphs and regression equations for translating the counts and averages into age estimates for each bone studied (i.e., femur, tibia, and fibula). Ahlqvist and Damsten (1969)

criticized this study, but Bouvier and Ubelaker (1977) demonstrated the reliability of Kerley's method. According to Stout (1989, p. 45), Kerley and Ubelaker's (1978; Table 8–19) modified regression equations are the most reliable available histological aging methods when the ages obtained from them are averaged together.

Stout (1989) also referred to his osteon-based, age-estimation studies of the rib and clavicle, which, in contrast to the long bones of the leg, are not weight-bearing. This factor is significant because a problem in using weight-bearing bones is that mechanical stresses can lead to bone remodeling (e.g., Frost, 1987a). With regard to the rib, at least two cross sections should be prepared from the middle third of the sixth rib and all osteons and osteon fragments counted together to achieve "total visible osteon density." Stout (1989, p. 47) offered the following regression equation to predict age:

$$y = 2.87351x - 12.3490$$

where y = age in years, x = total visible osteon density ($r = 0.68244$).

Subsequently, Stout and Paine (1992, p. 112)—who used a combined density (number/mm^2) of intact and fragmentary secondary osteons to produce the variable "osteon population density"—developed a different equation for predicting age based on the rib as well as equations for predicting age based on the clavicle alone and on the rib and clavicle together: "The data were made linear by using natural log (L_n) age as the dependent variable." The sample size was 40 individuals and the age range 13–62 years; $s = 12.9$ and $t_{0.05} = 2.0244$. For the rib, the predicting equation is

$$L_nY = 2.343 \pm 0.050877X_r$$

where $s_{yx} = L_n0.231$, $\bar{x} = 18.03$, $s_x^2 = 51.696$, $r^2 = 0.7211$.

For the clavicle, the equation is

$$L_nY = 2.216 + 0.070280X_c$$

where $s_{yx} = L_n0.239$, $\bar{x} = 14.86$, $s_x^2 = 26.256$, $r^2 = 0.6989$.

For the rib and clavicle together, the equation is

$$L_nY = 2.195 + 0.029904X_r + 0.035430X_c$$

where $s_{yx} = L_n0.209$, $r^2 = 0.7762$. If both bones are available, Stout and Paine suggest that the latter regression equation is the most accurate predictor of age.

Upon testing this predictor of age using the clavicle against a sample of individuals of known age from Spitalfriedhof St. Johann near Basel, Switzerland, Stout et al. (1996) discovered that the mean histologically predicted age and reported age differed by 5.5 years. Among other contributing factors, they attributed this statistically significant difference to decreasing reliability of the method for older individuals. In order to take this problem into account, Stout et al. (1996, p. 141) offered a new formula for predicting age using the clavicle:

$$L_n\text{age} = 2.033 + 0.085 \text{ (OPD)}$$

where OPD is the osteon population density.

Frost (1987b) also developed an algorithm for estimating missing osteons in fields of secondary osteon populations. The impetus was to try to deal with the fact that "as the number of secondary osteons increases in a given diaphyseal cross section, new ones can begin to remove all microscopic evidence of older ones. . . . [Thus] increasing numbers of observed osteons tend to become progressively smaller fractions of all osteons created in that domain" (Frost, 1987b, p. 239). Stout and Paine (1994, p. 123) tested Frost's algorithm on an autopsy sample ($N = 44$ ribs) and found that "[e]stimates of activation frequency . . . and bone remodeling rate . . . using the new algorithm are in reasonable agreement with age-matched tetracycline-based values." Therefore, they concluded, Frost's algorithm can be applied to archaeologically derived osteological material. Inasmuch as application of this algorithm requires determining mean osteonal cross-sectional area, mean cross-sectional diameter of intact osteons, intact and fragmentary osteon density (from which osteon population density can then be calculated), and accumulated osteon creations, the pursuit of this analysis would be limited to labs geared to such tasks. The interested student should therefore turn to these two publications for explication of the algorithm and procedure.

Clearly, the application of histological analyses to the determination of age at death of individuals is a potentially viable pursuit. As Frost (1987a) points out, however, many factors are known and/or suspected to affect osteon creation (Table 8–20). But, as he (p. 237) elaborates, before one can even deal adequately with these sources of potential error in estimation, one must know

. . . the normal drift patterns, rates, and durations at the diaphyseal level(s) the section(s) came from and for the bone involved. That means atlases are needed for those properties in standard normal bones, but such atlases do not exist yet because there has been little apparent need or usefulness for this kind

Table 8–20   Examples of Elements that Could
Affect Osteon Formation, Number, etc.

Actual age, mean tissue age, life span, sex, taxon

Anemias, Paget's disease

Bone growth, modeling patterns, hormones, skeletal
  maturation, regional acceleration

Diet, nutrition, vitamins, electrolyte imbalance

Drugs, toxic agents, radiation

Genetic (structural) disorders

Infectious/systemic disease, tumors

Metabolic disorders, alkalosis, acidosis

Mechanical strain, usage, disuse (acute), paralysis

Regional trauma, microdamage

Source: Modified from Frost (1987a).

of information. Certainly there has been little
incentive for expending the labor and time
needed to obtain such information. The bones
studied could be the tibia, femur, rib, a vertebra, humerus, parietal bone, radius, and/or
pelvis, and in such work an accounting should
be made for the effects of chronological age,
mechanical usage, species, and locations in
the bone itself. . . . A future task consists of
constructing such atlases.

More recently, Robling and Stout (2000)
endorsed Frost's vision of establishing reference samples and broadened his "wish list" to
include consideration of (1) sexual dimorphism (e.g., potential osteon size difference);
(2) physical activity (e.g., increased/decreased
mechanical loading due to level of techno-
logical development, topography of landscape,
and procurement of food); (3) population
variation (e.g., in metabolism and/or dynamics
of bone remodeling and turnover rates); and
(4) pathological conditions [e.g., metabolic
disturbances (see Chapter 1), fractures, infections, circulatory complications]. Hopefully,
future osteologists will undertake these tasks.

Chapter 9

Differentially Expressed Morphological Character States

Nonmetric Variation, Sex Determination, Populations, and a Discussion of Biodistance Analysis

Overview

The study of nonmetric variation, or the delineation of nonmetric traits, is the descriptive counterpart to analyzing measurements of teeth and bones. Thus, studies on nonmetric variation tend to focus on recording differences and similarities between individuals in details of skeletal morphology. The impetus for these studies is twofold: first, to try to describe groups or populations by the relative frequencies of expression of the traits recorded and, second, to try to use similarity in frequency of trait expression to determine how closely or distantly related to each other groups so defined are.

The underlying rationale is that if morphology has a genetic basis, morphologically more similar groups must be more closely related to each other. Conversely, the less similar groups are, the less closely related to each other they must be. Yet, in order to calculate the relative frequencies of trait expression in a group or population [which is necessary for calculating biodistance—how "close" or "far" from each other human groups are (see Pietrusewsky, 2000; Reed, 2006)], that group or population must first be defined or delineated, which it often is but on the basis of nonmorphological information (usually cultural, linguistic, archaeological, textual, situational, or some other nonmorphological attribute). Degrees of closeness, based upon the comparative relative frequencies of traits in different groups, are calculated using various statistical approaches, among which the Mahalanobis D^2 distance has been the most popular (see Pietrusewsky, 2000, and Van Vark and Schaafsma, 1992, for overviews of quantitative analyses). Although the trait lists brought to these analyses have been augmented over the years, the features scrutinized and recorded—dental and nondental—tend to be the same from one sample to the next. A suggested criterion in the choice of "good" nonmetric traits is that they should be "resistant to environmental stress" (Saunders, 1989, p. 96). But how does one decide which features are "good" or "not so good"? This is a particularly important question in light of recent research on development, stress proteins, and cell membrane physical states that underscores the hierarchical as well as individual nature of stress-induced response at the cellular and molecular levels (see Maresca and Schwartz, 2006, and references therein).

During the nineteenth century, most studies on what we now call nonmetric variation dealt with one of two subjects. Some sought to delineate distinguishing features of "races" (see review in Schwartz, 1999b). More ambitious investigations sought to interpret features in which non-European populations differed from the European ideal. These studies, however, tended to be cloaked in the guise of a purported evolutionary perspective, in which at least some group or groups among our own species were seen as the pinnacle of morphological change. For those individuals or groups of individuals who were found to differ from this "ideal," the variant features of the former were interpreted as retained primitive features or as *atavisms* (evolutionary reversals to more primitive states).

Early twentieth century studies began to concentrate on the developmental aspects of nonmetric traits. By the 1950s, interest shifted to the presumed genetic basis of nonmetric variation (e.g., summarized in Grüneberg, 1963). In the 1960s, the focus was on the general applicability of the study of nonmetric variation and the delineation of the differential distribution of nonmetrically variable traits to the problem of distinguishing populations from one another (e.g., Berry, 1968). Debate continues as to which, if any, study of nonmetric traits reveals more population-specific information than metric analysis of skeletal variation (Saunders, 1989).

Nonmetrically variable traits are often identified as features whose representation from one individual to the next may differ. Character states are recorded, for instance, as present/absent, open/closed, more/fewer (than an agreed upon "expected" number), or within/outside of suture. Typical examples of categories of nonmetric variation are (1) "foramen spinosum: open or closed" or "third trochanter on femur: present or absent"; (2) "frontal (metopic) suture: fully persistent, partially expressed, or totally obliterated" or

"talar calcaneal facet: single, pinched, double, or anterior facet absent"; (3) "infraorbital foramen: single or multiple"; and (4) "mastoid foramen: sutural or exsutural (i.e., in the mastoid region)." Most frequently in textbooks, the topic of nonmetric variation is discussed in its own chapter or section and thus set apart from what in reality constitute other expressions of nonmetric morphological variation. As I hope you will come to appreciate, features whose expression may differ between the sexes (i.e., features related to sexual dimorphism)—which, in turn, are often presented in a separate chapter—also constitute nonmetric morphological variation.

There is still debate as to what the best way is for recording and evaluating nonmetric variables (as traditionally identified). For example, how does one take into account a trait that, in one individual, might be expressed only on one side of the body (*unilaterally*) but, in another individual, may be expressed *bilaterally*? Some osteologists advocate recording a trait as present regardless of whether it is unilaterally or bilaterally expressed in an individual. In this case, the frequency of representation of the feature is a ratio based on the number of individuals with the trait relative to the total number of individuals in the sample. Other osteologists insist on counting the number of variant traits on each side of the body—in which case, the frequency of a feature is a ratio based on the number of individuals with the trait on a given side of the body relative to the total number of individuals for which data on that side of the body can be collected. There is lack of agreement as to which approach is "better" or more accurate, primarily because it is still not known whether trait expression is truly side-dependent (i.e., genetically based and correlated with a specific side of the body) or essentially random (see Saunders, 1978, 1989). Further complicating such analyses is the suggestion that changes in frequency of trait expression—specifically, a shift from unilateral to bilateral—may increase with age (e.g., Saunders, 1978). But even if true for some traits (e.g., "accessory facets" or "bony bridging"), it would be unlikely to be true for all traits (e.g., multiple foramina, whose presence results from the formation of bone around soft tissue structures) classified as "nonmetric variants."

Considering the Bases of Nondental, Nonmetric Variation

On a general level, there is a well-established tradition of how nonmetric traits are categorized and recorded (see Table 9–1). With regard to the nondental skeleton (teeth are dealt with separately below), one kind or category of nonmetric trait recognized in the literature is characterized by the proliferation of bone into, or in response to the action of, soft tissue structures. Such a trait is often referred to as **hyperostotic**. Examples include (1) a **trochlear spur** (a small spit of bone in the medial orbital wall above and behind the posterior lacrimal crest), which may result from either mineralization of a ligament that is ultimately associated with the superior oblique muscle (one of the muscles that move the eye) or the muscle pulling on the bone via

Table 9–1   Examples of Nonmetric, Nondental, Non-Sex-Related Traits

Hyperostotic (i.e., increased/extended bone growth)
Division/bridging of foramina/canals
   Optic canal
   Hypoglossal (anterior condylar) canal
   Infraorbital foramen
   Jugular foramen
   Transverse foramen (vertebral)
   Bridging of neural canal of first cervical vertebra (lateral or posterior)
Mineralization of connective tissue/deep fascia
   Complete supraorbital foramen
   Extension of lacrimal hamulus across inferior border of lacrimal fossa (in greatest degree of expression, excludes maxilla from inferior border of lacrimal fossa)
   Pterygospinous bridge (between lateral pterygoid plate and sphenoid spine, lies medial to foramen ovale)
   Pterygobasal spur/bridge (spur lies lateral to foramen spinosum and ovale, bridge is enlarged spur-to-lateral pterygoid plate)
   Clinoid bridge (between anterior and middle clinoid processes or between anterior and posterior clinoid processes)
   Mylohyoid bridge (internal, mandible; spans mylohyoid groove; variable in expanse)
   Enlargement of posterior lacrimal crest
   Infraorbital canal length > infraorbital groove length
Response to soft tissue
   Trochlear spur (medial orbital wall)
   Extension of vaginal process (postglenoid plate/sheath of styloid process)
   Highest nuchal line
   Distension of maxillary tuberosity
   Enlarged hamulus (medial pterygoid plate)
   Shape/size of posterior nasal spine
   Shape/size of genial tubercles (mandible)
   Precondylar tubercle
   Third trochanter (femur)
   Peroneal tubercle (calcaneus)
Accelerated closure/union
   Craniostenosis/cranial synostosis [e.g., of squamous portion of temporal and adjacent parietal bones (i.e., squamoparietal synostosis), sagittal suture (i.e., scaphocephaly), coronal (with/without sphenoparietal synostosis)]
   Reduction in lumbar vertebrae via incorporation into sacrum

Table 9–1   (*continued*)

Excessive bone deposition of unknown origin/etiology
  Palatal torus
  Mandibular torus
Excessive bone deposition of apparent (non–soft tissue) known origin/etiology
  Exostoses—for example, auditory torus (induced by cold/hydrostatic pressure)[1]
  Osteomas
  Osteochondromas
Supernumerary elements: additional centers of ossification
  Ossicles [at lambda, bregma, asterion, parietal notch, pterion (i.e., epipteric bone); in coronal, sagittal,
    lambdoid sutures; not including Inca bone]
  Os japonicum (inferior portion of zygoma, delineated superiorly by transversozygomatic suture)
Supernumerary elements: aggressive preossification differentiation
  Vertebrae
  Digits
  Cervical ribs
Hypostotic (i.e., less extended/truncated bone growth)
Incomplete coalescence of elements: cranial sutures
  Frontal (metopic) suture (resulting in metopism)
  Transverse occipital (mendosal) sutures/fissures (complete patency yields transverse occipital suture,
    delineating Inca bone inferiorly)
  Infraorbital suture
  Squamomastoid suture (extending from parietal notch, temporal bone)
  Transversozygomatic suture (yielding os japonicum)
Incomplete coalescence of elements: postcranial
  Pinched/doubled occipital condyle
  Spina bifida (lumbosacral, occulta or severe)
  Sacral segments (separate)
Incomplete mineralization of soft tissue precursor
  Foramen of Huschke (dehiscence in floor of tympanic plate)
  Septal aperture (olecranon fossa of humerus)
  Infraorbital groove length > infraorbital canal length
  Enlarged foramen lacerum
  Incomplete carotid canal (petrosal, internally and/or externally)
  Emarginate/(vastus) notched patella
Incomplete preossification development
  Sternal aperture (incomplete coalescence of cartilaginous sternal bars)
  Pharyngeal fossa (midline depression/pit in basiocciput externally; possible vestige of pharyngeal
    pouch)
  Anterior facial cleft (lip/palate)
Foramina
Absent
  Posterior ethmoid
  Mastoid
  Foramen spinosum (?)
  Zygomaticofacial (?)
Atypically present
  Parietal
  Zygomaticofacial (?)
  Supraspinous (scapula)
  Foramen in clavicle (for supraclavicular nerve)

Table 9–1 (*continued*)

Multiple
  Infraorbital
  Anterior ethmoid
  Posterior ethmoid
  Supraorbital/frontal
  Frontal process
  Nasal bone (superior or inferior/marginal)
  Zygomaticofacial
  Mastoid
  Palatine
Position
  Infraorbital high on frontal (i.e., frontal foramen)
  Mastoid (exsutural, high on mastoid region)
  Anterior ethmoid (exsutural)
  Posterior ethmoid (exsutural)
  Groove for middle temporal artery (restricted to external surface of parietal bone)
Articular facet—configuration and number
  Calcaneus (talar): single, pinched (symmetrically, asymmetrically), subdivided (anterior, middle), anterior absent
  Talus (calcaneal): single, pinched (symmetrically, asymmetrically), subdivided (anterior, middle), anterior absent
  Trochlear surface (olecranon process of ulna): single, pinched (asymmetrically, from medial margin), subdivided (superior, inferior)
  Intermetatarsal (proximal end; single, subdivided, truncated)
  Intermetacarpal (proximal end; single, subdivided, truncated)
  Accessory occipital condyle (uni-, bilateral)
  "Squatting facets"[1]
Other
  Bifid anterior nasal spine
  Coincidence/separation of dacryon and lacrimale (i.e., degree of approximation of superior extents of posterior and anterior lacrimal crests, reflecting openness/closedness of lacrimal fossa superiorly)
  Position of base of medial pterygoid plate relative to width/sides of basisphenoid
  Position of base/alae of vomer relative to spheno-occipital synchondrosis
  Degree of extension of maxilla medially beyond superior margin of infraorbital foramen
  Distension/reflection of inferior orbital margin
  Degree of depression of region at/below infraorbital foramen (general region of canine fossa)
  Course/configuration of zygomaticomaxillary suture (e.g., arced, straight on the diagonal, "cornered" or "reclining checkmark")
  Course/configuration of zygomaticotemporal suture [e.g., arced, straight on the diagonal, stepped (short, long)]
  Palate/dental arcade shape (e.g., "V," "U")
  Degree of alveolar prognathism
  Degree of anterior projection/curving back of zygoma (extreme anterior projection increases relative zygomatic arch length and produces angularity or "cornering" of zygomatic arch at region of maxillary tuberosity)
  Degree of lateral flaring of zygomatic arch (produces large, triangular or small, narrow temporal fossa)
  Orientation of articular fossa (directly lateral or diagonally forward)
  Jugular foramen right/left asymmetry (size, configuration, degree of "pocketing")
  Jugular foramen orientation (anteriorly, vertically, posteriorly angled)
  Carotid foramen orientation (anteriorly, vertically, posteriorly angled)
  Angularity/"filling out" of parietal notch

---

[1] Although these features are traditionally included in lists of nonmetric traits, it must be borne in mind that they result from culturally emphasized behaviors. As such, this kind of feature, while perhaps providing insights into cultural activities, does not contribute to a systematic analysis of relationships.

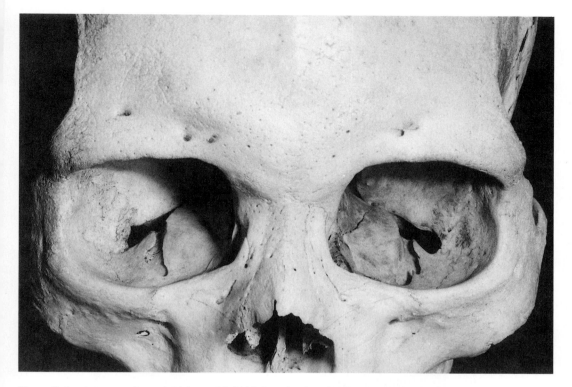

Figure 9–1   Accessory frontal (*right* and *left sides*) and infraorbital (only *right side*) foramina; also note, for example, differences in length of infraorbital groove (longer on the *left*) and development of sharp, ledgelike posterior lacrimal crests (postcontact?, Alaska).

its ligament; (2) a **third trochanter** on the femur, which represents an enlargement of the gluteal tuberosity, onto which the lower part of the gluteus maximus muscle inserts; (3) **bridging of cranial canals** (e.g., **optic** or **hypoglossal canals**); (4) **doubling of cranial and vertebral foramina** [e.g., **accessory frontal** or **infraorbital foramina** (Figure 9–1) and **accessory cervical transverse process foramen**, respectively]; (5) **excessive ossification** along the "roof" of the infraorbital groove (which leads to a relative lengthening of the infraorbital canal; see Figure 10–28); (6) **reduction** of number of **lumbar vertebrae via incorporation into the sacrum** (Figure 9–2); and (7) development of an **osteochondroma** (Figure 9–3). One might also reasonably suggest that **muscle scars** or markings, including those of less frequent appearance (e.g., **highest nuchal line**) as well as well-delineated, thickened bands or tori of bone [e.g., **palatal torus** (see Figure 7–16) or **auditory torus** (Figure 9–4)], should be regarded as hyperostotic characters, even if their precise etiology is, at present, unknown, as in the case of palatal tori; auditory tori, however, may be induced by regular bouts of aquatic activity, such as diving, in which the coldness of the water causes the membrane lining the ectotympanic tube to shrink, inducing bone formation (e.g., Katayama, 1988).

Ossenberg (1976) suggested that early and complete closure or **synostosis** of the squamoparietal suture between the squamous portion of the temporal and the parietals—as especially noted in the posterior extent of the

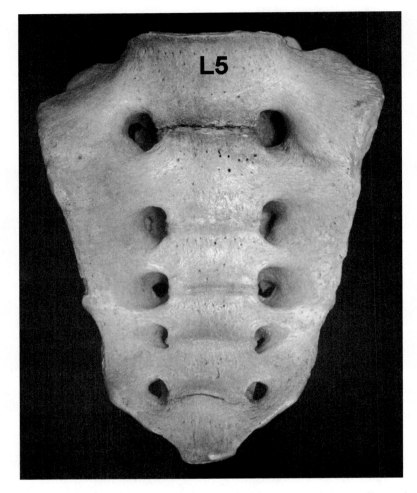

Figure 9–2   Sacrum with L5 partially fused to it (University of Pittsburgh teaching collection).

squamoparietal suture—should be categorized as a nonmetric variant. If synostosis is recognized as a nonmetric variant, it could be thought of as a hyperostotic trait because premature or accelerated closure represents atypical ossification into the connective tissue of the sutural zone. But if one particular example of premature or accelerated closure of a cranial suture (i.e., **cranial** *synostosis* or **craniostenosis**) constitutes a nonmetric variant, then do not all other forms or expressions of cranial synostoses? The question, which is applicable to all decisions regarding what is or is not the variant, is "How has the 'norm' been determined?"

Squamoparietal synostosis neither is associated with nor causes noticeable cranial alteration. However, other synostoses are or do—for example, premature closure of the sagittal suture appears to be correlated with the development of an atypically elongate cranium as well as, perhaps, with deflection of the frontal forward and/or of the occipital inferiorly and the development of an elevation along the posterior portion of the synostosed suture (producing **scaphocephaly**);

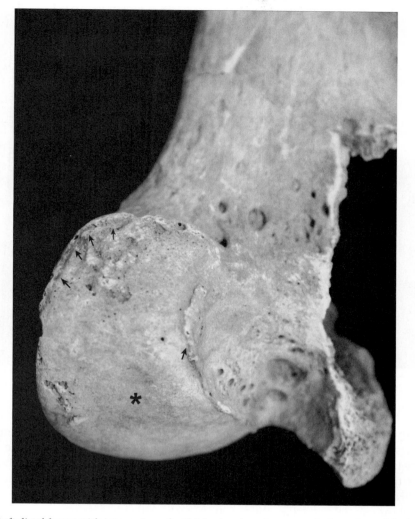

Figure 9–3   Left distal femur with impression of a tibial osteochondroma on the lateral condyle (*asterisk*); also note osteophytes along margin of insertion of articular capsule diagnostic of arthritis (*arrows*) (prehistoric, Pennsylvania).

and premature synostosis of the coronal and sphenoparietal sutures is associated with a **postbregmatic depression**. Since the latter forms of cranial synostosis change skull shape, the osteologist encountering one or the other might not be able to distinguish between a group that had a "preferred" skull shape (which would have been caused by a particular craniostenosis) and a group in which cranial synostosis was simply one of its

distinguishing developmental features (one would predict, however, that the development of the feature preceded its being maintained in the group).

Supernumerary or extra **structures** constitute another potential source of variation or deviation from the typical skeletal pattern. Often listed in this category are the variably small to medium-sized islands of bone [**ossicles** or **wormian bones** (Figure 9–5)] that may

Figure 9–4   Auditory torus (exostosis) occluding acoustic meatus (postcontact?, Chile).

develop within the territories of sutures—that is, along the sagittal, lambdoidal, coronal, and sphenoparietal sutures, at *asterion* (i.e., the juncture of the occipital bone, petromastoid region of the temporal bone, and parietal bone), and at the parietal notch (i.e., the region of transition between the squamosal and mastoid sutural regions of the temporal bone, into which the "corner" of the parietal nestles). Details of ossicles are presented elsewhere in this text, with the descriptions of individual cranial bones. It should be noted here, however, that, although the presence and frequencies of sutural ossicles within skeletal populations have long been recorded, the etiology of these supernumerary structures is still debated (e.g., see Ossenberg, 1970). Perhaps confounding an understanding of the subject is the sometime identification of the sutures that delineate these islands of bone as

supernumerary sutures (e.g., Ossenberg, 1970; Saunders, 1989). To be sure, the presence of extra "pieces" of cranial vault bone necessitates the presence of additional sutural joints. However, since the existence of sutural contacts depends first on the growth of membranous bones from their initial centers of ossification, it would seem that focus should be on the ontogeny of the ossicle itself, not on the suture that comes to surround it.

Ossenberg (1969) found in the 22 Native North American populations she studied that, although frequency of posterior and lateral cranial vault ossicles differed among groups, there was a correlation within each population between the frequencies of posterior and lateral ossicle expression (Pearson's r_s = .609, $p < .01$). In undeformed crania, posterior and lateral cranial vault ossicles occurred with equal frequency. However, in skulls that were

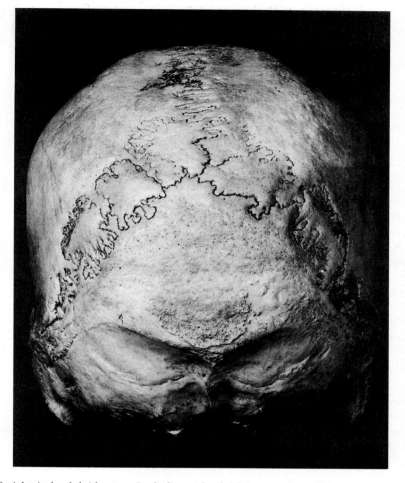

Figure 9–5   Ossicles in lambdoid suture (including at lambda) (recent, Australia).

posteriorly deformed (e.g., in infants who were swaddled to a cradleboard) compared to undeformed skulls from the same population, Ossenberg (1970) noted a higher incidence of posterior ossicles but a decrease in the frequency of lateral ossicles; specifically, posterior ossicles occurred three times more frequently than lateral ossicles. Ossenberg (p. 366) interpreted this as indicating "that inhibition of the normal growth rate produces stress which either encourages the formation of supernumerary sutures, or delays their closure (or both); while acceleration of the normal growth rate either inhibits their

formation or speeds up their obliteration (or both)." Nevertheless, as suggested above, it might be more fruitful to investigate ossicle formation in terms of factors that lead to the development of supernumerary centers of ossification along the margins of expanding cranial vault bones. One might, for instance, ask the question, "Do differential rates of cranial vault bone expansion create regions that would be depauperate of (presumptive) bone if additional centers of ossification did not arise?"

A totally different kind of supernumerary element has been noted in the zygomatic region. This so-called **os japonicum** of the zygoma presumably derives from the appearance of an extra ossification center; a "normal" zygoma otherwise develops from a single center. In its full-blown state, the os japonicum is distinguished as the inferior portion of a zygoma in which a transverse suture (**transversozygomatic suture**) subdivides a

Table 9–2   Examples of Nonmetric Dental Traits

Elaboration/proliferation of enamel
   Shovel-shaped upper central incisors [e.g., marked with lingual tubercle, faint margocristae (marginal crests)]
   Shovel-shaped upper lateral incisors [e.g., marked with lingual tubercle, faint margocristae (marginal crests)]
   "Distal accessory ridge" on upper/lower canines
   Enamel extends beyond neck [possible on all teeth; on molars, enamel "tongue" may also develop between buccal roots (i.e., enamel extension)]
   Enamel pearl (isolated enamel nodule below neck, often in cleft of roots)
Cusp, etc., elaboration/proliferation
   "Carabelli's cusp" (the protostyle of other mammals)
   Metastylid on lower molars (especially on M_1 = so-called cusp 7)
   Protostylid on lower molars (especially on M_1)
   Enlargement of lingual tubercle on P_1 (large—could be identified as a metaconid)
   "Twinning" of P_2 metaconid
   "Twinning" of lower molar hypoconulid (especially on M_1, entoconulid, so-called cusp 6)
Supernumerary structures (i.e., polydontia/polygenesis)
Teeth
   Post-M3 "peglike" tooth (i.e., protoconid; uni- or bilaterally, upper or lower)
   Post-M3 "partial" tooth (protoconid + metaconid + talonid basin; uni- or bilaterally, upper or lower)
   Post-M3 tooth (complete; uni- or bilaterally, upper or lower)
   Post-M4 structures (complete; uni- or bilaterally, upper or lower)
   "Twinned" tooth (usually antemolar secondary teeth; twinned I^2 often in association with anterior facial cleft/cleft lip–palate)
Roots
   "Third" root on lower molars (may be thin or stout but conical/tapering)
   "Fourth" root on upper molars (above hypocone)
   "Split"/bifid tip of buccal root on upper premolars (especially on first)
Reduction in tooth number (i.e., agenesis/hypodontia)
   M3 absence (uni- or bilaterally, upper or lower)
   P2 absence (uni- or bilaterally, upper or lower; often associated with retention of dm2)
   I2 absence (uni- or bilaterally, upper or lower; in upper jaw, sometimes associated with anterior facial cleft)
Reduction in root number (i.e., "coalescence"/lack of separation)
   Upper premolars (partial, bifid tip, complete)
   Upper molars (mesiobuccal and lingual roots spanned by laminae; roots appressed to one another; coalescence partial, tips separate, complete)
   Lower molars [mesial and distal root coalesced buccally (forming "C" or reversed "C"); roots appressed to one another; coalescence partial, tips separate, complete]
Other
   "Carabelli's pit" on upper molars (may be in association with "Carabelli's cusp")
   Groove on internal surface of buccal root of upper premolar
   Winging of I^1 (uni- or bilaterally)
   Counterwinging of I^1 (uni- or bilaterally)

---

typically single zygomatic bone into upper and lower moieties. Sometimes this suture is complete in its horizontal course across the zygomatic bone; at other times, it may extend only a centimeter or so from a juncture with the zygomaticotemporal suture. In the literature, the expression of a truncated transversozygomatic suture is often identified as an incomplete suture, which, in some individuals, may come to extend fully across the zygomatic bone (e.g., Ossenberg, 1976). It is, however, developmentally more likely that a truncated transverse suture results from the partial obliteration of a suture whose presence was dictated by the appearance in the presumptive zygoma of a supernumerary ossification center.

In general, it is reasonable to think of the development of supernumerary elements as hyperostotic in nature. Included in this category, therefore, would be sutural ossicles (but not the special case of the Inca bone) as well as an os japonicum with a complete transversozygomatic suture. The variable obliteration of the transversozygomatic suture would result from the imposition of yet another kind of hyperostotic event—that is, the premature ossification of sutural connective tissue. **Supernumerary digits (polydactyly)**, **supernumerary vertebrae**, and **supernumerary cervical ribs**, for example, arise not from the development of excessive ossification of (or into) a normally extant structure but via the ossification of precursors that had arisen via hyperdifferentiation. For the sake of simplicity here, such supernumerary skeletal elements (which result from aggressive preossification differentiation) are included in Table 9–1 as a subset of hyperostotic nonmetric traits. [**Supernumerary molars** may be likened to supernumerary vertebrae (in that they arise via hyperdifferentiation of premineralized precursors; see Table 9–2).]

If one also takes into consideration features attributed to sexual dimorphism, the number of traits that can be identified as variants of hyperostotic origin increases. For example, consider such stereotypically male

features as a pronounced external occipital protuberance, enlarged malar and maxillary tuberosities, an elongate and stout mastoid process, a marked occipitomastoid crest, and a projecting ischial spine; flare and/or eversion of the margin of the gonial region of the mandible; robustness and/or rugosity of the linea aspera, temporal and nuchal lines, and the inferior surface of the zygomatic arch; or even protrusion of the anterior nasal spines (which grow either as a response to traction from the septopremaxillary ligament or as an infilling of bony matrix in the wake of an anteriorly expanding nasal capsule).

In opposition to hyperostotic features are **hypostotic** features, which result from the incomplete or arrested ossification of a structure or from the incomplete or arrested union of structures. Representative of hypostotic traits are, for example, (1) the **persistence** of certain **cranial sutures** or contacts that would otherwise close [e.g., the **frontal** or **metopic** suture (i.e., **metopism**; Figure 9–6), **infraorbital suture** (Figure 9–6), or **transverse occipital (mendosal) fissures**; a complete transverse occipital suture, which delineates a large occipital ossicle or **Inca bone** (which looks like a large, equilateral triangle, with its apex at lambda, its sides delineated by the right and left arms of the lambdoid suture that course down and away from lambda, and its base delineated by the transverse occipital suture; Figure 9–7)]; (2) the **persistence** of separate **sacral segments**; (3) **arrested ossification**, for example, of the floor of the auditory meatus (leaving a **foramen of Huschke**; Figure 9–8), of the medial border of the **foramen ovale**, or of the "roof" of the infraorbital canal [which increases the relative length of the **infraorbital groove** (e.g., Figure 9–1)], and of the petrosal along the length of the **carotid canal** (which can lead both to truncation of the petrosal bone and thus increase the size of the foramen lacerum and to patencies along the internal and external walls of the petrosal; Figure 9–8); (4) the **lack** or **incomplete fusion** of vertebral spinous processes [**lumbosacral spina bifida** (*occulta* or

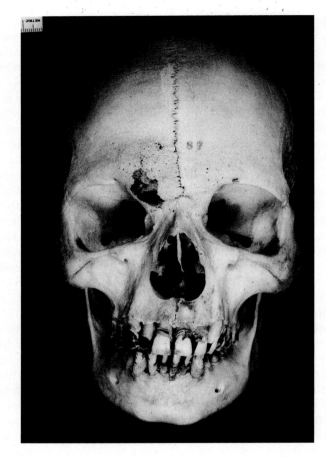

Figure 9–6   Persistence of frontal (metopic) and infraorbital sutures; also note, for example, abscessing of right frontal sinus, somewhat deviated vomer, three-dimensional calculus around exposed tooth roots, and betel nut staining on crowns of various teeth (recent, Indo-Pakistan).

severe)] (Figure 9–9); and (5) the presence of a **vastus notch** on the patella (Figure 9–10). Incomplete or arrested ossification would also result in the persistence of the patency identified in the olecranon fossa of the humerus as a **septal aperture**.

As there appear to be different developmental backgrounds to different hyperostotic traits, there may also be different etiologies for hypostotic traits. For example, a **sternal aperture** or patency in the body of the sternum (Figure 9–11) is sometimes identified as a hypostotic trait (Saunders, 1989). However, a sternal aperture does not arise in the same

manner as, for example, a septal aperture or a foramen of Huschke. Rather, a sternal aperture results when, during fetal development, the right and left cartilaginous bars that contribute to the formation of the presumptive sternum do not coalesce completely along the midline (see discussion of the sternum, Chapter 4). Because separate centers of ossification within the cartilaginous precursor of the sternum give rise to its various segments, ossification of a cartilaginous precursor in which the sternal bars have left a patency leads to the development (i.e., persistence) of a sternal aperture. Thus, a sternal aperture

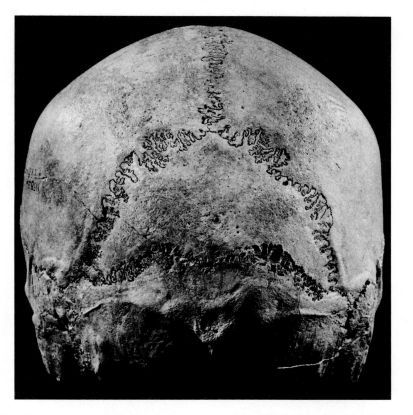

Figure 9–7   "Inca" bone; also note, for example, development of highest nuchal line (especially visible on *left side*) and ossicle near parietal notch (*right side*) (prehistoric, Pennsylvania).

exists because of disturbances in preossification development. Similarly, the persistence of a small centrally emplaced pit or **pharyngeal fossa** in the basiocciput would be a hypostotic trait that results from incomplete preossification development (see Chapter 1).

Anterior facial cleft (lip) and **cleft palate** are often discussed exclusively under the categories of congenital abnormalities and skeletal malformations. However, it might not be unreasonable to identify anterior facial cleft and cleft palate as hypostotic in nature, being analogous to a sternal aperture in the sense that these conditions result from the disruption of, or interference with, proper development of a preossification precursor. In the case of **anterior facial cleft**, the embryonic

maxillary isthmus, which normally maintains mesenchymal continuity between the maxillary and median nasal prominences, is severed (Andersen and Matthiessen, 1967). During normal development, separate ossification centers—one appearing in the median nasal prominence and the other in the maxillary prominence—give rise, respectively, to the premaxilla and maxilla; they eventually ossify across the maxillary isthmus, which then forms the roof of the infraorbital foramen (Andersen and Matthiessen, 1967). If, however, connection between median nasal and maxillary prominences is severed, the ossifying facial elements will be separated from one another, leading, in this case, to anterior facial cleft (Anderson and Matthiessen, 1967;

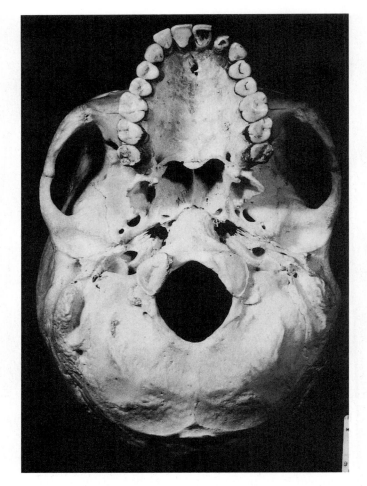

Figure 9–8 Examples of various "nonmetric" features: bony spur across incisive foramen, incomplete ossification of carotid canals, large foramina lacera, slight foramen of Huschke (*left side*), incompletely ossified styloid processes, notably asymmetrical jugular foramina, obliterated posterior condyloid canals, exsutural mastoid foramina (multiple on *left side*) (recent, India).

Schwartz, 1982). **Cleft palate** results from failure of the embryonically vertical and separate palatal shelves to hydrate sufficiently for them to elevate and meet at the midline to form the presumptive palate. Ossification of the separated palatal shelves yields a gap. The co-occurrence of anterior facial cleft and cleft palate involves the expression of two different, developmentally disruptive phenomena, both of which occur prior to the onset of ossification. Thus, these two varieties

of clefting would represent hypostotic traits that result from incomplete preossification development.

Sometimes cited as a hypostotic variant is the so-called **supraorbital notch**, which results from lack of ossification of soft tissue that bounds the inferior border of the **supraorbital foramen**. However, the presence of a supraorbital *foramen* is also often cited as a nonmetric variant, which would be identified as a hyperostotic trait. Regardless of which

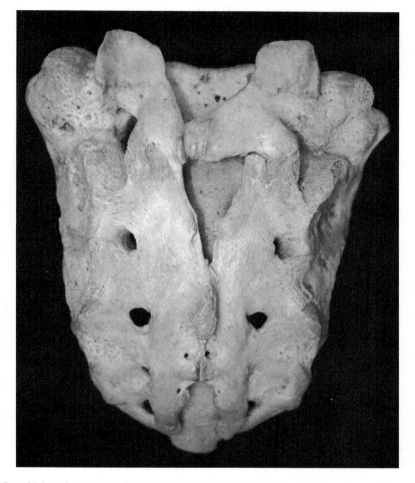

Figure 9–9   Fused L5 and sacrum with spina bifida occulta (University of Pittsburgh teaching collection).

character state—notch or foramen—is the true variant, one can usually find in a series of skulls examples that span the gamut of variation from "broad, shallow notch" to "deeper and more narrowly constrained notch" to "partial closure of notch" to "complete foramen." The particular variant of notch/foramen may be expressed either bilaterally or asymmetrically. In the context of a broad comparison among primates (including fossil hominids), however, even a cursory study of supraorbital notch versus foramen expression reveals that, if there is anything of note in the supraorbital margin, the most prevalent fea-

ture is a notch (Schwartz, unpublished data; see illustrations, e.g., in Hershkovitz, 1977; Schwartz and Tattersall, 2002, 2003, 2005). As such, it would appear that the supraorbital foramen in *Homo sapiens* represents the variant trait, and it is so listed in Table 9–1.

Variations in articular facet number and/or morphology have typically been taken as representing a unified category of nonmetric traits. It is, however, probably more appropriate to break up this category into ontogenetically meaningful sets. For example, one should single out the development on the occiput of **doubled** (or **twinned**) **condylar facets**

Figure 9–10   Left patella with vastus notch (University of Pittsburgh teaching collection).

and deal with this variant separately and as a potential hypostotic feature because, while various postcranial articular facets (e.g., the talar calcaneal facet) may be expressed in a doubled (twinned) form, the development of a doubled condylar facet is due to a disruption of the coalescence of two ontogenetically discrete elements. That is, the anterior part of an occipital condyle arises on the basiocciput, while the posterior portion arises on the lateral part of the occipital (see discussion of the cranium, Chapter 2). During growth, the lateral parts of the occipital coalesce medially with one another. Anteriorly, each lateral part unites with an arm of the basilar portion of the occipital, which, in most individuals, results in the formation of a pair of single, unified occipital condyles. Incomplete coalescence of the parts that contribute to the formation of a single condylar facet yields a disrupted facet. In contrast, disjunctions in

the surfaces of postcranial articular facets have different origins. A more substantial list of nonmetric variants of possible hypostotic origin is presented in Table 9–1.

Additional or **supernumerary** articular **facets** as well as variations in **number** and/or **morphology** of postcranial articular **facets** appear to represent other kinds of nonmetric traits: the former result from unusual or exaggerated contact between bones (Figure 9–12); the latter, although between articulating bones, are of unknown origin (Figure 9–13). Common examples of additional facets are the typically ovoid or elliptical **squatting facets** that develop when sustained bouts of extreme hyperflexion cause contact between the posterior distal femur just above the condyles and the posterior proximal tibia, the anterodistal margin of the tibia (in the midline) and the head of the talus, and the distal portions of the distal phalanges and the metatarsal heads

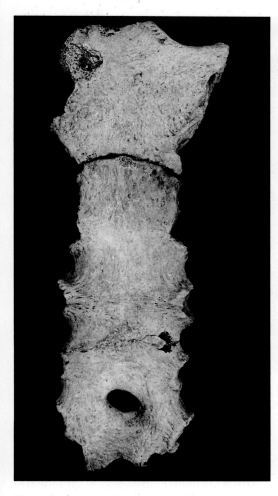

Figure 9–11   Sternal aperture (prehistoric, Pennsylvania).

(Figures 9–14 and 9–15). Uni- or bilateral development of **accessory occipital condyles** also results from unusual contact, in this case between the first cervical vertebra and the base of the skull lateral to the occipital condyle itself.

Although variation in number and morphology of postcranial articular facets is often illustrated in texts, their etiology and ontogeny remain unknown, in spite of the fact that one can make a case for their heritability (Saunders, 1978). An additional and curious aspect of such variation in articular facet morphology is that expression of a facet variant on one bone does not appear to be correlated with expression of an analogous (mirror-image) variant on the opposing articular facet. For example, a talar calcaneal facet may be subdivided into anterior and middle moieties, but the sister facet on the talus may be smoothly single (not even pinched or otherwise modified). Given their different paths of development, it is not surprising that an occipital condylar facet may be single but its sister facet on C1, pinched or variably divided (Schwartz, personal observations). Table 9–1 lists the general regions in which one typically finds variation in facet number and/or configuration. Details of facet variation are discussed in the descriptive section for relevant bones.

Distinguishing foramina that are "doubled" or otherwise subdivided by virtue of the development of a thin bony septum within their walls (as mentioned above in the context of hyperostotic traits) from other kinds of **secondary** or **accessory foramina** depends on the focus of the observer: is it on the foramen or on the soft tissue structure around which the foramen has formed? As Fazekas and Kósa (1978) point out, though, bone forms around soft tissue structures (e.g., neurovascular bundles). Thus, in the case, for example, of the infraorbital foramen, the difference between a foramen that is subdivided internally by a septum and a typically large and single infraorbital foramen in the vicinity of which lie one or more smaller foramina reflects ossification around infraorbital neurovascular bundles at different points of arborization. As such, one can offer the following explanations for typical variations in infraorbital foramen number: a single infraorbital foramen formed around an undivided neurovascular bundle; a septum within a foramen formed near the base of a furcation in a neurovascular bundle, or at least prior to extensive separation between branches; a primary foramen and one or more separate

Figure 9–12   Left scapula (*left*) with extra articular facet on acromion and coracoid (*arrow*) and a supras-capular foramen; left scapula (*right*) with additional suprascapular foramen (both prehistoric, Pennsylvania).

accessory foramina ossified around a more fully arborized neurovascular bundle. The problem that remains, however, is determining on a case-by-case basis whether the plane of the foramen remained constant relative to the point of arborization of the neuro-vascular bundle or whether it shifted relative to the point of arborization. That is, differences between individuals with a large single infraorbital foramen, those with a subdivided infraorbital foramen, and those with separate secondary foramina would seem to arise because either (1) the trunk was arborizing at different points along its length or (2) facial growth differed such that the neurovascular bundle was captured at different points along its length. In a broader context, therefore, the study of foraminal variation could serve as a window into differential rates of growth of parts of the skeleton in different populations.

Other oft-cited osteological nonmetric variants may be explained in terms of ossification around soft tissue structures and the relative location and/or path of these soft tissue structures. Consider, for the moment, the cases of **foraminal position/location/trajectory** as well the **presence** or **absence** of **grooves** or **foramina**. Typical examples of variants in location of cranial foramina are a foramen's being either *sutural* (i.e., created in a sutural joint between bones as a result of a nerve and/or blood vessel being captured there) or *exsutural* (the result of the same nerve and/or blood vessel coursing the membranous or cartilaginous precursor of a particular bone—e.g., a mastoid or anterior ethmoid foramen).

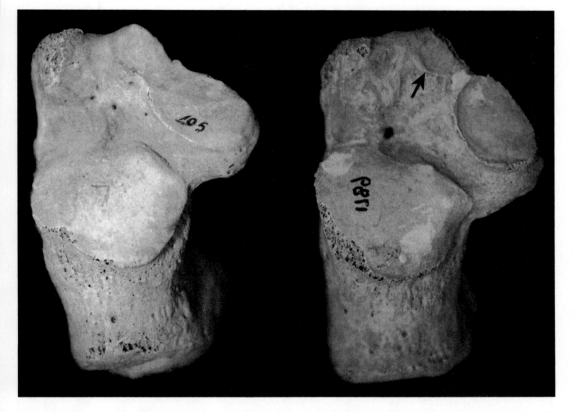

Figure 9–13 Left calcaneus (*left*) with slightly pinched, single anterior facet; left calcaneus (*right*) with divided anterior facet (*arrow*) (both prehistoric, Pennsylvania).

Displacement of nerve or vascular course/trajectory within the preossification matrix would explain the occasional appearance of a *clavicular foramen* (the clavicle ossifies around the supraclavicular nerve), the **presence** or **absence** of **grooves** (e.g., a groove made by the middle temporal artery), as well as the **presence** or **absence** of **foramina** [e.g., suprascapular (Figure 9–12), zygomaticofacial, or posterior ethmoid foramina]. In the case of absence of the posterior ethmoid foramen, the posterior ethmoid nerve may be lacking, while the posterior ethmoid artery pursues a different course into the anterior cranial fossa.

Taking note of and recording these kinds of traits is straightforward (see Table 9–1).

It is, however, still unclear for many of them which character state actually represents the variant. For example, agreement is lacking on the interpretation of polarity of presence versus absence of the zygomaticofacial foramen. Nonetheless, a survey of primates (Schwartz, 1997, and unpublished data; also see illustrations in Hershkovitz, 1977) reveals that the presence of this foramen is widespread, with foraminal size, number, and position (e.g., below/level with/above the infraorbital margin) being the variables of interest. It would thus seem that "absence of a zygomaticofacial foramen" and "more than one zygomaticofacial foramen" represent variants (or, more precisely, different character

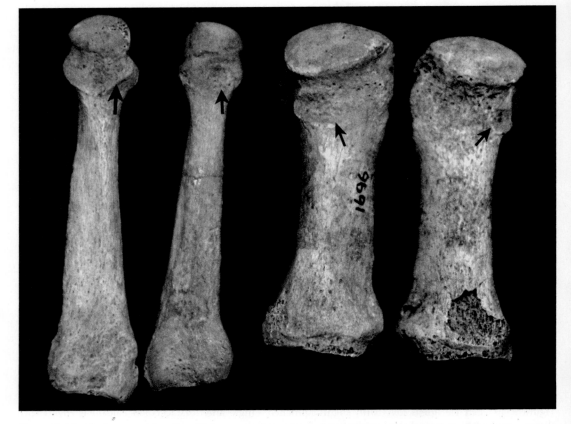

Figure 9–14   Various metatarsals with large squatting facets (*arrows*) induced by contact with proximal phalanges resulting from repeated hyperflexion (prehistoric, Pennsylvania).

states) in primates, including humans; they are listed as such in Table 9–1. In a study of 500 human crania representing a global sampling of populations, two foramina were often present and their position typically above the level of the infraorbital margin (Schwartz, 1997, and unpublished data).

In addition to the features discussed above, which represent a variety of categories of non-sex-attributed nonmetric traits, I have included under the section "Other" in Table 9–1 a series of different kinds of features that have either been studied as "racially revealing features" or not been studied at all. This section is not meant to be exhaustive by any means. Rather, as with the traits listed in

other sections of Table 9–1, these features are presented in order to provoke new approaches to the study of nonmetric skeletal variation.

The features discussed above are listed in Table 9–1; additional examples are included within trait categories.

Dental Nonmetric Traits

In general, categories of dental trait variation (Table 9–2) appear to parallel those of the nondental skeleton. That is, there appears to be an analogue of excessive growth leading to hyperostotic-like traits, as well as a dental counterpart of interrupted or diminished growth leading to hypostotic-like traits. As

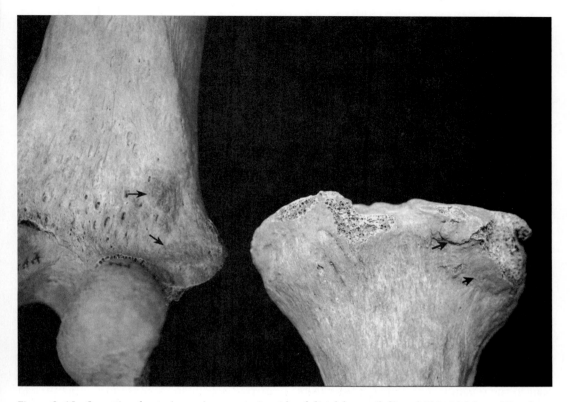

Figure 9–15   Squatting facets (*arrows*) on posterior side of distal femur (*left*) and tibia (*right*) resulting from repeated hyperflexion (prehistoric, Pennsylvania).

emphasized in Chapter 7, though, tooth development must be appreciated on its different levels: primary versus secondary (i.e., the level of cellular proliferation that leads to the differentiation of tooth classes and the number of primary teeth within a tooth class, with a secondary tooth being derived from the external dental epithelium of the tooth it eventually replaces).

Reduction in tooth number due to the nondevelopment of a tooth (**agenesis** or **hypodontia**) can occur among primary or secondary teeth. Clearly, if a primary tooth that is normally replaced by a secondary tooth does not develop, the secondary tooth will not develop either. What we see ultimately as agenesis of a primary tooth can, however, be affected in more than one way: failed

signal transduction pathways of molecular communication, inhibition of cellular proliferation, resorption of a tooth germ, or the obliteration of a presumptive tooth by a disruptive or traumatic assault on it or its environs. Permanent molar agenesis, which occurs at the end of the primary molar class (resulting, e.g., in "loss" of M3), may result from inhibition of cellular proliferation posteriorly ("truncation of the dental lamina") or resorption of a presumptive tooth germ (e.g., in marmosets, which are unique among primates in their *erupting* two, rather than three, molars in each quadrant of the jaw). Reduction in secondary tooth number would be affected by inhibition of cellular proliferation from the predecessor primary tooth, failure of interaction between the secondary dental lamina and neighboring mesenchyme, or resorption of the presumptive secondary tooth germ (see Ooë, 1969, for the latter). An interesting phenomenon, but one that occurs infrequently in humans, is, in the same individual, "loss" of M3, retention of dm2, and inhibition or resorption of P2. This reduces premolar number but maintains three functional, molariform teeth [and has been proposed as a mechanism for reduction of the molar class in various mammalian groups (Osborn, 1978)] (see Figure 7–10).

It has been suggested that fusion of (primary) tooth germs may be another way in which tooth number is reduced absolutely (e.g., see Berkovitz and Thomson, 1973, on fetal ferrets). However, because one cannot pursue a longitudinal study of the process, it is equally possible that presumptive tooth germs that appear to have coalesced may, instead, be presumptive tooth germs that failed to separate [i.e., their cell masses did not migrate sufficiently apart from one another for a zone of inhibition to intervene and thus lead to the separation of dental entities (e.g., see Lumsden, 1979)].

There appear to be two ways in which **polydontia** or **polygenesis** (the development of extra or **supernumerary teeth**) can arise. One is related to process (excessive cellular proliferation within a tooth class), while the other is more mechanistic (the splitting of a tooth germ). Although "extra teeth" are, indeed, produced in each case, the bases of supernumerary tooth development are very different from one another.

Supernumerary structures at the end of the molar series are by far the most commonly noted supernumerary dental features (see review by Schwartz, 1984). They range from being conical–crowned, single-rooted structures to two-cusped, one- or two-rooted teeth (with variable talon/talonid development) to full-blown teeth. These possible developmental states actually conform to a morphocline of tooth formation, culminating in a morphologically full-blown supernumerary tooth

(a "fourth" molar) that conforms to the size–shape gradient of the molar series. A second supernumerary molar (i.e., "fifth" molar or part thereof) will not be more morphologically complex than the supernumerary molar anterior to it and will also conform to this size–shape gradient. Although these examples constitute indirect evidence, it appears that a gradient of size and morphological complexity of supernumerary structures results from the maintenance of tooth-forming competence (at the molecular and cellular levels) of a posteriorly expanding molar tooth class (Schwartz, 1984). [Truncation of the molar tooth class, with, e.g., third molar agenesis can be appreciated as the converse of supernumerary molar development (see Figure 7–14)].

Supernumerary teeth can also arise in the middle of a tooth class, apparently by interstitial budding between two tooth germs that migrate apart from one another (see Schwartz, 1982, 1984). The latter explanation accommodates the supernumerary primary tooth germs that Ooë (1971) found in humans: in one specimen, a presumptive germ had begun to differentiate between the growing dm_1 and dm_2 and, in two other specimens, a supernumerary primary tooth germ had begun to differentiate between the enlarging di_1 and di_2. It is important to point out that these examples of interstitial supernumerary teeth are of primary teeth. A secondary supernumerary tooth can develop only if preceded by a primary supernumerary tooth.

The most common example of an extra tooth structure that can be explained by an externally induced splitting of a tooth germ is the "twinned" di^2 that sometimes accompanies anterior facial clefting (see review by Schwartz, 1984). Developmentally, the epithelial thickening of the presumptive di^2 arises in the maxillary prominence (presumptive maxilla) and migrates into the median nasal prominence (presumptive premaxilla) (Ooë, 1956, 1957). As the tooth germ migrates, it crosses the maxillary isthmus, which maintains mesenchymal continuity between these two prominences (Andersen and Matthiessen, 1967). Inasmuch as anterior facial clefting results from a severing of the maxillary isthmus, causing separation of the median nasal prominence (Andersen and Matthiessen, 1967), a migrating di^2 germ would be affected if it were in the region of the maxillary isthmus (Schwartz, 1982). Since another dental feature associated with anterior facial clefting is the absence of a lateral incisor and this phenomenon can be explained as the ablation of a tooth germ in the wake of the destruction of the maxillary isthmus, it seems reasonable to suggest that severing of the maxillary isthmus might disrupt and "split" a tooth germ (Schwartz, 1982). In general, it appears that more instances of "extra" teeth can be explained as resulting from excessive cellular proliferation within or at the end of a tooth class than from an external source or insult.

Given the above, I recommended that the term "supernumerary" should be used to refer to teeth that, like supernumerary digits (producing polydactyly), arise because of something intrinsic to the process of development and differentiation (e.g., excessive cellular proliferation). "Supernumerary" is then restricted to those situations in which the extra tooth (or teeth)—first, and at least, at the level of the primary tooth class—either fits into the normally present morphological/size gradient of the tooth class in which it occurs or adds another successional generation of teeth (e.g., a "third" dentition) (see Figure 7–15). The mechanical disruption of a developing tooth germ leading to its being cleaved in some way might better be called "twinning" (i.e., "twinned" structures are produced).

The Assessment of Skeletal Sexual Dimorphism

Although *Homo sapiens* is only weakly sexually dimorphic—in size and some morphology—compared to other large-bodied hominoids, males tend to be somewhat larger than females, and there are a few skeletal features that seem consistently to distinguish one sex from the other (Tables 9–3 to 9–7).

Since there is often a size difference between females and males, various studies have attempted to distinguish one from the other on the basis of measurements of skull and mandible (Table 9–3), skull and postcranial elements (Table 9–4), deciduous and permanent teeth (Table 9–5), as well as postcranial elements alone (Table 9–6). With males being on average somewhat larger than females, a generalization commonly employed in determining sex on the basis of skeletal morphology is that the cortical bone of males will be thicker and, overall, individual bones will be more massive and heavier. Because of the latter, muscle markings on a male's bones will be more pronounced and rugose. For example, in males, as the generalization goes, features such as the linea aspera, gluteal lines on the ilium, deltoid crest on the humerus, inferior margin of the zygomatic arch, nuchal region, mastoid process, temporal lines, and gonial angle of the mandible will tend to be more clearly delineated, marked, and/or distended (see Tables 9–7 and 9–8). Obviously, degree of physical activity will affect the robustness of bone and its attendant muscle markings. Therefore, females can display supposedly male attributes and vice versa.

Stereotypes also exist for analyzing sex-related differences between females and males in cranial, mandibular, and especially pelvic (including sacral) morphology. In the skull and mandible, for instance (see Figures 9–16 to 9–20), the female is supposed to have a more vertical frontal bone, higher and more rounded orbits lacking attendant supraorbital or glabellar thickening or distension, thinner zygomatic arches, a more obtuse gonial (mandibular) angle, and a less pronounced mental trigon (see Table 9–7). Because females may bear children, features of the articulated pelvis and its separate elements are supposed to reflect this aspect of reproduction. As such, the female pelvis (including

Table 9–3 The Skull and Mandible: Determining Sex via Discriminant Function Analysis of Measurements

Cranial Measurements	Based on White Americans						Based on African-Americans						Based on Japanese	
	Function						Function						Function	
	1	2	3	4	5	6	7	8	9	10	11	12	13	14
Maximum cranial length	3.107	3.400	1.800		1.236	9.875	9.222	3.895	3.533		2.111	2.867	1.000	1.000
Maximum cranial breadth	-4.643	-3.833	-1.783		-1.000		7.000	3.632	1.667		1.000		-0.062	0.221
Maximum cranial height	5.786	5.433	2.767			7.062	1.000	1.000	0.867					1.865
Basion–nasion length		-0.167	-0.100	10.714				-2.053	0.100	1.000		-0.100		
Bizygomatic breadth	14.821	12.200	6.300	16.381	3.291	19.062	31.111	12.947	8.700	19.389	4.963	12.367	1.257	1.095
Basion–prosthion length	1.000	-0.100		-1.000	-1.000	-1.000	5.889	1.368		2.778		-0.233		
Alveolare–nasion length	2.714	2.200		4.333	1.528	4.375	20.222	8.158		11.778		6.900		0.504
Maxilloalveolar breadth	-5.179		-6.571				-30.556		14.333					
Portion–mastoidale	6.071	5.367	2.833	14.810			47.111	19.947	14.367	23.667	8.037			
Sectioning point (> in male)	2,676.39	2,592.32	1,296.39	3,348.27	536.93	5,066.69	8,171.53	4,079.12	2,515.91	3,461.46	1,387.72	2,568.97	579.96	380.84
Percent of sample correctly identified	86.6	86.4	86.4	84.5	85.5	84.9	87.6	86.6	86.5	87.5	85.3	85.0	86.4	83.1

	Based on White Americans			Based on African-Americans			Based on Japanese
	Function			Function			Function
Mandibular Measurements	1	2	3	4	5	6	7
Symphyseal height	1.390	22.206	2.862	1.065	2.020	3.892	2.235
Mandibular body height		-30.265			-2.292		
Maximum projective length		1.000	2.540		2.606	10.568	
Mandibular body breadth			-1.000			-9.027	
Minimum ramus breadth			-5.954			-3.270	1.673
Maximum ramus breadth			1.483			1.000	
Coronoid process height	2.304	19.708	5.172	2.105	3.076	10.486	2.949
Bigonial breadth	1.000	7.360		1.000	1.000		1.000
Sectioning point (> in male)	287.43	1,960.05	524.79	265.74	549.82	1,628.79	388.53
Percent of sample correctly identified	83.2	85.9	84.1	84.8	86.9	86.5	85.6

	Based on Japanese			Based on African-Americans		
	Function					Function
Cranial and Mandibular Measurements Combined	1	2	3	4	5	6
Maximum cranial length	1.000	1.000	1.000	1.000	1.000	1.289
Maximum cranial height	2.614	2.519		2.560	2.271	-0.100
Bizygomatic breadth	0.996	0.586	0.785	1.084	1.391	1.489
Alveolar–nasion length						
Porion–mastoidale						4.289
Symphyseal height	2.364				2.708	-0.987
Maximum projective length						-0.544
Coronoid process height	2.055	2.713	1.981	2.604		3.478
Bigonial breadth		0.661	0.404			1.400
Sectioning point (≥ in male)	850.66	807.40	428.05	809.72	748.34	718.23
Percent of sample correctly identified	89.7	89.4	86.4	88.9	88.8	88.3

SOURCES: After Giles (1970) and Hanihara (1959).

Table 9–4   The Skull and Postcranium: Determining Sex via Discriminant Function Analysis of Measurements Based on Japanese

Measurement	Function						
	1	2	3	4	5	6	7
Maximum cranial length	1.000			1.000	1.000	1.000	1.000
Maximum cranial height		1.000	1.000				
Femur: physiological length	0.107	0.031	0.176	0.138		0.220	
Scapula: length of glenoid cavity	6.644	4.390		8.117	−5.586	−3.816	
Ischiopubic index	−5.050	−2.654	−3.281	−5.156			
Atlas: maximum breadth	2.678		2.090		2.152	2.491	2.124
Sectioning point (> in male)	299.18	117.11	142.12	157.76	233.09	194.55	494.36
Percent of sample correctly identified	99.0	98.8	96.4	98.6	98.8	97.4	92.5

SOURCE: Based on Giles (1970).

Table 9–5   Discriminant Function Equations for Determining Sex Using Deciduous and Permanent Teeth, where >0 Indicates Male and <0 Indicates Female

1. 1.091 (bl rt di^1) + 1.500 (bl rt di^2) + 0.654 (bl lft dm^2) − 1.489 (bl lft dc^1) + 1.640 (md rt dc_1) − 20.342
2. 1.899 (bl rt di^2) + 1.174 (bl lft dm^2) − 1.750 (bl lft dc^1) + 1.653 (md rt dc_1) − 20.138
3. 1.625 (bl rt di^2) + 1.239 (bl rt dc^1) + 1.135 (bl lft dm^2) − 1.141 (bl lft dc^1) − 18.564
4. 2.084 (bl rt di^2) + 1.688 (bl lft dm^2) − 1.353 (bl lft dc^1) − 18.425
5. 3.079 (md rt dc_1) − 18.861
6. 0.574 (bl rt di^2) + 0.393 (bl lft dm^2) − 0.371 (bl lft dc^1) + 1.521 (bl lft M^1) − 21.314
7. 2.049 (md rt dc_1) + 0.887 (md lft M_1) − 0.516 (bl lft M_1) − 16.872

SOURCE: Modified from DeVito and Saunders (1990).
Abbreviations: bl, buccolingual; md, mesiodistal; rt, right; lft, left.

sacrum) is thought of as being broader; having a more vacuous pelvic inlet with less protrusion into its realm of, for instance, the ischial spine and distal sacrum and coccyx; having an obtuse subpubic angle; and having relatively longer pubic bones, deeper and more laterally flared ilia, and more obtuse and open greater sciatic notches (see Figures 9–20 to 9–27 and Table 9–8). Phenice (1969) has also identified an indentation on the inferior margin of the inferior pubic ramus as characteristic of females (in part because it has the effect of increasing the spread of the subpubic angle).

Inasmuch as generalizations of female/male differences are based largely on European material, it is not surprising that they do not hold up uniformly across all groups of humans, present and past. The sex or degree of sexualization (Acsádi and Nemeskéri, 1970) of an individual is therefore often determined as an average of features coded as female, male, or indeterminate. In order to take into consideration different factors that could produce female versus male features, Acsádi and Nemeskéri (1970) introduced a weighted scheme of sex determination. Here, features are scored as follows: +2, hypermasculine; +1, masculine; 0, indeterminate; −1, feminine; −2, hyperfeminine. (There are stereotypes, imaginary or real, of what constitutes "hyper" versus "regular" versus "who knows?" It is, however, perhaps more meaningful to try to determine these relative states of sexualization

Table 9–6   The Postcranium: Determining Sex via Discriminant Function Analysis of Measurements

Measurement	Based on White Americans				Based on African-Americans				Based on Japanese			
	Function				Function				Function			
	1 (Right)	2 (Left)	3	4	5	6	7	8	9 (Right)	10 (Left)	11 (Right)	12 (Left)
Femur: physiological length	1.000	1.000			0.070	1.000	1.000	1.980			1.000	1.000
Femoral head: maximum diameter	30.234	30.716			58.140	31.400	16.530				9.854	9.351
Femur: least transverse diameter	−3.535	−12.643									11.988	8.369
Femur: max. bicondylar breadth	20.004	17.565									4.127	3.575
Ischial length				0.607	16.250	11.120	6.100	1.000				
Pubic length				−0.054	−63.640	−34.470	−13.800	−1.390				
Sciatic notch width			−0.115	−0.099								
Acetabulosciatic breadth			−0.182	−0.134								
Acetabulum–innominate line length			0.828	0.451								
Anterior iliac spine–auricular surface			0.517	0.325								
Humerus: maximum length					2.680	2.450			1.000	1.000		
Humerus: biepicondylar width					27.680	16.240			8.726	6.198		
Clavicle: maximum length					16.090							
Humerus: midshaft circumference									7.394	3.221		
Sectioning point (> in male)	3,040.32	2,656.51	9.20	7.00	4,099.00	1,953.00	665.00	68.00	1,189.51	804.28	1,431.82	1,277.83
Percent of sample correctly identified	94.4	94.3	93.1	96.5	98.5	97.5	96.9	93.5	92.9	93.6	96.2	95.9

Table 9–6 (*continued*)

Measurement	Based on Japanese Function											
	13 (Right)	14 (Left)	15 (Right)	16 (Left)	17 (Right)	18 (Left)	19 (Right)	20 (Right)	21 (Right)	22 (Left)	23 (Left)	24 (Left)
Radius: maximum length	1.000	1.000										
Radius: midshaft circumference	1.917	1.273										
Radius: head circumference	2.991	3.163										
Radius: maximum distal breadth	9.126	7.711										
Ulna: maximum length			1.000	1.000								
Ulna: transverse shaft diameter			8.068	6.501								
Ulna: maximum capitulum diameter			5.551	2.881								
Tibia: length					1.000	1.000						
Tibia: maximum midshaft anterior–posterior diameter					4.264	2.954						
Tibia: minimum shaft circumference					7.544	5.605						
Tibia: distal epiphyseal breadth					12.213	10.212						
Scapular height							1.000	1.000	1.000	1.000	1.000	1.000
Scapular spine: length							6.335	1.899		1.929	1.846	
Scapula: length of glenoid cavity							12.664	11.922	10.940	6.949	7.107	6.800
Scapula: breadth of glenoid cavity							10.991			2.120		
Scapular breadth									1.350		1.494	
Sectioning point (> in male)	763.92	696.97	441.54	370.25	1802.10	1494.54	1660.16	782.10	634.75	669.79	611.03	508.35
Percent of sample correctly identified	96.7	97.0	88.9	90.5	95.7	95.3	96.8	96.0	95.6	94.8	94.7	94.1

SOURCE: Based on Giles (1970).

Table 9–7   Features of the Skull and Mandible that May Differ between Females and Males (Based on the Stereotype)

Feature	Female	Male
Overall	Anatomical details, muscle marking, and lines less marked and smoother; bone thinner	Details more marked, bone thicker
Skull	More gracile and rounded, smaller, lighter (avg. 595 g)	Frontal more sloping, larger, heavier (avg. 795 g)
Frontal eminences	Moderate to marked, raised to rounded	Weak to undistinguished, low to absent
Parietal eminences	Moderate to marked, rounded to pointed	Weak to undistinguished, low to absent
Facial skeleton	Narrower, smaller	Especially zygoma, more rugose and massive; broader
Zygoma	Surface low, smaller, more arced, contours less defined	Surface higher and thicker, tubercle and marginal process marked
Zygomatic arch	Typically thin, moderate; weakly scarred inferiorly	Heavy, thick; muscle scarred inferiorly
Supraciliary arch	Trace to moderate	Moderate to extraordinary (toral)
Glabellar region	Flat to moderately swollen	Moderate to prominent swelling
Orbits	Large relative to face; rounded to circular; sit high on face; sharp, thin superior edge	Smaller relative to face; square to rectangular; sit low on face; blunt, thick superior edge
Nasal margin	Less clearly delineated	More crisply demarcated
Nasal bones	Relatively smaller, less protrusive	Relatively larger, more protrusive
Anterior nasal spine	Smaller, thinner	Larger, bulkier
Alveolar margin (facial aspect)	Impressions of roots faint	Impressions of roots more marked
Palate	Shorter, rounder, flatter	Broader, longer, more vaulted
Cranial base	Flatter, less marked	More rounded, marked
Occipital bone (squamous part)	Smooth to traces of nuchal lines	Marked to roughened nuchal lines and occipital crest
Mastoid process	Generally narrower, pointier (low/narrow to high/pointed)	Generally more massive, broader, stubbier, (high/massive to broad/stubby/low)
External occipital protuberance	Smooth to weak	Marked to massive (studlike)
Mandible	Short, narrow, low, gracile, lightweight	Long, broad, high, robust, heavy
Mandibular (goneal) angle	Obtuse (>125°), rounded	More acute (<125°), rectangular
Goneal region	Surface smoother	Edge and surface of masseteric tuberosity more marked
Mental trigon	Rounded, smooth to somewhat delineated	Pronounced protuberances, protruding triangle or inverted "T"
Depth between incisors and mental trigon	Relatively short	Relatively long
Mandibular condyle	Smaller	Larger

Table 9–8   Features of the Articulated Pelvis, Os Coxa, and Sacrum that May Differ between Females and Males (Based on the Stereotype)

Feature	Female	Male
Complete pelvis	More lightly built, less muscle scarring	More robust with more muscle scarring
Pelvic aperture	Broader, more "lima bean"- to ellipse-shaped	Narrower, more "heart-shaped"
Subpubic angle	More obtuse (80°–85°), rounded, more "U"-shaped	More acute (50°–60°), narrow, more "V"-shaped
Ischiopubic ramus	More gracile, tapers toward pubic symphysis, roughened, edge everted	Deeper, flatter anterior surface
Sacrum (shape)	Broad triangle (broader superiorly more severe taper inferiorly), appears shorter	Narrow triangle (narrower superiorly, long taper inferiorly), appears more elongate
Sacrum (curvature)	Less pronounced and intrusive into pelvic aperture	More pronounced and intrusive into aperture
Lumbosacral articular facet	$<^1/_3$ of superior width, alae appear relatively large	$>^1/_3$ of superior width, alae appear relatively small
Sacroiliac articulation (auricular facet)	Extends to second segment	Extends to third segment
Ilium (blade)	More flared laterally, wide, low	More vertical, higher, narrower
Ilium (crest)	Less rugose, less sinuous path, anteriorly curve medially directed	More rugose, more pronounced outer lip, more sinuous path
Iliac auricular surface	Raised, narrow	Depressed, wide
Preauricular sulcus	Wider, deeper and shallow	Absent or more narrow
Postauricular area	Thin, smooth	Thick, rough
Postauricular center	Delineated (e.g., knob, bar)	Undistinguished, thick, rough
Postauricular groove	Common	Rare
Postauricular space	Wide, "loose"	Narrow, tight
Greater sciatic notch	"U"-shaped, broader, shallower, more open, more obtuse angle	"V"-shaped, narrower, deeper, more closed, more acute angle
Ischial spines	Shorter, less intrusive into pelvic aperture	More prominent and intrusive into pelvic aperture
Acetabulum	Smaller, oriented anterolaterally, diameter < distance from anterior edge to pubic symphysis	Larger, laterally oriented, diameter = approx. distance from anterior edge to symphysis
Obturator foramen	Edges sharp, triangular, low, wide	Blunter edges, longer vertical axis, oval
Pubic tubercle	Blunter, thicker, farther from pubic symphysis	Pointier, closer to symphysis

for each sample studied.) Some features are weighted twice as heavily as others (see Tables 9–7 and 9–8 for details of features). Characters that Acsádi and Nemeskéri (1970, p. 89) weight more heavily ($w = 2$) include the relative states of development of the glabella, supraorbital region, mastoid process, external occipital protuberance, orbital shape, supra-orbital margin, mental trigon, subpubic angle, greater sciatic notch, ischiopubic index, and diameter of femoral head. Other features ($w = 1$) include the frontal and parietal eminences, nuchal region, zygomatic arch, malar surface, body of mandible, gonial (mandibular) angle, mandibular head, articulated pelvis os coxa, sacrum, linea aspera, obturator foramen, and cranial bone thickness. Degree of sexualization is calculated using the following formula:

$$M = \frac{\Sigma wx}{\Sigma w}$$

where M is the mean value of the degree of sexualization, x is the score given a feature, and w is the weight of that feature.

Interestingly, as far as archaeologically and often forensically derived skeletal material is concerned, the only features of sexual dimorphism noted for the adult skeleton that appear to distinguish female from male neonates and children include degree of protrusion of the mental trigon, greater sciatic notch angle, and degree of sigmoidal curvature of the iliac crest (Schutkowksi, 1993; see Table 9–8). [Of these, the only feature that appears to be distinctive of our own species, *Homo sapiens*, is the development of a true "chin" (cf. Schwartz and Tattersall, 2000; Stringer et al., 1984).] Thus, it is not surprising that an Indo-European standard for distinguishing females from males would not necessarily be relevant to all groups of humans.

Nonmetric Variation, Sexual Dimorphism, and Populations

The expression of nonmetrically variable traits has been linked to sex, size, and developmental differences between individuals. The situation is complicated further by the fact that, in some instances, sex and size may be correlated, as indeed size and robustness often are. For example, with males being on average (even slightly) larger and more robust skeletally than females, males might be expected to exhibit higher frequencies of hyperostotic traits. Females, on the other hand, might be expected to exhibit hypostotic variants. A realization of this generalization is dependent on the degree to which sex and skeletal robustness are actually correlated within a given population.

Ultimately, however, all so-called nonmetric features represent differentially expressed as-

pects of an individual's skeletal and dental morphology, regardless of whether these features are significant at the level of the species, a population within a species, a sex within a population, or a specific individual within a population. In some cases, it may not even be possible to determine (or at least hypothesize) anything significant about a "variant" feature: it is just there; it is noise. Not every feature is significant, nor, if so, is every feature significant at the same level (e.g., species, population, or subset of population).

Perhaps we have been guided too long by the history of a discipline in which, for example, comparative anatomists and phrenologists sought to find features in teeth and bone that might be correlated with external differences between sexes or "races." But in reality, it is still a matter of debate just how many of the features that have been catalogued over the decades as being distinctive of one "race" or population or one or the other sex are truly distinctive of or unique to any, rather than being merely descriptive (and descriptive of more than one). All too often, our analyses are formulated and channeled by our inherited history, received wisdom, and expectations. We bring into our supposedly objective analysis a partially answered question. We might not think that this is the case. But even just knowing where a particular skeleton or skeletal population came from—temporally and/or geographically—or with what external trappings it may have been associated (e.g., burial goods, grave construction, type of dwelling) or just calling a collection of skeletons representatives of a population can provide an unconscious bias: certain people are supposed to be characterized by certain features or traits, or at least by different frequencies of different traits.

If, however, an osteologist were presented with a skeletal collection of unknown origin (and, thus, of unknown "racial" and sexual composition), she or he would have to start from scratch. Is this assemblage a sampling of individuals from the same species? (This

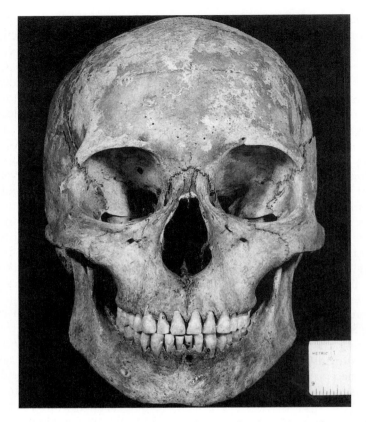

Figure 9–16   Skull exhibiting "mostly female" features (see text for discussion); also note dehiscence exposing right C^1 root and apical abscess below left I$_1$ (prehistoric, Alaska).

might appear to be an unnecessary question, but it is one that has to be broached, even if unconsciously.) If so, are all individuals from the same population? And are differences between females and males discernible?

In order to deal with this nested set of questions, the osteologist must first sort out the differential representation of morphological character states in the sample: Which features are shared by all individuals? Which by only some individuals? Then comes the task of determining whether any features were truly distinctive of a group or subgroups and, if so, whether within these hypothesized groups there were character states that could be attributed to sexual dimorphism.

By "features that might be distinctive of a group," I do not mean "features that are merely descriptive of a group" or "features other than those potentially due to sexual dimorphism that are expressed only in some individuals of a group." The former—"features that are merely descriptive of a group"—if typical of *Homo sapiens* (e.g., the development of a chin), cannot then also be distinctive of any subset of *Homo sapiens*. The latter—"features other than those potentially due to sexual dimorphism that are expressed only in some individuals of a group"—would not be distinctive of any subset of *Homo sapiens*, precisely because they are not expressed in all individuals. Furthermore, because these latter

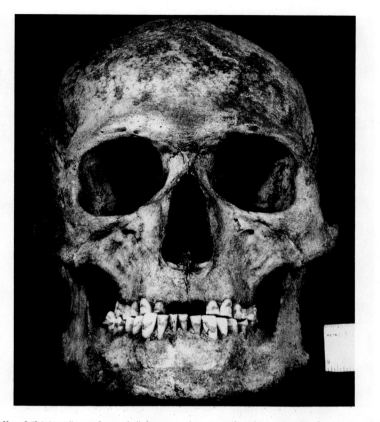

Figure 9–17 Skull exhibiting "mostly male" features (see text for discussion); also note varying degrees of calculus buildup on teeth (prehistoric, Alaska).

features would not be present in all members of a group and would thus not be distinctive of that group, their frequencies could not be determined until a group had been defined (on the basis of other characters or attributes, whether biological, cultural, linguistic, or material). In practice, a "population" is typically assumed on the basis of nonmorphological criteria and then "defined" on the basis of the relative frequencies or expressions of certain morphological traits—morphological traits which, in turn, have already been decided as being reflective of populational differences.

But this is circular. How can one know beforehand what features will be distinctive of any group? Furthermore, how can one claim to have delineated a group if, for example,

only 80% or 17% of its presumed members have a certain feature? This question applies to any group, whether a population or one of the sexes.

That so-called racial features fall under the aegis of nonmetric variation is self-evident. Referring to these features as population markers, or by some similar phrase, does not alter the possibility that subsets of *Homo sapiens*, indeed, may have been and/or are distinguishable from one another by certain traits that are unique to each "group." If we were discussing the systematics of a nonhuman animal—such as the Malagasy primate *Eulemur fulvus*—we would not be constrained by social or political overtones were we to find (as is the case) that subspecies are distin-

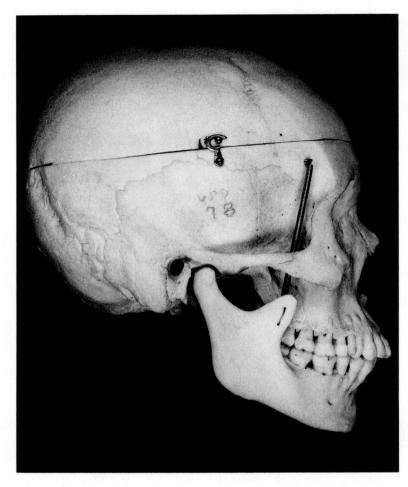

Figure 9–18   Skull exhibiting "mostly female" features (see text for discussion); also note fenestration exposing P¹ root and linear enamel hypoplasia (especially visible on lower teeth) (provenience unknown, University of Pittsburgh Dental School collection).

guishable on the basis of discrete and unique morphologies (Tattersall and Schwartz, 1991).

Problems arise in discussing the delineation of subsets of *Homo sapiens*, in part because of potential racist implications and in part because of the history underlying the study of "races" (see historical review in Schwartz, 1987, 1999b). In the course of forensic investigations involving personal identifications, "racial identity" becomes a matter of legal importance—which, then, gives a certain (although perhaps unreasonable)

credence to the assumption that skeletally expressed "racial" or populational differences will coincide with our individual and personal visual perceptions of how "we" differ from other humans and how "they" differ from "us." Indeed, the history of racial stereotyping on the skeletal level has often been that of trying to *find* traits "in the bones" that reflect how one divvied up the "races" on the basis of external appearance (however perceived or defined) (see review in Schwartz, 1999b; also Blumenbach, 1775 and 1795).

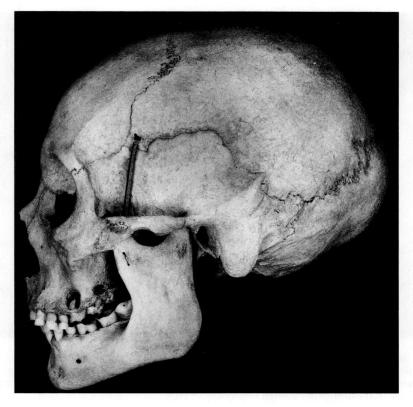

Figure 9–19   Skull exhibiting "mostly male" features (see text for discussion); also note severe dental attrition, alveolar remodeling, and apical abscesses in upper jaw (provenience unknown, University of Pittsburgh Dental School collection).

Stereotypes of different groups of humans give the appearance of portraying significant distinguishing features. For example, Inuits have been characterized as having narrow and unprojecting nasal bridges as well as flat, vertical, and very broad faces, while native Africans are supposed to have wide and moderately projecting nasal bridges as well as slender, narrow faces with rounded foreheads and alveolar prognathism (e.g., see review by St. Hoyme and İşcan, 1989; also Table 9–9). But a broader comparison reveals, for example, that not only Arctic populations but also other Native Americans as well as Europeans develop narrow nasal bridges and that southern Asians and Melanesians have narrow "African" faces (Table 9–9). Clearly,

nasal bridge and facial widths, respectively, distinguish by uniqueness neither an Eskimo nor an African. In fact, it would require broader comparisons among hominids and, preferably, among hominoids (the group that includes humans and their fossil relatives as well as apes and their fossil relatives) to provide the perspective necessary to judge whether the feature "narrow nasal bridge" that is *descriptive* of (probably at best most) Inuits, Native Americans, and Europeans *uniquely* characterizes them as a group among *Homo sapiens* or if "narrow face" *uniquely* delineates Africans, southern Asians, and Melanesians as a potential group within our species.

The question that must be raised about research into the delineation of subsets of

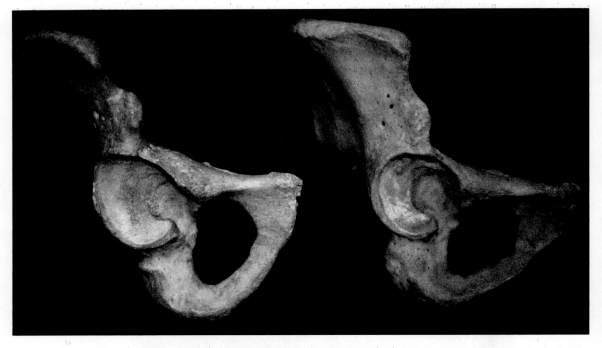

Figure 9–20 Male (*left*) and female (*right*) os coxae. Note especially differences in size and orientation of the acetabulum, depth of the ilium, shape of the pubic tubercle, length of the superior pubic ramus, obtuseness versus acuteness of the subpubic angle, length of and degree of indentation (*arrow*) in the inferior pubic ramus, and shape of the obturator foramen (University of Pittsburgh teaching collection).

Homo sapiens is, "What is the purpose of the inquiry?" Surely, if one is investigating "the peopling of the New World" or "the origin of modern human populational diversity," one is actually faced with a phylogenetic and systematic problem—albeit below the species level—that should be dealt with in the same way and with the same rigor and objectivity as one would approach the phylogeny and systematics of, for example, a group of potentially related species (Schwartz, 1995, 1999a).

Consider, for the moment, the oft-cited trait category "number of cusps on the first lower molar." Some apparently "natural" groups of humans have been characterized as having five cusps and other groups as having four cusps on M_1 (e.g., Turner, 1984). Indeed, these characterizations may be descriptively accurate. However, are these features really of equal significance in delineating each group on the basis of uniqueness? A survey of fossil hominids and extant and fossil apes reveals that the common condition among hominoids is the development of five cusps on M_1 (e.g., Schwartz, 1986). Thus, while it may be true that some human groups possess five-cusped M_1s, this feature is not unique to any of them (Schwartz, 1995; Schwartz and Brauer, 1990). In contrast, it is among that subset of *Homo sapiens* with four-cusped M_1s that we find the distinctive—and potentially systematically significant—variant (character state) of the category "lower first molar cusp number."

In other words, although a description may be accurate, phylogenetic significance—which is the only thing that counts in determining relatedness, whether in large-scale systematic analyses or measures of biodistance commonly practiced in bioarchaeology (e.g., Larsen, 1997)

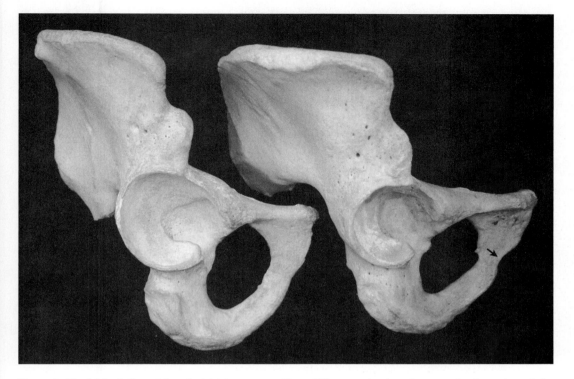

Figure 9–21   Male (*left*) and female (*right*) os coxae. Note differences in size, shape, and orientation of the ilium as well as features in Figure 9–20; also note exposure in this view of the wide greater sciatic notch in the female (University of Pittsburgh teaching collection).

—can only be hypothesized on the basis of the relative uniqueness with which a biological feature or features is shared: that is, *uniquely shared* either by individuals who are then hypothesized as constituting a biological group or by groups (each delineated by its own unique feature or features) that are then hypothesized as being related (Reed, 2006; Schwartz, 1995, 1999a). It is only *after* a "group" has been hypothesized on the basis of at least one biological feature that all individuals uniquely share (to the exclusion of all other *Homo sapiens*) that one can set to the task of describing individual variation and, eventually, compile frequency distributions of (incompletely shared) features within that group.

From this perspective, the problem underlying the common claim (e.g., Carbonell, 1963) that Native Americans are derived from an Asian group by virtue of their common possession of shovel-shaped upper incisors becomes obvious. While it may be true that Asians and Native Americans exhibit a relatively high frequency of shovel-shaped upper incisors, so too do Africans and Indians (e.g., see Figure 9–28), as well as most fossil hominids, orangutans and their fossil relatives, and even to some extent chimpanzees (e.g., Schwartz, 1988, 1997; Schwartz and Tattersall, 2002, 2003, 2005). The presence of shovel-shaped upper incisors among hominids, at least, is thus rather ubiquitous and certainly not distinctive of any subset of *Homo sapiens*. To the contrary, the *lack* of shoveled upper incisors [as often seen in Europeans (cf. Carbonell, 1963; Schwartz, unpublished data; Scott, 1972)] emerges as

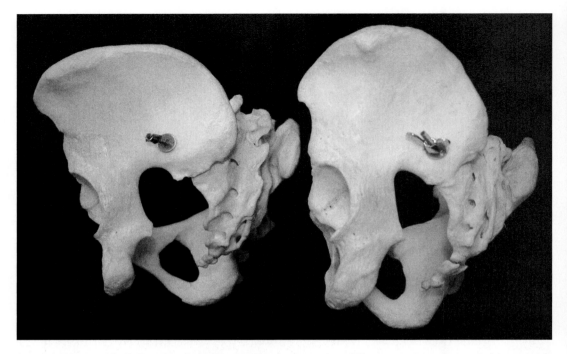

Figure 9–22 Female (*left*) and male (*right*) articulated pelves. Note differences in size, shape, and orientation of the ilium; degree of openness of the greater sciatic notch; size and orientation of the acetabulum; and inward curvature of the sacrum (University of Pittsburgh teaching collection).

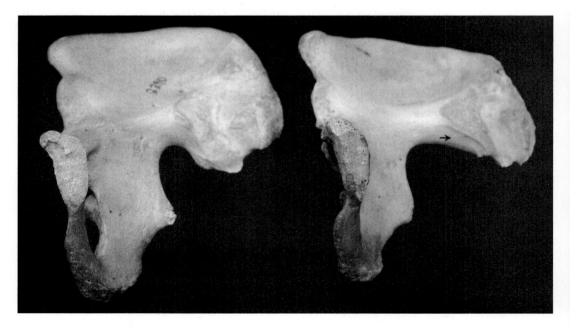

Figure 9–23 Male (*left*) and female (*right*) os coxae (internal view). Note differences in shapes of the pubic tubercle and inferior pubic ramus as well as in features in Figure 9–22; also note presence (*arrow*) versus absence of preauricular sulcus (University of Pittsburgh teaching collection).

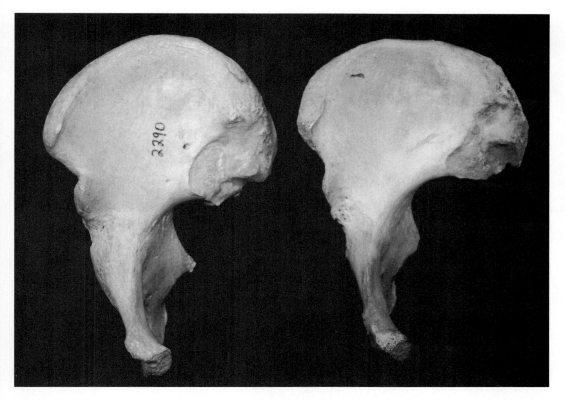

Figure 9–24   Male (*left*) and female (*right*) os coxae (superior view). Note differences in size, shape, and orientation of the ilium; openness of the greater sciatic notch; length of superior pubic ramus; shape of the pubic tubercle; and medial protrusion of the ischial spine (University of Pittsburgh teaching collection).

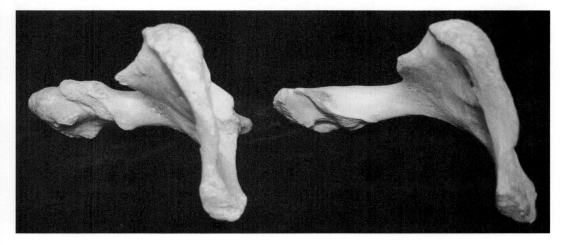

Figure 9–25   Male (*left*) and female (*right*) os coxae (posterior view, ischial tuberosity is on the *left*). Note differences in size, shape, and orientation of the ilium; length of ischium; and shape of ischial tuberosity (University of Pittsburgh teaching collection).

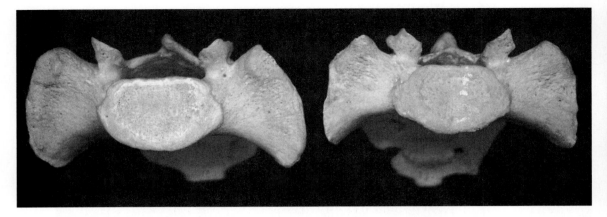

Figure 9–26   Male (*left*) and female (*right*) sacra (superior view). Note differences in the relative sizes of the articular surface and the alae, as well as of the inward curvature of the inferior end (University of Pittsburgh teaching collection).

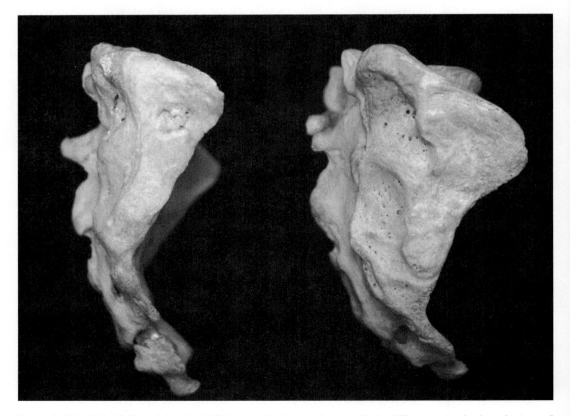

Figure 9–27   Male (*left*) and female (*right*) sacra (right lateral view). Note differences in the length, inward curvature of the inferior end, and extension inferiorly of the auricular facet (University of Pittsburgh teaching collection).

Table 9–9   Distribution of Cranial Features among Various Humans

Population	Nasal Bones	Nasal Bridge (W)	Nasal Bridge (Proj)	Prognathism	Subnasal Margin	Anterior Nasal Spine	Nasal Aperture (H/W)	Zygoma (W/Shape)	Bizygomatic/ Bifrontal (W)	Chin
S. Pacific	Reduced			Alveolar/dental	Deep pits					Rounded
Inuit		Narrow	Least	Face vert/very flat	?Distinct	Prominent	High/average	br/lg malar tubercle	Greatest/widest face	?Projecting
Native American		Narrow	?Least	Face vertical	?Distinct	Prominent	?High/average	br/lg malar tubercle	≤Greatest	?Projecting
European		Narrow	Greatest	Upper face/mid-face	Distinct	Prominent	High/average	Slender/triangular	Medium	Projecting
African		Wide/flat	Medium	Alveolar/dental	Dep/smooth/round	Short/blunt	Low/average	Slender/triangular	Smallest	Intermediate
Melanesian		Wide/flat	?Least	?Face vertical	?Distinct	?Prominent	Low/average	br/lg malar tubercle	Smallest	Rounded
Asia: North			Least	Face vertical	Distinct	Prominent	High/average	br/lg malar tubercle	Smallest	Projecting
Asia: South			Least	Face vertical	Distinct	Prominent	Low/average	br/lg malar tubercle	≤Indistinct	Rounded

SOURCE: Based on St. Hoyme and İşcan (1989).

Abbreviations: W, width; Proj, projection; H, height; dep, depressed; br, broad; lg, large.

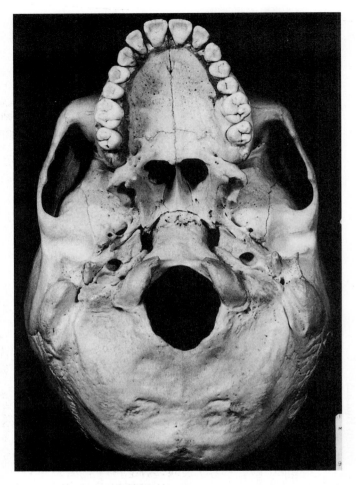

Figure 9–28   Example of shovel-shaped upper incisors; also note, for example, lack of definitive hypocones on both M2s, shape of palate, arc of zygomatic arch, bilateral expression of foramen of Vesalius, bridging of right foramen ovale, size of foramen lacerum, patent right and occluded left posterior condyloid canals; M3s are still unerupted and the spheno-occipital synchondrosis is not fully closed (recent, India).

the potentially phylogenetically significant character state.

On a broader level of inquiry, any feature of bone or tooth can be of potential significance in delineating a subset or group of *Homo sapiens*. But the degree of inclusiveness of that group—from the smallest possible subset to a group comprising many small groups defined on the basis of other features unique to them—will depend on the degree to

which the potentially significant variant is distributed within the species.

Rather than reiterating or seeking additional features to corroborate established stereotypes, it would seem more productive to approach the topic of human populational diversity in a different way (Reed, 2006; Schwartz, 1995). That is, instead of first deciding what the groups are—on the basis, for example, of biogeography or linguistic or cultural attributes or such "morphological" features as "skin color"—and then trying to find skeletal morphologies that appear to distinguish some individuals of one "group" from some individuals of another, it would be methodologically logical to begin by analyzing an array of different morphologies whose subsequently determined distributions of relative uniqueness within the species might yield a nested hierarchy of potential groupings. Ultimately, after perhaps many trials (during which characters that started off "looking good" were discarded because they varied too widely to be "characteristic" of more than a few individuals), one might be fortunate to delineate uniquely shared features that distinguish subsets of *Homo sapiens* (of whatever size or inclusivity) from one another. If the investigator chooses to identify these distinctive features as those of "race" or "population," that is her or his prerogative. In the end, some "groups" might conform to "expectation," but other emergent groupings or their hypothesized relationships to one another might be totally unexpected. At the level of inquiry into the potential subspecific systematics of our own species, however, the issue is—or should be—devoid of negative connotations.

The same theoretical and practical concerns should be brought to bear on determining an individual's sex from osteological remains. For although stereotypes of skeletally expressed traits of "maleness" and "femaleness" abound in the literature, operationally the determination of an individual's sex is achieved as an average or a weighted average of the total number of traits over the range of their degree of development. In some groups (e.g., northern Asians and Arctic Native Americans), females present more features of "maleness" (e.g., supraorbital, glabellar, zygomatic, and mandibular thickening or distension) than would be expected from the perspective of an Indo-European standard of comparison. [St. Hoyme and İşcan (1989) give other examples of the problem of separating the

analysis of sexual dimorphism from that of populational differences (e.g., chin shape).] Obviously in such cases, the total number of "traditionally" sexually diagnostic or revealing features would be diminished noticeably. On the other hand, a few pelvic and mandibular features attributable to sexual dimorphism appear reliable in distinguishing females from males in any population of *Homo sapiens* (e.g., Schutkowski, 1993; Schwartz and Houghton, unpublished).

Although the obscuring of cranial clues to the sexual identity of females in populations such as Arctic Native Americans may be viewed as a hindrance to analysis—when the goal of the analysis is to discriminate females from males—what this situation actually reflects is that certain nonmetric traits are expressed more equally in both sexes of some populations than they are in females and males of other groups. Since all potentially sexually dimorphic traits will not be expressed similarly —with equal intensity, clarity, or frequency —among different populations, it might be worthwhile to deal with the determination of sex as a subset of the overall study of the differential expression of skeletal features (i.e., of nonmetric variation). Thus, if skewed representations of certain morphologies do indeed reflect populational differences within our species, then sexual differences within each population (each, as suggested above, defined by its own unique feature or suite of features) would be delineated by other morphologies (Schwartz, 1995). It is the responsibility of the investigator to approach each sample without preconceptions about what she or he will delineate as diagnostic at either the populational or sexually dimorphic level of the analysis.

A Discussion of Metric Approaches to Biodistance Analysis

In addition to nonmetric features, metric approaches have been used to try to unravel the "genetic" relationships of individuals and groups. In fact, according to Pietrusewsky

(2000, p. 384), "there is some consensus . . . that biological relatedness as measured by biological distances based on metric data reflects genetic similarity overall." This is an interesting twist to the argument since the fundamental question is, "Does overall genetic similarity actually reflect closeness of relatedness?" (e.g., Schwartz, 2005a). In spite of a general consensus on this assumption, the answer is actually "no," or at least "probably not and not in all cases" (Schwartz, 2005b).

Although a measurement (traditionally linear but recently, because of digitizing units such as the Microscribe 3D digitizer, three-dimensional) or an index calculated from measurements is a metric representation of size (length, height, or breadth) or shape, it is referred to as a "trait." Typically, metric traits are contrasted with nonmetric traits in that they are considered "continuously variable" [as appears to be the case when one arranges people or ears of corn from smallest to largest (see Morgan, 1916)], whereas nonmetric features are considered to be "discrete," "discontinuous," or "quasi-continuous." As with interpreting a chronological ordering of seemingly similar fossils as representing gradual change or the sudden emergence of novelty (see Morgan, 1903, and Schwartz, 2006), the "reality" of a distinction between metrically "continuous" and nonmetrically "discontinuous" depends, I suggest, on one's a priori perspective.

Larsen (1997, pp. 331–332) defended the viability of biodistance analysis (whether based on nonmetric morphology or shape and size) in "identify[ing] patterns of biological relatedness between and within populations on the grounds that it "is not simply a modern attempt at racial typology." Rather, he suggested that it "reveal[s] continuities and discontinuities that are valuable for showing how past populations were structured and for interpreting key biological trends when viewed in a temporal or spatial perspective, including disease history, activity patterns, dietary changes, and other param-

eters." While an ideal to strive for, there are two questions that need to be addressed.

In a very accessible overview of metric approaches to biodistance analyses, Pietrusewsky (2000) summarizes the statistical methods that have been used: for example, principal components analysis (PCA), discriminant (canonical) analysis, Q- and R-mode analyses, Euclidean distance, and Mahalanobis' generalized distance or D^2 (which he considers the only realistic metric measure of biological distance), which is based on the squared Euclidean distance. Stepwise discriminant analysis (in which the first transformed variables account for most of the variance between groups and subsequent variables the residual variation) has been used to assign a specimen to a group and to evaluate group distance. With regard to the former, a specimen is assigned to a group on the basis of the discriminant scores it receives, which reflects the probability of group membership as well as the likelihood of group membership in terms of the average variability of all groups analyzed. Many statistical programs available to perform these tests (e.g., the commercially available BMDP, SAS, SYSTAT, and NTSYS) can also cross-check group assignment by assessing the probability of a specimen being misclassified (e.g., by running the statistics on a subsample of the original).

While discriminant analysis is based on criteria that maximize differences among groups, PCA attempts to delineate common patterns of variation among groups via investigation of their shared underlying factors (represented along x, y, and z axes). Two programs that rely not on discriminant analysis but PCA and Euclidean distances to assign skulls of unknown "identity" are CRANID2 and POSCON, which incorporate Howells' (1989) extensive craniometric data. Howells' (1995), however, prefers discriminant analysis as the basis for using Euclidean distances to place skulls into his craniometric world "map."

The one element that most analyses require, however, is the a priori assignment of specimens to groups (Pietrusewsky, 2000, p. 384). That is, one must first decide that certain skulls are, for example, "Melanesian," "Australian Aborigine," "Northern Asian," etc., before running the analyses. And this series of assumptions, then, leads back to the problem discussed above with regard to the a priori identification of specimens belonging to certain "populations" on the basis of non-biological criteria: how do you know that the specimens you are working with constitute a "group"? This problem is further exacerbated by the fact that, even if one assumes that a priori group assignment of "known" specimens is accurate and, thus, the discriminant or other functions that one calculates are reliable indicators of group membership, subsequent assignment of "known" specimens to these predetermined groups is not always successful (see examples in Pietrusewsky, 2000).

This, however, is not necessarily surprising. Among the earliest uses of multivariate statistics to assign specimens to groups of known identity was Giles and Elliot's (1963) delineation between human females and males via discriminant analysis. While statistically viable, a relatively reliable method of discriminating between females and males (or assigning specimens to one sex or the other) does not, however, warrant application of the same approach to determining membership by a specimen of either sex in a group (subset) of *Homo sapiens*. Indeed, differences between human females and males (and the features that allow allocation of specimens to either group) transcend differences between possible groups within the species. Although the approach to both problems is the same—one must first decide which specimens are female and which are male or, for example, which are Ainu and which are Jomon—investigation of the former is not equivalent to determining the latter. Rather, while an individual's sex is biological (and thus should be reflected in discretely female versus male morphologies), being Ainu or Jomon or being Melanesian or Australasian may not (at least not necessarily)

be reflected in one's morphology. As argued above, investigating the latter is a systematic problem and must be addressed as such.

In turn, this question leads to another regarding another assumption of metric biodistance analysis. As Pietrusewsky (2000, p. 383) admits for cluster analysis and clustering algorithms and as is clearly true of other multivariate statistical analyses, "numerical (quantitative) methods for grouping [are] usually based on overall morphological (phenetic) similarity where each character, when assigned some numerical value, is considered to have equal weight." Regardless of the statistical approach, if the comparison yields results in terms of degrees of difference or similarity, it is only assessing "overall similarity." Yet, as I argued above with regard to interpreting nonmetric traits and as is appreciated by systematists of nonhumans, the demonstration of overall similarity does not alone constitute a demonstration of relatedness. Rather, as with morphology, one must tease apart the relative meaning of delineated similarity: is the similarity unique to the few specimens or samples being compared, or is it also demonstrable in the context of a wider comparison among specimens or samples (see the discussion of St. Hoyme and İşcan, 1989, above); if so, what does this mean metrically? The issue aside of whether the suites of measurements typically used in biodistance analyses (e.g., see review by Pietrusewsky, 2000; also Howells, 1989, 1995) are biologically relevant, the problem this question delineates is obvious when linear measurements are used because differently shaped crania, skeletal elements, or parts thereof can have the same maximum lengths or breadths. But the problem does not disappear when comparing three-dimensional shapes if one does not discriminate between the more widely and the more restrictively shared configurations. However, in order to determine how widely or restrictively any feature is shared by some number of our species, we must look beyond *Homo sapiens* to other hominoids, if not all catarrhines (Schwartz, 1995)—and it is for this reason that while determining sex may be accomplished via a within-species comparison, determining group membership cannot. In short, there is no methodological, theoretical, or philosophical reason why the same logic should not apply to nonmetric (i.e., morphological) and at least three-dimensional metric (i.e., shape) comparisons.

Mind you, the thrust of this intellectual exercise is not to discourage attempts to use metric—particularly three-dimensional—data in hypothesizing theories of relatedness. Rather, it is to point out that overall comparison for its own sake is not necessarily a reflection of closeness of relatedness (nor should it be).

*   *   *

In the preceding discussion, I have attempted to articulate a systematically rigorous approach to questions of which humans cannot seem to rid themselves: "Who are you?" "Who are they?" "Who am I?" It would perhaps be less unsettling if we could compile a list of traits that will always delineate populations as some features seem consistently to distinguish between the sexes. The challenge at hand, though, is to acknowledge that this will likely never happen, that even in the twenty-first century we still have much to learn about the shortcomings of skeletal analysis (and thus also about forensic and bioarchaeological analyses), and that even with all the sophisticated technology at our disposal, there is still much to refine in method and theory. Recording nonmetric traits or taking measurements for their own sake or only within the confines of a site report does not go beyond the level of isolated description. Description is only the beginning of a meaningful analysis of within- and between-population nonmetric variation.

Chapter 10

Pathology
Disease, Trauma, and Stress

Overview

The topics covered in this chapter—disease, trauma, and stress—can be characterized as assaults on an individual's body, in contrast to conditions often referred to as pathological (discussed elsewhere) that result from developmental, endocrine, or nutritional disturbances or imbalances (see Chapter 1, on bone, and chapters on individual bones). But while one can discuss disease and trauma separately, and both apart from stress, it is equally obvious that one can impact or lead to another—for example, infectious disease can arise secondarily from a traumatic or stress-related assault on the body. The problem with identifying any potential pathological agent is that it may not leave its mark on bone, even if (and perhaps especially because) it is lethal. In reality, the ability of the forensic osteologist or pathologist to identify any particular assault on bone is dependent on the injured individual's ability to survive the assault (e.g., Wood et al., 1992).

Of course, this note of negativity can be partially modified. Certain types of trauma will leave instantaneous signs on bone and teeth—for example, intentional mutilation or modification, complete/compound fracture, weapon wounds. But because the effects of other kinds of trauma—such as stress or fatigue fractures—as well as those of occupational stress and various infectious diseases accumulate over years, they may only be detectable if the individual lives longer. Thus, the number of potential assaults on an individual's body is greater than the number that *might* leave telltale signs on bone or tooth, which in turn is greater than the number of assaults that actually *do* leave their imprints on hard tissue.

Compounding potential difficulties in correctly assessing or even being able to assess certain kinds of assaults on the skeleton is the taphonomic history of the skeleton itself. For example, was it intentionally buried or had it lain exposed to the elements? If buried, was burial primary or secondary? What climatic/environmental conditions confronted it? What were the pH and other characteristics (e.g., wet, saline, peaty) of the surrounding soil? If excavated, washed, and/or cleaned, what techniques and implements were employed? Any of these could alter, if not obliterate, surface detail or even destroy fragmentary bone.

Care must be taken to distinguish between postmortem damage and perimortem assaults (Figures 10–1 to 10–5). On a gross level, postmortem breakage or puncturing of excavated bone, as well as of enamel and dentine, is often indicated by a difference in color between the edge created by the break and the rest of the bone or tooth structure. The broken surface (i.e., newly exposed bone or tooth surfaces) will usually be lighter in color. Postmortem breakage or impact also tends to produce fragments with edges that are unnaturally angular or serrated. Weathering (e.g., exposure to freezing and thawing, wetness and aridity) as well as acidic conditions can lead to loss of outer layers of cortical bone; teeth can split and enamel flake off. Weathering can create a multilayered, terraced effect on diaphyseal cortical bone and expose spongy bone beneath the thinner compact bone of epiphyseal regions; in addition to breakage, weathering can lead to exposure of tooth roots through thin alveolar bone, producing an artifact that might be mistaken for abscessing. Conditions that lead to rapid demineralization

but slow remineralization of bone (e.g., acidic clays and sandy soils) produce light, friable bone. Until the osteologist gains experience in detecting postmortem effects on bone and tooth, she or he should be suspicious of attributes such as "lack of symmetry," "irregularity," "angularity," or "sharpness" (of edges, corners). Even plant root marks etched on the surfaces of bones tend to look "unnatural."

The discrimination of one infectious disease from another—or any malady from an array of maladies that might leave seemingly similar clues—is referred to as *differential diagnosis*. The process of differential diagnosis requires a systematic elimination of the least and then less likely infections or other possible causes of the features observed. Since some diseases may affect the skeleton randomly (e.g., fungal infections), while other diseases will differently affect specific sites on bones, individual bones, and/or groups of bones, recording this information constitutes an initial step in differential diagnosis (Ortner, 2003c).

But while the clinical diagnosis of disease can achieve a high degree of certainty (e.g., from early, observable signs to precise laboratory tests), diagnosing disease from bony clues

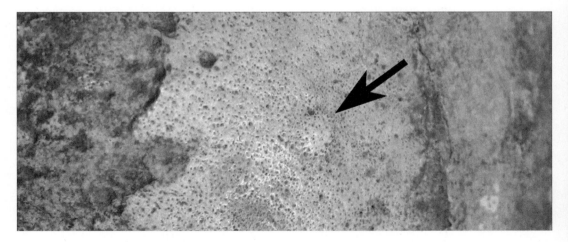

Figure 10–1   Postmortem damage to skull. Note how the flaked-off layer of bone creates a terracing effect, adjacent to the sclerotic outer layer of bone, which is affected by cribra crania (cranial hyperostosis) and bears a small osteoma (*arrow*) (prehistoric, Pennsylvania).

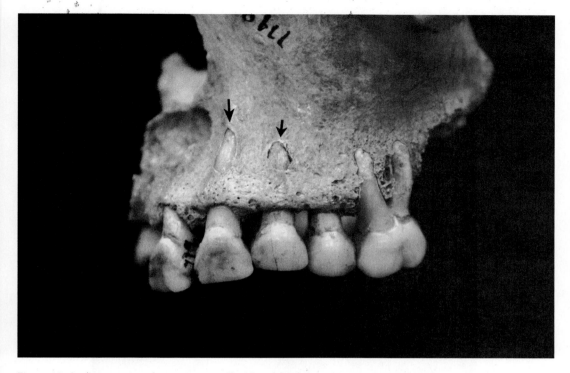

Figure 10–2   Postmortem damage to maxilla. Note false fenestrations exposing tooth roots (*arrows*); also note large abscess inferior to the nasal aperture and calculus on many crowns (prehistoric, Pennsylvania).

can be highly inaccurate and often seemingly subjective. Ultimately, and for whatever reason (e.g., state of preservation of the skeletal material, the parts of the skeleton preserved, the age and/or sex of the individual, how advanced the disease was at the time of death), even a careful differential diagnosis might not be able to whittle down the choices to fewer than two or three alternative diseases or conditions. A choice between alternatives might then be suggested by criteria such as age, sex (including, e.g., sex-related immune response), and/or geography as well as by cultural, behavioral, or other inferred and reconstructed aspects of the population [e.g., populational density (tuberculosis is associated with crowding), agriculture (certain fungal infections, e.g., blastomycosis, are soil-borne and may be expressed in one sex more than the other depending on division of labor), types

of domesticated animals maintained (certain bacteria, e.g., *Brucella*, can be transmitted across species and may be expressed in one sex more than the other depending on division of labor)] (Ortner, 2003e; Schwartz, 1995).

It is often difficult (even for the experienced osteologist) to decipher and apply with confidence the terminology, descriptions, and even illustrations in available reference works. Since this is not a paleopathology textbook, no attempt is made to be all-inclusive in the description or illustration of disease [for that the reader can turn to Aufderheide and Rodríguez-Martín (1998), Ortner and Putschar (1981), Ortner (2003c), and Rothschild and Martin (1993)]. The information provided here is meant primarily to be introductory and the discussion is intended to focus thought on approach and method and to provoke continued research in this field. Details of most of

Figure 10–3   Postmortem damage to skull. Note light-colored bone exposed where breaks occurred; also note various fields of healed lesions (*arrows*) (prehistoric, Pennsylvania).

the diseases and conditions discussed below are summarized in Table 10–1. The accompanying figures show examples of types of lesions central to the diagnosis of diseases: lytic (Figure 10–6), resorptive (Figure 10–7), proliferative (Figure 10–7), and erosive (Figure 10–8).

Ortner (2003c, p. 49) recommends an initial evaluation

to determine whether the abnormality consists of (1) a solitary abnormality in which there is a single focus for the pathology, (2) a bilateral multifocal abnormality in which defective bone tissue is apparent in two or more sites within a skeleton, (3) a randomly distributed multifocal skeletal pathology, (4) a diffuse abnormal reduction of bone mass that is distributed throughout the skeleton although not necessarily equally in all areas, and (5) a local or generalized disturbance of normal size or shape in bone that may be accompanied by abnormal quality of bone tissue.

Infectious Diseases of Bone

Although an infection may lead to death, it may not detectably alter the bone. As Kelley (1989) summarized, the following infectious diseases can potentially produce telltale lesions on bone: actinomycosis, blastomycosis,

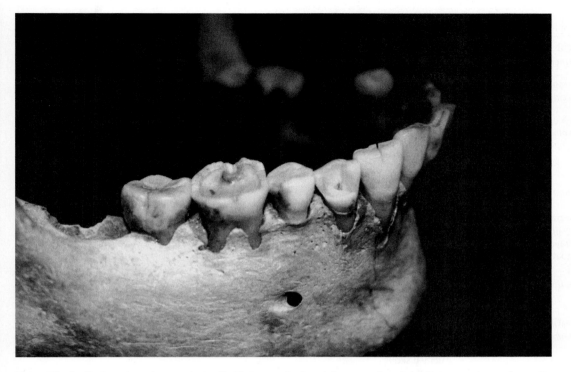

Figure 10–4   Postmortem damage to teeth. Note angular break in enamel and dentine exposing pulp cavity of right P_1 (compare the "clean" look of the exposed dentine compared to darker dentine on other teeth that was exposed through tooth wear); also note the broadly "V"-shaped occlusal wear of right I_1 and I_2, which reflects repeated anterior tooth use (prehistoric, Pennsylvania).

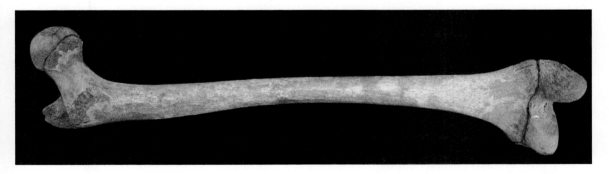

Figure 10–5   Postmortem damage of femur. Note irregular, lighter-colored patches of flaked cortical bone surrounded by darker, intact cortical bone; also note unfused epiphyses (prehistoric, Pennsylvania).

brucellosis, coccidioidomycosis, cryptococcosis, echinococcosis, histoplasmosis, leprosy, osteomyelitis (suppurative/pyogenic and non-suppurative/nonpyogenic), periostitis, polio-myelitis, smallpox, sporotrichosis, treponemal infections (treponematosis) [syphilis (endemic/nonvenereal and venereal) and yaws], tuberculosis, and typhoid spine. Of course, since the diagnosis of disease derives from studying the effects of known, identified diseases, it is

Table 10–1 Summary of Major Diseases that Impact the Skeleton (not Including Osteomyelitis or Tumors, See Text for Fuller Description)

Disease	Features	Skull	Mandible	Ribs
Fungal				
Actinomycosis	Min. reactive new bone, many small resorptive foci		1° "lumpy jaw"	3° resorbed
Blastomycosis	Lytic lesions, soil-borne, inhaled, N. America	3°		2°
Coccidioidomycosis (San Jaoquin Valley fever, valley fever)	Inhaled/abrasions; 1° males, resorptive lesions, bony prominences, Meso- and S. America			x
Cryptococcosis	Lytic lesions, bony prominences, 1° Europe	1°		
Histoplasmosis	Murky lytic lesions, <1 yr or adult, Mississippi/Ohio Valleys			
Sporotrichosis	Thorn punctures; 1° males; hematogenous; periosteal lesions; latitudes 50° N, S	4° cranium, face		5° (including clavicle)
Bacterial (other than osteomyelitis)				
Brucellosis (Mediterranean, goat's milk fever)	Multifocal lytic lesions; min. reactive bone formation; dog, goat, pig, cow; 2° leads to osteomyelitis, suppurative arthritis			
Viral				
Smallpox	1° infants, children (bone); destruction of metaphyses; separation of epiphyses; no sequestrae; primary periostitis			
Parasitic				
Echinococcosis	Tapeworm; dog, pig, sheep, cow; erosive, expansive lesions; monostotic; calcified sacs			
Granulomatous (*unknown etiology*)				
Sarcoidosis	Youth/ young adult; lesions multiple, lytic, round, 1° distal epiphysis, 1° bilateral; no reactive bone formartion; similar to lepromatous leprosy	2° cranium, especially nasal bones		
Joint				
Ankylosing spondylitis (spondylarthropathy)	Syndesmophyte formation at vertebral margin, symmetrical; few peripheral joints (pauciarticular); ankylosis; young individuals; 3 males: 1 female			Costovertebral joints
Osteoarthritis	1° diarthrodial and weight-bearing joints			
Spondylosis deformans	Perpendicular osteophytes, diarthrodial joints do not ankylose			
Rheumatoid arthritis	Symmetrical, polyarticular (virtually all peripheral diarthrodial joints, no ankylosis), periarticular loss of trabecular bone			
DISH (diffuse idiopathic skeletal hyperostosis)	Typically males >50 yr; no reactive sclerosis or erosion; ossification at sites of insertion of ligaments, tendons, joint capsules	x	x	x
CPDD (calcium pyrophosphate deposition disease) (crystalline arthritis)	Often precedes osteoarthritis; primary CPDD correlated with aging (30–75 yr); pauciarticular, subchondral erosions not crisp; trabecular bone remains dense; no reactive bone formation; zygoapophyseal, costovertebral joints spared			

Abbreviations: x, common site of disease; 1°, 2°, 3°, etc., typical frequency/order of sites of disease; PIP, proximal interphalangeal; DIP, distal interphalangeal; MP, metacarpophalangeal; min., minimally.

Vertebrae	Pelvis	Leg	Arm	Hand	Foot
2° neural arches, centra (thoracic, lumbar)					
1° centra, 2° neural arch		3° tibia		3°	3°
Centra, neural arches, spines 1° C1/2		Long bones	Long bones	Diaphyses	Diaphyses
		1° diaphyses, cortical thickening, rarified areas	Same as leg		
5°		1° knee and tibia	3° ulna, radius	2°	2°
1° centra, resorptive lesions	1° sacroiliac joint, resorptive lesions	2° diaphyses, periostitis	Same as leg		
		3° knee joint	1° elbow joint, including radius, bilaterally	2° wrist joint	2° ankle joint
Centra (may lead to gibbus)	x	Long bone metaphyses	Same as leg		
2° centra (not disc or pedicle), ≥1 involved (not necessarily contiguous)	2°	2° tubular bones	Same as leg	1° fingers	1° toes
Zygoapophyseal joints, centra squared up					
Zygoapophyses	x		x	Especially carpal–metacarpal, PIP, DIP	x
Lipping of centra					
1° juncture of C1/2 (when spine involved)			1° ulnar styloid	1° carpal, metacarpal–phalangeal, and PIP joints	1° metatarsal–phalangeal, PIP and ankle joints
1° anterior spinal ligament (distinct from centrum)	x	x	x	x	x
Calcification of nucleus pulposus or annulus fibrosus		Joints at cartilage margin	Joints at cartilage margin (shoulder, radiocarpal, elbow)	Sclerotic margins, cysts communicating with surface, MP, PIP, DIP joints; subchondral erosion 1° MP, PIP, DIP	Subchondral erosion 1° PIP, DIP; ankle, midtarsal

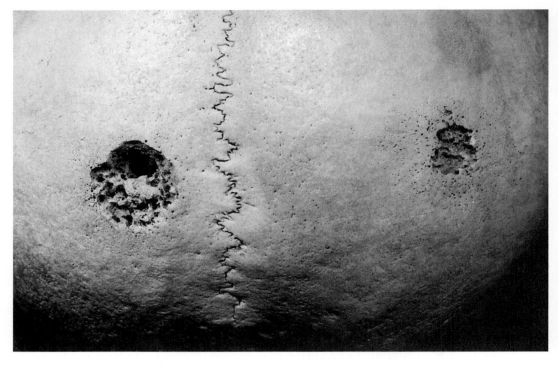

Figure 10–6   Examples of lytic lesions on cranial vault (postcontact, Siberia).

always possible that lesions on the skeletons of past populations represent a now extinct or unknown infectious disease.

More importantly, as recognized by Kelley (1989) and others (e.g., Buikstra and Ubelaker, 1994; Mensforth et al., 1978), what is still lacking in the field of paleopathology is a precise terminology shared even by experts. Mensforth et al. (1978) suggest describing lesions as "active," "healed," "unremodeled," or "remodeled." Kelley (1989) distinguishes lesions as being "resorptive," "lytic," "proliferative," "periostitic," "fused," "healed," or "unhealed." Clearly, there is more information in delineating, for example, resorptive from lytic lesions and less in distinguishing only between remodeled and unremodeled states. But there is also information to be had in describing the characteristics of the bone that surrounds a lesion (e.g., see Buikstra, 1976; Rothschild and Martin, 1993).

Infection leads to inflammation of tissues. **Osteitis** refers to inflammation of bone, which can result from various insults to the skeleton. Acute bacterial infections are generally indicated by pus formation and, in bone, are reflected in the development of *abscesses* (lacunae or pockets caused by bone death or necrosis) and drainage holes (*cloacae*). Viral infections, which usually do not lead to pus formation, typically do not leave such bony clues.

Inflammation of the periosteum leads to **periostitis** (Figure 10–9); inflammation of the endosteum leads to **endostitis**. The normal surface morphology of cortical bone is disturbed such that the reactive, unhealed bone appears more porous and lamellar. Healed periosteal lesions are denser, less porous, and more sclerotic (i.e., thicker, denser) but often retain their distinctiveness from surrounding, undisturbed cortical bone. Periosteal inflammation

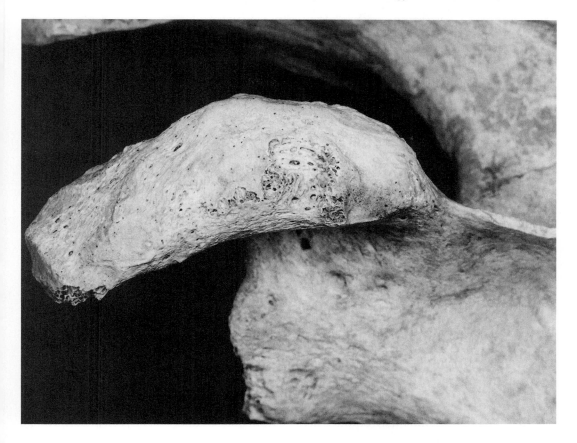

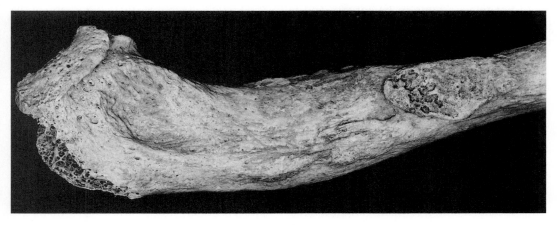

Figure 10–7   (*Top*) Coracoid process of left scapula with resorptive lesion surrounded by sclerotic, remodeled bone; (*bottom*) inferior view of right clavicle displaying proliferative lesion of acromial end, lateral to which is a healed, resorptive lesion and then reactive bone (prehistoric, Pennsylvania).

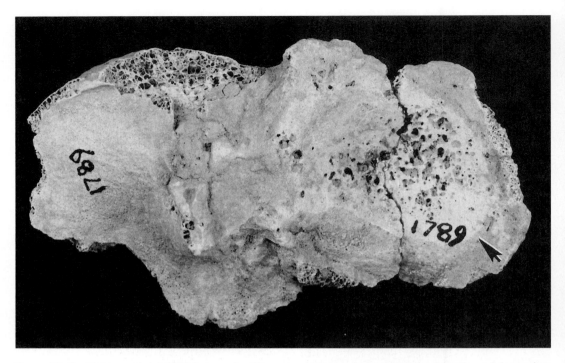

Figure 10–8   Right proximal tibial fragment with erosive lesion and eburnation on lateral condylar surface (*arrow*) (prehistoric, Pennsylvania).

due to infection or trauma may produce **primary periostitis**. Primary periostitis typically does not affect a bone in its entirety and is thus localized or unevenly distributed. Often, the affected regions are thickened or appear to have an additional layer applied or adherent to them. **Secondary periostitis** arises in conjunction with or as a result of some other disease (e.g., osteomyelitis), which can spread to or otherwise affect the periosteum. Thus, for example, **secondary periostitis** in concert with osteomyelitis is distinguished from **primary periostitis** by its association with features consistent with osteomyelitis (i.e., abscesses, involucra, sequestra, and cloacae). One should attempt to distinguish between primary and secondary periostitis.

Infection of bone (**osteomyelitis**) may be due to the introduction of bacteria locally (i.e., from an adjacent wound or infected soft tissue) or via the bloodstream (*hematogenously*)

from a site of infection elsewhere in the body. *Hematogenous osteomyelitis* typically affects one (and much less frequently two or more) of the long bones, especially the femur and tibia (Ortner and Putschar, 1981). Since it is transported by the vascular system, hematogenous osteomyelitis arises in the medullary cavity of a bone; in young individuals, it tends to localize in the metaphyseal regions of growth. Infection spreads through the medullary cavity, destroying spongy bone. Cortical bone death eventually occurs because pus and other exudates expand within marrow spaces and eventually cut off the blood supply. Loss of blood supply can lead to bone death (*necrosis*) and eventually the production of islands of dead cortical bone (*sequestra*, s. *sequestrum*). Often, a sequestrum is surrounded by thickened, hard or hard-looking (*sclerotic*), heavily vascularized bone (identified as *involucrum*, pl. *involucra*), which the periosteum produces

Figure 10–9   Left and right tibiae (*left* and *right*) with (apparently primary) periostitis (see text for discussion) (prehistoric, Pennsylvania).

in response to the infection. Ultimately, expanding pus may erupt to the surface of the bone through an opening (*cloaca*) in the involucrum (Figures 10–10 and 10–11). Often, a sequestrum is found in the cloaca's aperture. In extreme cases, the entire diaphysis may become a sequestrum that is surrounded by a sheathlike involucrum perforated by one or more cloacae.

Osteomyelitis may be suspected in tubular bones in which the diaphysis appears unnaturally and asymmetrically swollen or expanded along its length. This would constitute the site of the involucrum. In addition, one metaphyseal region may appear unusually swollen and the other unusually thinned. The bone's external surface can also be reactive and variably textured (e.g., nodular, stringy, fibrous, layered—rather than relatively smooth as in normal cortical bone), as well as thickened and even harder and more compact than usual.

It may be difficult, however, to diagnose osteomyelitis arising from infection spreading to underlying bone from soft tissue because the infection may not invade the medullary cavity of the bone. Such osteomyelitis would,

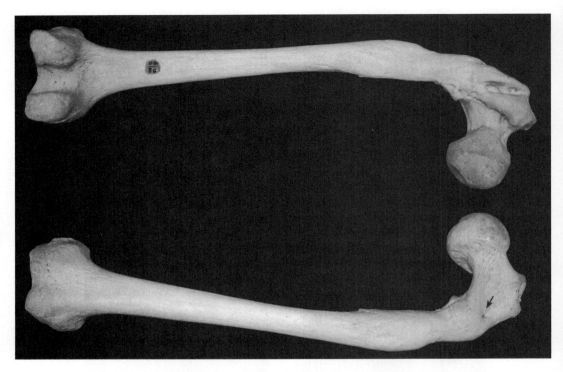

Figure 10–10 Left femur (*top*, posterior view; *bottom*, anterior view) with healed fracture (probably spiral and compound) and evidence of infection (osteomyelitis); note drainage hole (*arrow*).

therefore, be more localized and not necessarily accompanied by the typical signs of the infection (abscess, involucrum, sequestrum, cloaca). Furthermore, such localized osteomyelitis may not be distinguishable from primary periostitis and what sclerosing had occurred because of the infection might have been obliterated.

Osteomyelitis can also be caused by various chronic infections that are neither pyogenic nor even bacterial in origin. These nonpyogenic infections are called **granulomas** because their mechanism of healing often involves an overgrowth of tissue by granulation, which causes a general swollen appearance or a series of swellings (e.g., Figures 10–3 and 10–12; "oma" refers to "tumor," which means "swelling"). Typical granulomas are *tuberculosis (tubercular osteomyelitis), leprosy, syphilis (syphilitic osteomyelitis), yaws,* and *fungal (mycotic) infection.* The most common overall similarities between granulomatous infections and pyogenic bacterial infections leading to osteomyelitis are the development of lesions, abscesslike lesions (*gumma*), and even abscesses. When abscesses are present, however, they often lack the large sequestra typical of pyogenic osteomyelitis. Sometimes the drainage material itself may become calcified. Although cortical bone may be altered (e.g., through either healing of lesions or various avenues of cortical thickening or cortical erosion and destruction), osteitic and/or periostitic reactive bone is not always associated with the disease.

Tuberculosis, which is caused by a bacterium and transmitted between humans via the respiratory system, most frequently affects children and subadults (Kelley, 1989; Rothschild and Martin, 1993). In its skeletal

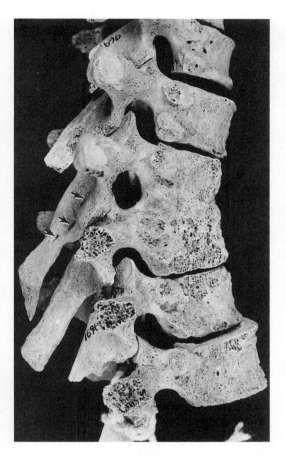

Figure 10–11   Osteomyelitis in thoracic vertebrae; note small cloacae (*arrows*) in the spine of the upper of the two ankylosed vertebrae (prehistoric, Pennsylvania). (Tuberculosis would be distinguished by the development of abscesses anteriorly in thoracic vertebrae.)

osteomyelitis, traumatic arthritis, rheumatoid arthritis, malignancy, typhoid spine, sarcoidosis, actinomycosis, blastomycosis, coccidioidomycosis, Paget's disease (osteitis deformans), osteochondritis, and neuroarthropathies. Compression or crush fractures can also create what might appear to be tuberculosis-affected vertebrae.

Features specific to tuberculosis are erosion of bone (i.e., destruction of cortical as well as spongy bone, leading sometimes to the excavation of cavities in the trabecular bone of articular regions; Figure 10–13) and absence of bone regeneration. The stereotypic picture of spinal tuberculosis is unabated destruction and erosion of vertebral bodies, leading to their eventual collapse. Typically, however, the maximum number of vertebrae involved is four and transverse processes, neural arches, and vertebral spines are infrequently affected. Again, typically, bone does not regenerate and affected adjacent vertebrae rarely fuse (*ankylose*). Visually as well as radiographically in lateral view, collapsed vertebral bodies are wedge-shaped, tapering toward their ventral (anterior) sides (for an example, see Fractures and Trauma, later). In severe cases of collapse, the vertebral column can become kyphotic with an angular (rather than curved) anterior deflection in the sagittal plane (producing gibbus). But although tuberculosis may result in kyphosis, kyphosis is not associated exclusively with tuberculosis. As summarized by Morse (1978), gross similarities between tuberculosis and pyogenic osteomyelitis can be found in the development of cold abscesses (with cloacae) posteriorly in the cervical region and anteriorly in the lower thoracic and lumbar regions. Tubercular abscesses are characterized by large surface apertures and downward-coursing erosive penetration.

The hip joint (acetabulum and/or femoral head) is next most frequently affected by tuberculosis, followed by involvement of the knee, foot, and elbow joints. Much less frequently affected are, in adults, the sternum, ribs, shoulder joint, and pelvic (sacroiliac)

manifestation, tuberculosis is typically thought of as affecting the vertebral column. The reason for this is that, as difficult as it is to diagnose spinal tuberculosis incontrovertibly, the effects of tuberculosis elsewhere in the skeleton are more difficult to discriminate from other possible causes. Even the diagnosis of spinal tuberculosis is not necessarily a simple matter. For example, Morse (1978, p. 45) points out the number of diseases that must be eliminated from consideration prior to diagnosing tuberculosis: chronic pyogenic

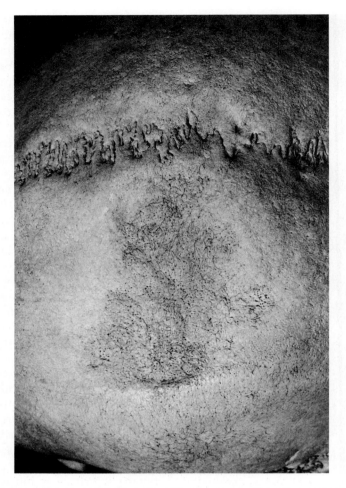

Figure 10–12 Healed granulomatous infection of cranium; note overall swollen appearance as well as more localized swellings (postcontact, Siberia).

joint and, in children, the tubular bones of the hands and feet. Cold abscesses in the pelvic region result from tubercular infection. In general, the affected region is characterized by erosion and perforation of cortical bone and destruction of subchondral/subarticular as well as trabecular bone (sometimes with development of cavities) but without noticeable development of reactive bone.

Brucellosis is a bacterial (bacillary) infection that secondarily produces osteomyelitis and suppurative arthritis (spondylitis). Also known as undulant, Mediterranean, or goat's

milk fever, dogs, goats, pigs, and cattle can transmit the disease to humans. Lumbar centra and the sacroiliac joint are most frequently affected. Typically, lesions are multiple and produce cavities and, in vertebrae, may extend the length of a centrum and/or pass through an intervertebral space into an adjacent centrum. Its lesions have been described as either lytic or resorptive. Infection may secondarily affect the diaphyses of leg and arm bones and provoke periostitis. The spread of brucellosis is likely correlated with the domestication of herding animals

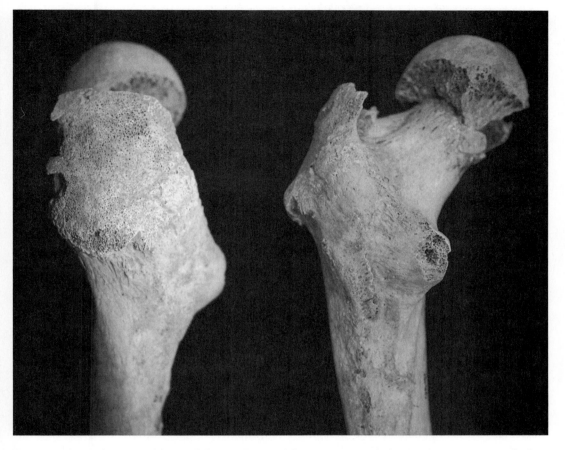

Figure 10–13   Left proximal femur (*left*, lateral view; *right*, posterior view) showing apparent signs of tuberculosis (note destruction of cortical bone) (prehistoric, Pennsylvania).

and the use of the dog in herding (Ortner, 2003a; Schwartz, 1995).

Syphilis (both endemic and venereal) and **yaws** (*treponematoses*, s. *treponematosis*) are caused by a treponemal spirochete that invades the host through either the skin or superficial mucous membranes. Both syphilis and yaws can lead to periostitis and osteitis of the long bones (often involving the tibia) as well as of the skull, with the frontal and parietal often being the first cranial bones affected. In long bones, cortical thickening and diminution of the medullary cavity may occur. However, unlike in pyogenic osteomyelitis, sequestra usually do not form.

Periostitic layering on the tibia can create the so-called *saber shin*, which is characteristically bowed anteroposteriorly because bone remodeling thins the anterior margin (making it convex) and thickens the posterior surface (making it concave). Although typically thought of as a feature of syphilis, saber shin may develop in cases of tertiary yaws.

Syphilis is distinguished from yaws by its effects on the skull. Although in both diseases bone surrounding the nasal cavity may be destroyed, in syphilis destruction also typically subsumes the nasal septum, spreading down and throughout the hard palate or up and into the nasal bones. Lesions of tabular

Figure 10–14   "Worm-eaten" appearance of cranial bone characteristic of advanced syphilis (postcontact, Siberia).

cranial bones—the frontal and parietal, in particular—are more characteristic of syphilis and take the form of a central zone of destruction surrounded by an often irregular perimeter of reactive bone. Over time, as the process of lesion healing and eruption of new lesions proceeds, the cranium's sclerotic surface takes on a mottled, sculpted appearance, producing the "worm-eaten" look so characteristic of crania in advanced cases of syphilis (e.g., Ortner and Putschar, 1981; Figure 10–14).

Although venereal syphilis can be transmitted across the placenta from mother to fetus, it and endemic syphilis are most frequently seen in postpubescent individuals. In contrast, yaws is typically contracted and expressed in young individuals. However, yaws may go undiagnosed because its initial lesions may heal without leaving clues. In terms of biogeography, the preantibiotic distributions of yaws and endemic syphilis also differ. Yaws was prevalent in populations throughout the tropics, whereas endemic syphilis was restricted to temperate and subtropical regions of the non-European Old World. The spread of venereal syphilis has traditionally been linked to European colonization (e.g., Baker and Armelagos, 1988).

Leprosy is caused by a bacterium that primarily infects nonosseous tissues, but it may spread to the skeleton. Ortner and Putschar (1981) suggest that, although the distribution of leprosy used to be all but global (exclusive of the Arctic), it was probably introduced into the New World after European contact and in conjunction with European colonization. Its spread was slow because, in spite of being an infectious disease, it is not very contagious.

Ortner and Putschar (1981) delineate two skeletal manifestations of leprosy: (1) *lepromatous osteomyelitis and subperiosteal periostitis* and (2) *neurotrophic bone* as well as *lesions* to and *degenerative arthritis* of the *weight-bearing joints of the foot and ankle*. The effect of lepromatous osteomyelitis on the lower face and nasal region is like that of syphilis and yaws in destroying the nasal septum, hard palate, and nasal bones; but it can also lead to widening of the nasal aperture and destruction of the nasal conchae. In contrast to syphilis, in leprosy the flat cranial bones are not involved. If subperiosteal periostitis develops in the long bones, it is characterized by thin, longitudinally oriented striations.

Neurotrophic bone changes, caused by loss of sensory nerve function, can lead to the loss of fingers and toes. The process typically begins in the hand: segments are slowly resorbed distoproximally; metacarpals may also be affected. In foot bones, the process proceeds from the metatarsal–proximal phalangeal region to the terminal phalanges. Although changes may resemble those seen in rheumatoid arthritis, which also affects the hands and feet, there are no arthritic changes. Because of sensory nerve death, the affected individual is vulnerable to injury, which can lead to infection and, in turn, to ordinary osteomyelitis and septic arthritis.

Fungal (*mycotic*) **infections** that may ultimately affect the skeleton include *actinomycosis, blastomycosis, coccidioidomycosis, cryptococcosis, histoplasmosis*, and *sporotrichosis*. [For the purposes of discussion here, actinomycosis, which Ortner and Putschar (1981) attribute to a higher bacterium rather than a fungus, and sporotrichosis, which Kelley (1989) does not include with fungal infections but Ortner and Putschar (1981) and Rothschild and Martin (1993) do, will be grouped as fungal infections.] All but sporotrichosis, which is introduced via thorn punctures and spreads hematogenously (i.e., through the bloodstream), are inhaled, often infecting the lungs first. Rothschild and Martin (1993, p. 73) identify the focus of fungal infection on joints as the articular surface of a bone that is typically undercut by marginal erosion. Ortner and Putschar (1981)

suggest that, because the bony lesions produced by these fungal infections are indistinguishable from one another morphologically, biogeographic distribution is a primary criterion for diagnosis. Rothschild and Martin (1993; see also Buikstra, 1976) do, however, itemize some morphological characteristics that appear to be distinctive of each kind of fungal infection. Because of the potential ambiguities in diagnosing fungal infections as well as possible confusion with the diagnosis of tuberculosis and other infections, the student is advised to consult Table 10–1 in addition to the brief descriptions provided below; further detail and illustration are provided by Ortner (2003b).

As a class, fungal (mycotic) infections affect the skeleton in relatively low frequencies. Thus, although globally distributed, the incidence of bony lesions due to **actinomycosis** is quite low. Clues to its diagnosis are (1) (especially), the development of "lumpy jaw" and (2) the presence of many small lesions emanating from the periosteum, which often creates a "worm-eaten" look due not only to bone resorption but also to an increase in vascularization of the affected region. [The literature is contradictory on the incidence of reactive new bone formation: e.g., Ortner and Putschar (1981) illustrate and describe significant periosteal reactive bone deposition in specimens diagnosed as having actinomycosis, whereas Buikstra (1976) states that bone proliferation in association with actinomycosis occurs infrequently. The difference here, however, seems to be a matter of identifying periostitis versus endostitis (i.e., sclerotic bone proliferation originating in the endosteum, as seen in pyogenic osteomyelitis). Clearly, in the case of the mandible, periosteal inflammation leading to periostitis would produce a "lumpy jaw."] After the mandible, the next most frequently affected regions of the skeleton are the thoracic and lumber vertebrae (neural arches and centra alike) and then the ribs, which may show areas of resorption. Vertebral collapse is rare.

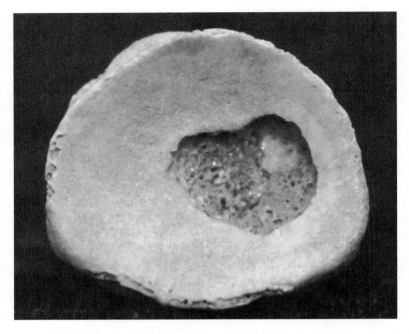

Figure 10–15   Proximal end of proximal phalanx of first pedal digit showing lytic lesion; such a lesion in this joint is diagnostic of blastomycosis (prehistoric, Pennsylvania).

Blastomycosis, which is found primarily in North America (especially the Ohio and Mississippi Valleys), affects males five times more frequently than females. Because it is soil-borne, its skeletal presence is likely associated with the development of agriculture (e.g., Kelley, 1989). Blastomycosis typically produces lytic lesions (with crisp perimeters) and most frequently affects vertebral bodies, sometimes also involving neural arches and adjacent ribs. Bones of the skull, hands, and feet (Figures 10–15 and 10–16), as well as the tibia may subsequently be affected. Contradictory diagnoses suggest that destruction of vertebral bodies and intervertebral discs, leading to collapse and kyphosis, is either characteristic of blastomycosis or rarely associated with the infection.

Coccidioidomycosis (San Joaquin Valley or feveralley fever) is found in the North American Southwest and areas of Meso- and South America. Seen more frequently in males than in females, it may be contracted through inhalation or a cut or scratch; its lesions have been described as either lytic or resorptive. The vertebral column is most frequently involved, with centra and neural arches being affected (in contrast to tuberculosis, in which the neural arches are not involved). (If vertebral and intervertebral disc collapse are associated with kyphosis, the specifics of vertebral involvement are important criteria in distinguishing between coccidioidomycosis and tuberculosis.) Infection often spreads from the vertebral column to the ribs. Thereafter, parts of the skull and articular/ epiphyseal regions of long bones are at risk because coccidioidomycosis characteristically attacks bony prominences and protuberances. In the hands and feet, however, diaphyses are typically infected.

Cryptococcosis, which occurs primarily in Europe and is neither sex- nor age-specific, also tends to lodge in bony prominences. It typically produces lytic lesions, and although any bone in the body is potentially vulnerable,

Figure 10–16 Left calcaneus with postmortem damage as well as identifiable lytic lesions diagnostic of fungal infection (probably blastomycosis) (prehistoric, Pennsylvania).

it frequently affects the cranium and the vertebrae closest to it.

A fungal infection common in the Ohio and Mississippi Valleys that only rarely shows up skeletally is **histoplasmosis**. It is typically found in adults and infants less than 1 year old and primarily affects the arm- and leg-bone diaphyses and cranial bones. Multiple lytic lesions may be associated with cortical thickening as well as with areas of rarified bone.

Sporotrichosis may on rare occasions affect the skeleton. Primarily found in individuals living in regions between latitudes 50° north and south and predominantly affecting males, it is usually introduced via thorn punctures and spreads hematogenously. Lesions are typically periosteal and occur most frequently in the tibia, followed by bones of the hands and feet, bones of the face and skull, and then the ribs, clavicle, and vertebrae.

Echinococcosis, which is the only parasitic disease in humans that produces significant bony change, appears to have arisen regionally throughout Europe in conjunction with the domestication of herding animals and the use of dogs to control them. Tapeworm larvae are transmitted to humans from dogs, cattle, sheep, or pigs. It can result in the ossification of cysts and/or the development of expansive, erosive lesions. Commonly affected are the pelvis, vertebral centra [possibly leading to collapse and gibbus (compare with tuberculosis)], and, because the larvae are blood-borne, the metaphyseal regions of the long bones of the arms and legs.

A viral disease that can affect the skeleton (of infants and small children but not adults) is **smallpox**, which used to be global in its distribution. Diagnostic clues include the settling of the infection in joints and the destruction of metaphyses, leading at times to the separation of epiphyses. Particularly telling is that (1) the elbow (the primary locus of the infection) is affected bilaterally and (2) in addition to the humerus and ulna, the radius (which usually is not involved in other infectious diseases) is affected. After the elbow joint, the wrist and then the ankle and knee are most frequently involved. Periostitis is commonly associated with smallpox. If abscessing occurs, sequestra typically do not form.

Sarcoidosis is a granulomatous collagen-vascular disease whose cause remains unknown. It is characterized by multiple round, lytic lesions that typically occur bilaterally and are not accompanied by reactive bone formation. Primarily affected are the fingers and toes and thereafter the tubular bones of the leg and arm, vertebral centra (multiple but not necessarily contiguous), and the bones of the pelvis and skull (especially the nasal bones). Children and young adults are most frequently affected. Sarcoidosis is most similar to lepromatous leprosy.

Joint Diseases

A cursory review of the literature on archaeologically derived skeletal material reveals that *osteoarthritis* (i.e., arthritis impacting bone) is arguably the most frequently diagnosed malady of the postcranium. But although changes of joint surface topography may be associated with osteoarthritis, not all joint surface changes are the result of osteoarthritis.

In spite of its suffix "itis," arthritis is not an inflammatory disease (e.g., Ortner and Putschar, 1981). Inflammation of joints may be a complication of osteoarthritis, but it is not an attribute of it. Consequently, "osteoarthrosis" has been put forth as a more accurate term of reference (see Rothschild and Martin, 1993). Another misconception about osteoarthritis is that while it may not be an inflammatory disease, it is a degenerative disease. Hence the term "degenerative osteoarthritis" has enjoyed widespread use in the literature, either as a misnomer for an array of joint diseases or in an attempt to distinguish among various kinds of joint disease.

But if osteoarthritis is not what it has typically been assumed to be, what is it?

An apparently obvious reason why osteoarthritis has been thought of as being either inflammatory or degenerative in nature is that it is associated with morphological changes in joint surfaces: normal joint surface topography is noticeably altered or destroyed, and these abnormal-looking joint surfaces often bear spike- or sheetlike outgrowths along their perimeters. Seemingly similar changes in joint surface morphology with seemingly similar attendant growths also appear to characterize other diseases that are truly inflammatory or degenerative. But differences within these apparent similarities do exist.

Osteoarthritis develops as a result of interruption of or interference with normal joint function and stability, which is brought about primarily by injury to the joint cartilage and the bone beneath the cartilage. Ortner (2003d) distinguishes two categories of osteoarthritis: *primary osteoarthritis* (which results later in life from, e.g., biomechanical stress and trauma) and *secondary osteoarthritis* (which occurs earlier in life in abnormally formed or pathological joints). The more stabilized the joint (e.g., ankle), the less likely it is to suffer osteoarthritic change. The less stabilized the joint (e.g., knee), the more likely it is that osteoarthritis will develop in it. And as it turns out, the knee is the joint most frequently affected by osteoarthritis.

Osteoarthritis occurs because the rate at which the articular cartilage and attendant subchondral bone is destroyed is faster than the rate of repair (Rothschild and Martin, 1993). Subchondral bone modification results in (1) sclerosis in the area of injury (often because of healing of minute fractures caused by the injury) and (2) the outgrowth of bony spicules or spurs (*osteophytes*) around the perimeter of the joint cartilage. These bony outgrowths are thus not the result of ossification of soft connective tissue. In fact, the attachment areas of tendons, ligaments, and joint capsules remain unaffected in osteoarthritis (although mineralization of soft tissue may characterize other forms of "arthritis"). Over time, remodeling of subchondral bone can lead to its becoming not only denser but also polished [*eburnated* (Figure 10–8)] and even grooved as bone-on-bone contact increases with the destruction of intervening joint cartilage (in advanced cases of osteoarthritis, the knee joint is often eburnated and grooved). Although cysts may develop in subchondral bone, it neither erodes nor becomes resorbed. In contrast to other joint diseases, fusion or *ankylosis* of adjacent bony elements is not a feature of osteoarthritis (Ortner and

Putschar, 1981). Thus, sclerotic subchondral bone, perhaps with cysts, and osteophytes are the primary clues to diagnosing osteoarthritis.

Although osteophytes may develop in various places on vertebrae (creating "lipping" on centra), strictly speaking "osteoarthritis" should be diagnosed only when diarthrodial/synovial joints exhibit subchondral bone remodeling and osteophytosis (Rothschild and Martin, 1993). Technically, lipping of vertebral bodies should be identified as **spondylosis deformans** (Figure 10–17). Only if the zygoapophyseal joints (which are diarthrodial/synovial joints between articulating superior

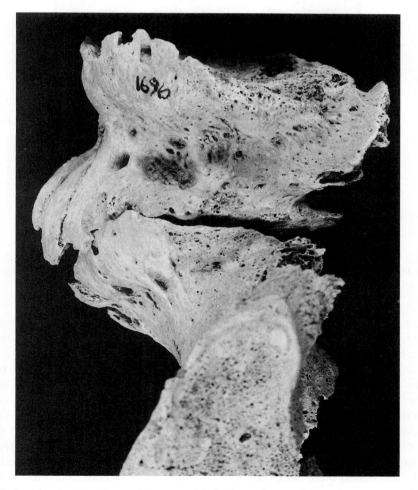

Figure 10–17   Last lumbar vertebra and sacrum with horizontal osteophytes characteristic of spondylosis deformans (prehistoric, Pennsylvania).

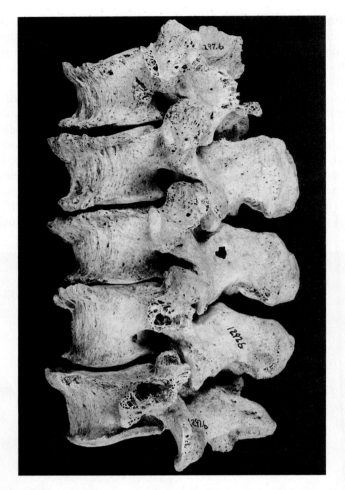

Figure 10–18 Horizontal osteophytic growths on T12 to L4 (same individual as in Figure 10–17 with spondylosis deformans), illustrating lack of involvement/ankylosis of zygoapophyseal joints (which would be involved in arthritis).

and inferior articular surfaces of vertebrae) bear osteophytes can one identify osteoarthritis in the vertebral column. Thus, in general, the recognition of osteoarthritis rests almost entirely on changes in diarthrodial joints throughout the skeleton, not on general changes in the vertebral column.

In **spondylosis deformans**, osteophytes grow from the superior and/or inferior margins of a vertebral body and eventually extend across the intervertebral space toward an adjacent centrum (Figures 10–17 and 10–18). An

important diagnostic feature of spondylosis deformans is that osteophytes grow somewhat parallel to the vertebral end plate and project out from (often at a right angle to) the vertebral body; they then curve toward neighboring vertebrae (Rothschild and Martin, 1993, p. 121). In advanced cases of spondylosis deformans, adjacent osteophytes often ankylose. Thus, spondylosis deformans is characterized by the uniting of vertebral bodies via their ankylosing osteophytes and not, as in *ankylosing spondylitis*, directly between bones in zygoapophyseal or costovertebral joints (Figure 10–18). In spondylosis deformans, cervical and lumbar vertebrae are more frequently affected than thoracic vertebrae. (In diffuse idiopathic skeletal hyperostosis—which can also affect the spine and has often been mistakenly identified as spondylosis deformans—thoracic vertebrae are most frequently affected.) Also, in spondylosis deformans, the intervertebral disc cartilage may be extruded (*herniated*) either through a portion of the perimeter of the vertebral body (through the annulus fibrosus, the fibrous ring that bounds the pulpy center of an intervertebral disc) or, as seen on bone, into the vertebral body's end plate, often near its center. The herniated nodule of cartilage is a *Schmorl's node*, and the depression it leaves in the end plate (typically also identified as a Schmorl's node) can look like an erosive (lytic) lesion. The size of such a depression can vary, depending on how large the Schmorl's node becomes. The disc cartilage may herniate superiorly and inferiorly into the end plates of adjacent vertebrae.

Ankylosing spondylitis is probably the best-known (and most frequently diagnosed) example of the class of inflammatory *arthritides* (the plural of arthritis) known as **spondyloarthropathies**. In spondyloarthropathy, there is a tendency for erosion, reactive bone growth, and fusion of adjacent bones as well as for mineralization of attachment areas of soft connective tissue (Rothschild and Martin, 1993). Osteophytelike spurs that arise as a result of ossification of soft connective tissue (e.g., ligaments, tendons, annulus fibrosus) are identified as *syndesmophytes*. Ankylosing spondylitis is characterized by involvement of the vertebral column and sacroiliac joints. Diagnostic changes include erosion followed by fusion of the zygoapophyseal joints; erosion of the superior and inferior anterior margins of vertebral bodies, which eventually causes the typically concave surface to become straighter, thus squaring up the vertebral body; vertically oriented syndesmophytes (rather than horizontal

osteophytes) that arise from the margins of vertebral bodies (because they originate in the annulus fibrosus between vertebrae) and form bony links connecting adjacent vertebral bodies; uniform or symmetrical formation of syndesmophytes around the perimeters of the vertebral bodies; erosion and/or fusion of the sacroiliac joint; and erosion and fusion of zygoapophyseal (costovertebral) joints (Figure 10–19) (Rothschild and Martin, 1993).

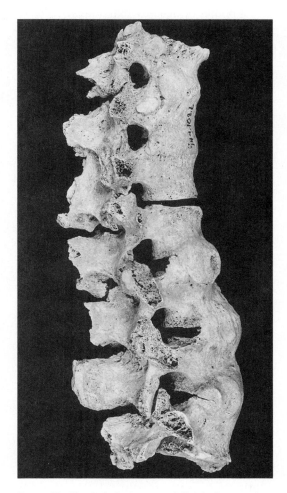

Figure 10–19   Ankylosing spondylitis in lower spine; note squared-off appearance of vertebral bodies, regular spacing of vertebrae, and uniform formation of vertically oriented syndesmophytes bridging various vertebrae (prehistoric, Pennsylvania).

Syndesmophyte formation (*syndesmophytosis*) typically proceeds along the vertebral column in an inferosuperior direction (i.e., from the lumbar region, through the thoracic vertebrae, and eventually up through the cervical vertebrae). Since syndesmophytes form continuously around the margins of adjacent vertebrae, the bridged intervertebral space thickens, becoming reminiscent of the segmental ring of a stalk of bamboo. In advanced ankylosing spondylitis, with a number of vertebral bodies bridged by marginal syndesmophytes, the configuration is often described as *bamboo spine*. Ankylosing spondylitis occurs in males three times more frequently than in females.

Rheumatoid arthritis is the best-known example of a class of arthritides called **polyarticular erosive arthritis**. Aside from affecting many joints (i.e., being poly- rather than pauciarticular), rheumatoid arthritis—which occurs three times more frequently in females than males—is characteristically expressed bilaterally and symmetrically and most commonly in adults (although there is a juvenile form of the disease). Because it is an erosive form of arthritis, lesions expose trabecular bone as smoothly remodeled fields within cortical bone (Figure 10–20). A field of trabecular bone produced by rheumatoid arthritis will be at the same level as, and appear to be continuous with, surrounding cortical bone. In contrast, a field of trabecular bone exposed by postmortem damage or weathering will lie below the level of the cortical bone, creating a terraced effect.

Typical of rheumatoid arthritis, lesions begin to resorb bone "at the 'bare area' or cartilage margin [which is] the region . . . between the subchondral bone (that is covered with cartilage) and the point of insertion of the joint capsule into the bone" (Rothschild and Martin, 1993, p. 93). Diagnostically, these lesions are not associated with sclerotic or reactive bone formation. Over time, the joint is destroyed and the surrounding bone becomes markedly porotic (on the surface and subchondrally). Bone remodeling across the joint space can lead to ankylosis. Often occurring in the hand, remodeled articular surfaces cause a deviation of the bones medially (i.e., toward the ulnar side).

Most frequently affected in rheumatoid arthritis are carpal, metacarpophalangeal, metatarsophalangeal, proximal interphalangeal (on both hands and feet), and ankle joints (Ortner and Putschar, 1981; Rothschild and Martin, 1993). Joints of the hand tend to be involved first and more often than those of the feet. Also frequently involved is the ulnar styloid process, but distal interphalangeal joints are not affected as often as proximal interphalangeal joints. Essentially, though, any peripheral joint (e.g., shoulder, hip, elbow, knee) can be affected by rheumatoid arthritis. Also typically diagnostic of rheumatoid arthritis is the lack of involvement of most of the vertebral column and the sacroiliac joint. If, however, the vertebral column is involved, the first and second cervical vertebrae are most frequently affected.

A form of crystalline arthritis with characteristics that can variably be misinterpreted as having been caused by osteoarthritis, rheumatoid arthritis, or ankylosing spondylitis is **calcium pyrophosphate deposition disease (CPDD)**, which can arise from both mechanical problems and inflammation (Resnick and Niwayama, 1988; Rothschild and Martin, 1993). Secondary CPDD is associated with aging, typically affecting individuals older than 30. Details may differ between "varieties" of CPDD, but they all involve bone formation (e.g., subchondral bone, wall-like extrusions around the perimeter of articular surfaces, or calcification across joint spaces).

Although calcification of the annulus fibrosus and squaring up of the anterior faces of vertebral bodies can be confused with ankylosing spondylitis, in CPDD, zygoapophyseal and costovertebral joints are unaffected. Calcification of the pulpy center (nucleus pulposus) of intervertebral discs can also occur in CPDD. There may be erosion in some forms of CPDD, but these lesions are less crisply delineated around their perimeters

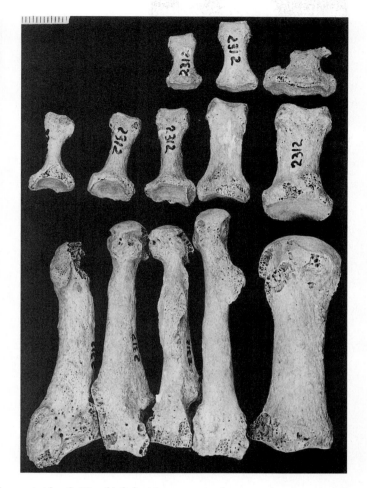

Figure 10–20 Rheumatoid arthritis of left foot bones; note, for example, characteristic exposure of cancellous bone (see text for discussion); also note squatting facets on metatarsal heads (prehistoric, Pennsylvania).

than the lytic and well-defined lesions of ankylosing spondylitis and rheumatoid arthritis; CPDD lesions do not affect the trabecular bone, as in rheumatoid arthritis. Erosion, which may simply destroy an articular surface, typically affects only a few joints (i.e., is pauciarticular) and is not accompanied by reactive bone formation. If cysts form, they are huge, not typically small cysts as in osteoarthritis; they communicate with (i.e., open upon) the articular surface and have sclerotic margins. Some individuals with CPDD also develop "holes" in bones, particularly in bones of the metacarpophalangeal

and wrist joints. In distinguishing CPDD from osteoarthritis, ankylosing spondylitis, and rheumatoid arthritis, one should also assess the number and distribution of joints involved (see Table 10–1).

Diffuse Idiopathic Skeletal Hyperostosis

Diffuse idiopathic skeletal hyperostosis (DISH) is easily mistaken for spondylitis deformans. But once its characteristics are understood, DISH can be distinguished from the latter (see Rothschild and Martin, 1993).

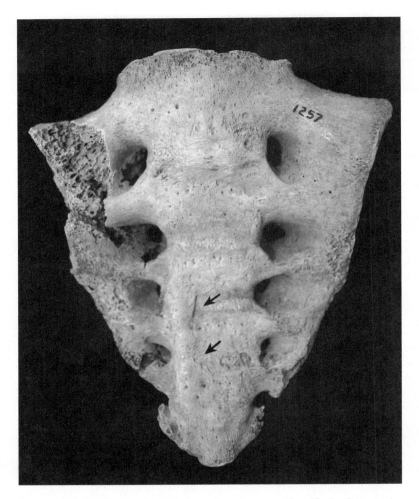

Figure 10–21   Sacrum with DISH; note "dripping wax" appearance of ossified soft tissue (*arrows*); also note fused coccyx (prehistoric, Pennsylvania).

DISH is typically found in old individuals and affects males twice as often as females (Ortner, 2003d). As the word "hyperostosis" indicates, a major feature of DISH is aggressive ossification of soft tissue; in particular, the ligaments that run alongside and in parallel with the vertebral column mineralize (which come to look like dripping wax; Figure 10–21). Thus, zygoapophyseal and sacroiliac joints are not affected. Most frequently, the anterior longitudinal spinal ligaments and the region of the thoracic vertebrae are involved. In the thoracic region, ossification is asymmetrical, affecting the side not in association with the aorta (thus ossification typically occurs in the right anterior longitudinal spinal ligament). If lumbar vertebrae are involved, both anterior ligaments ossify and motion is impeded. Because vertebrae become linked as the ligament that courses alongside them mineralizes, there is a space (visually or radiographically detectable) between the ossified ligamentous band and the vertebral bodies. In contrast, the osteophytes and syndesmophytes that form, respectively, in spondylitis deformans and ankylosing spondylitis are confluent with the vertebral bodies.

Although most frequently observed in the vertebral column, DISH can affect any site of soft connective tissue attachment (tendon, ligament, joint capsule). In these cases, DISH is often noted as a spur at the locus of tendon or ligament insertion or, in the case of the iliac crest and ischial tuberosity, "pelvic whiskering" along these expansive surfaces (Figure 10–22).

Since many "nonmetric variations" are hyperostotic in nature, one should be careful not to conflate these "features" with DISH (if one can indeed distinguish between them).

Tumors

A tumor is the result of cellular proliferation. If the growth of the cell mass is confined in location and/or limited in size, the tumor is probably **benign**. If growth is unlimited—in

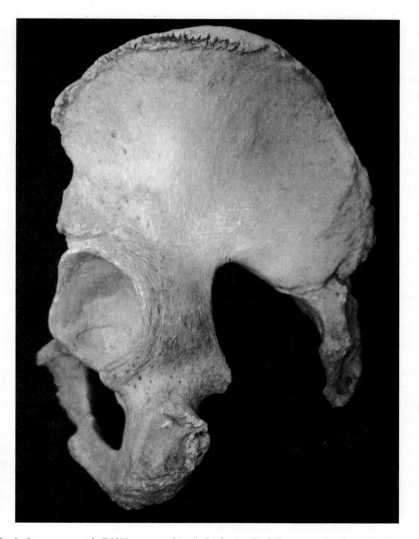

Figure 10–22   Left os coxa with DISH, as noted in "whiskering" of iliac crest (prehistoric, Pennsylvania).

size as well as location—it is most likely **malignant** (cancerous). Tumors specific to bone are **primary (skeletal) tumors.** *Metastatic* tumors (which develop via the spread of cells from a tumor that arose in soft tissue) are **secondary (skeletal) tumors.** Whether benign or malignant, primary skeletal tumors typically arise in individuals who are still actively growing. Primary skeletal tumors do not occur as frequently as secondary skeletal tumors (Ortner and Putschar, 1981).

A benign bony (osteoblastic) tumor is an **osteoma (pl. osteomata).** A common osteoma is typically expressed as a small, dense, buttonlike, low mound growing on the surface of one of the flat cranial bones (Figure 10–1). The perimeter of the base of this kind of osteoma is incised, making it look as if it had been "stuck" onto the bone. An individual may develop multiple cranial osteomata. On occasion, an osteoma may grow to c. 2 cm in diameter. Ortner and Putschar (1981) also list exostoses of the external auditory (acoustic) meatus (*meatal tori*)—which Saunders (1989) interprets as hyperostotic nonmetric variants—as examples of osteomata. Osteomata may proliferate within the paranasal sinuses as well. An **osteoid osteoma** develops intraosseously; it has an unossified center and should be suspected as lying within or under a localized area of markedly thickened cortical bone. Its existence can be confirmed radiographically.

Tumors associated with bone may also arise from cartilage. A common benign, cartilage-derived (chondroblastic) tumor is a **chondroma**, which typically develops from (exogenous) epiphyseal plate cartilage in the metaphyseal region of tubular bones. The most common chondroma, an **enchondroma** (i.e., singular as well as plural, although "enchondromata" is also a plural form), occurs intraosseously and most frequently affects the small tubular bones of the hand (Rothschild and Martin, 1993). Enchondromata can arise in females and males alike and often occur in older children through middle-aged adults (Ortner and Putschar, 1981). **Ollier's disease** is the development in early childhood of multiple enchondromata. The presence of an enchondroma is detected radiographically.

Another common benign chondroblastic tumor is an **osteochondroma**, which, like a chondroma, is an outgrowth of epiphyseal plate cartilage and which, like cartilage-derived bone in general, ossifies endochondrally. Osteochondromal growth ceases when epiphysis and diaphysis fuse. The distal femur and proximal tibia are most frequently affected, but osteochondromata may arise in any postcranial bone. Also referred to as a *cartilaginous exostosis* (Ortner and Putschar, 1981), an osteochondroma is distinguished from a "true" exostosis by the persistence of a cartilaginous cap (Rothschild and Martin, 1993), which is reflected osteologically in the irregular surface of the osteochondroma's "tip" (somewhat like a metaphyseal surface). The presence of an osteochondroma on one bone may be reflected in the articular surface of its partner as a shallow depression in, or flattening of, a curved surface (e.g., on a femoral condyle).

Tumors can also arise from the periosteum. The most common is a **fibroma** (fibrous cortical defect), which can affect the alveolar margins of the jaws as well as postcranial tubular bones (Resnick and Niwayama, 1988). A fibroma may be a single- or multichambered lesion lying just beneath the cortex. Its inner surface is lined with sclerotic bone. Although a fibroma may be suspected beneath a thin cortex, its presence must be verified radiographically. Although typically benign, a fibroma can become malignant. Rather than being cut off from the medullary cavity of the affected bone by its sclerotic lining, it may penetrate and produce an expansive lesion.

Malignant tumors affecting bone are relatively rare. The two most common forms are **osteosarcomata** and **chondrosarcomata**. Like benign tumors, osteo- and chondrosarcomata are typically associated with metaphyseal regions of actively growing bones. Osteosarcomata are typically intraosseous (most frequently arising in the knee joint and twice as frequently in males than females) and may be evidenced as destructive lytic lesions or growths of sclerotic, internally amorphous bone with exotic surface patterns. In response to the destruction of bone, the periosteum may produce new bone around the area of the lesion. Sometimes an osteosarcoma may develop on the cortex of the bone. Benign chondroblastic tumors as well as, potentially, any cartilage can give rise to chondrosarcomata. Although it may be difficult on skeletal material to differentiate between an osteosarcoma and an ossified chondrosarcoma, the latter tends to produce a more nodular tumor with a more structured, trabeculated interior.

Cancer can also spread and metastasize to bone from soft tissue. Tumor cells migrate through the bloodstream to the bone marrow, where they multiply, eventually replacing bone marrow and destroying spongy and eventually cortical bone. The axial skeleton is most frequently affected, but the femur, ribs, sternum, and humerus are often involved as well. As characterized by Ortner and Putschar (1981, p. 393), "fast growing tumors are mainly osteolytic, while slow growing ones elicit an osteoblastic response." Lytic lesions are the more common; they are often circular or ovoid in outline and may appear "punched out." Osteoblastic response may result in sclerotic lesions or various forms of hyperostotic bone growth (e.g., nodular or radiant). Of all carcinomas, breast and prostate cancers most frequently metastasize to bone. Metastatic breast cancer produces lytic lesions, while metastatic prostate cancer produces an osteoblastic response.

Bone Necrosis

Bone death or **necrosis** (Figure 10–23) results ultimately from loss of blood supply to a bone or region of bone. As discussed above with regard to pyogenic osteomyelitis and the formation of sequestra, infection can also lead to bone death. So too, for example, can trauma (such as dislocation) or fracture, vitamin D–deficiency rickets, rheumatoid arthritis, and some forms of metastatic carcinoma (Rothschild and Martin, 1993; see Figure 10–23). Necrosis frequently occurs in the hip joint (especially in the femoral head, which is a site of stress and potential trauma). As bone necrosis sets in, the affected bone or portion thereof diminishes in size and robustness (especially in cortical thickness), which can lead to fracture and articular epiphyseal displacement.

Fractures and Trauma

Undue or unusual stress can cause a bone to fracture because the force of the assault outpaces the rate at which the bone can respond to the stress (i.e., by remodeling and forming a callus to strengthen the bone in the region of stress). Since bone remodeling entails osteoclastic action prior to bone deposition, stressed bone becomes even more vulnerable to fracture.

Stress can be due to injury (e.g., a fall or wound), carrying or lifting heavy loads, assuming an awkward movement or position, and undergoing an invasive medical procedure. In the cases of injury or a medical procedure, the effect of the assault can cause an immediate response in the bone or bones involved. Types of stress that might be produced in a traumatic injury would be *compression, twisting* or *torsion, bending,* and *shearing;* in surgery, sawing and drilling, for example, leave their marks.

Collapsed vertebrae often result from *compression* (Figure 10–24). A blow to the head will also compress bone and create a depression fracture (Figure 10–25). Sometimes the blow is sufficiently severe that the bone of the inner table becomes detached (Figure 10–26). Because the force of the blow radiates outward from the point of impact, creating a conical field (as when a pellet hits plate glass), the affected area of the inner table is often larger than the

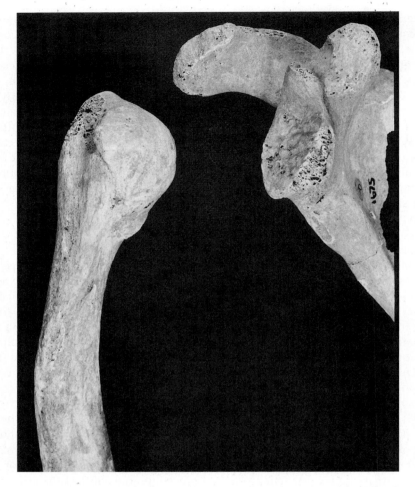

Figure 10–23   Necrosis of right shoulder joint; note, for example, alteration of scapular glenoid fossa and diminution in size and robustness of humerus (prehistoric, Pennsylvania).

affected surface area of the outer table. In long bones, compression along the long axis of the bone (e.g., the impact of a fall from a great height) can cause the diaphysis to bulge, with potential cracking of cortical bone.

Twisting (*torsion, spiral*) fractures are typically noted in long bones (Figure 10–10). The twist or unnatural rotation of part of a bone around its long axis can cause a break that is characteristically obliquely oriented (i.e., one side of the break is longer than the opposite side). Thus, a common feature of a spiral frac-

ture is that one end of each part of the severed bone looks pointed and the plane of the break along the thickness of the cortical bone looks "edgelike." One must use caution in identifying a spiral fracture as pre- rather than postmortem in origin; for example, taphonomic factors can produce a break that looks like a spiral. Such concern is particularly germane to animal bone analyses when human (at least, hominid) activity is suspected as the cause of the spiral fracture (e.g., twisting a bone in order to access marrow).

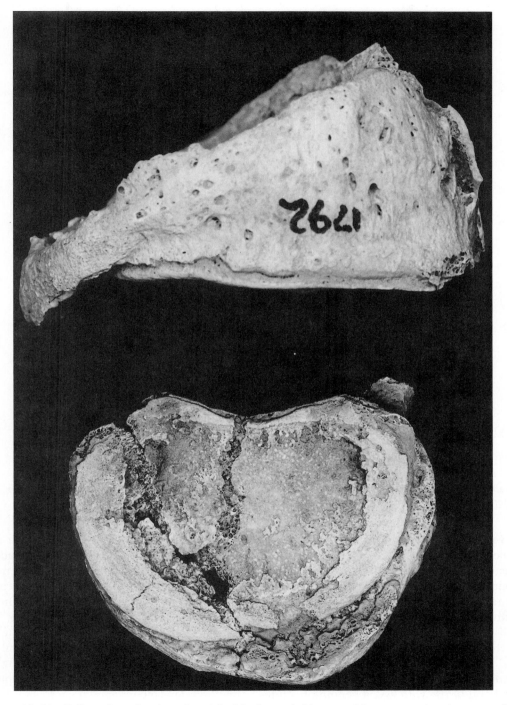

Figure 10–24   Collapsed, wedge-shaped vertebral body, probably caused by compression (as opposed to tuberculosis); (*top*) left lateral and (*bottom*) inferior views (prehistoric, Pennsylvania).

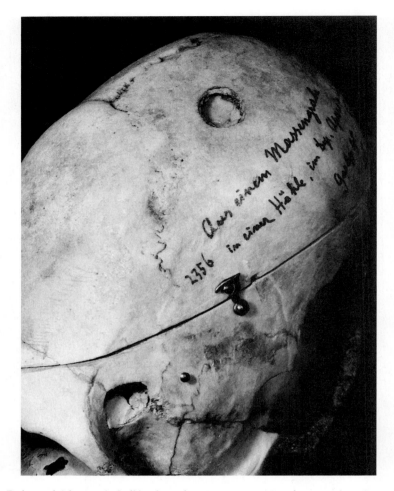

Figure 10–25 Deformed (elongate) skull with perforating compression fractures (one in parietal bone, the other near pterion); note possible healed compression fracture in coronal suture (prehistoric?, Peru).

Falling from a great height can also produce a bending fracture. Since any antemortem fracture affects green rather than dry bone, bone will "give" until it finally snaps. The more "give" a bone has—as in younger individuals—the more likely it is that a break will go only partway through the side being stretched; the break then often courses up and down the shaft of the bone producing a *greenstick fracture*. As bone becomes more rigid and brittle with age, a bending fracture is more likely to "pop out" a wedge-shaped piece of bone. Reminiscent of the spread of impact in a compression fracture, the wedge of displaced bone is narrowest on the side of the bone being flexed and broadest on the side being stretched. *Shearing fractures* —which result from opposing forces across a bone—may not be distinguishable from bending fractures.

The effects of various kinds of stress (e.g., opposing muscle action) can accumulate or increase over time, eventually leading to microfractures and then, perhaps, complete fractures [an incomplete fracture is sometimes called an *infraction*, while a complete break is referred to as a *fracture* (Merbs, 1989; Ortner and Putschar, 1981)]. The fracture of a

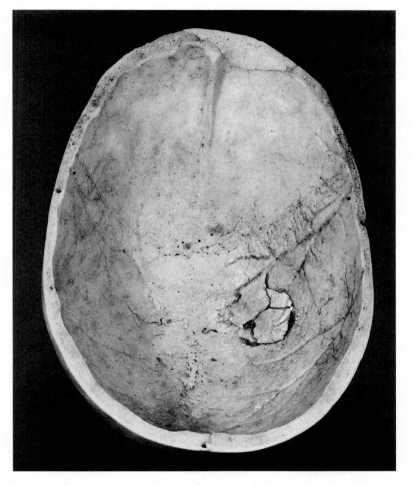

Figure 10–26   Interior of calvaria of individual in Figure 10–25 illustrating large area of bone dislodged (larger than area of impact).

neural arch from a vertebral body is referred to as *spondylolysis*, but the degree of separation can range from a minor severing to a marked chiasm and can be manifested either uni- or bilaterally. [Stress fractures leading to separation of the superior and inferior articular processes are particularly common in L5 and have been of special interest in archaeologically derived Arctic North American populations (e.g., Lester and Shapiro, 1968; Merbs, 1989).] Fractures resulting from the accumulation of stress rather than from a single stressful event are often referred to as *fatigue fractures*.

If bone is already weakened because of some preexisting condition (e.g., rheumatoid arthritis, leprosy, necrosis), the amount and kind of stress that can lead to fracture will be appreciably less than for healthy bone. Fractures in bones already affected by some pathology are *pathological fractures* (Merbs, 1989).

Since, in all fractures, the edges of the break become necrotic due to disruption of the blood supply, they do not participate in the healing process. If, however, the edges of the broken bone are not separated from one another even slightly, reestablishment of blood supply and eventual union of the fracture may be impeded, leaving an *ununited fracture*, whose ends may develop opposing articular surfaces. If the broken ends become too separated for healing to set in, a *displaced fracture* develops (Merbs, 1989). For proper healing to occur, the break must be "filled by a highly vascular fibrous callus containing focal avascular regions of fibrocartilage" (Rothschild and Martin, 1993, p. 57). Although new bone deposition occurs endosteally, ossification of the fibrocartilaginous "plug" is also necessary for complete union of the

ends of the broken bone. Eventually, normal cellular activities and features of osteogenesis characterize the region of the break. If the periosteum is also torn as a result of the fracture, its osteogenic potential will be activated and an external bony callus will form. If the fracture is profound enough to expose the break (to the outside)—constituting a *compound fracture*—secondary infection can set in, leaving evidence of both the break and the infection (Figures 10–9 and 10–27). [A fracture that is not exposed to the outside is a *simple* or

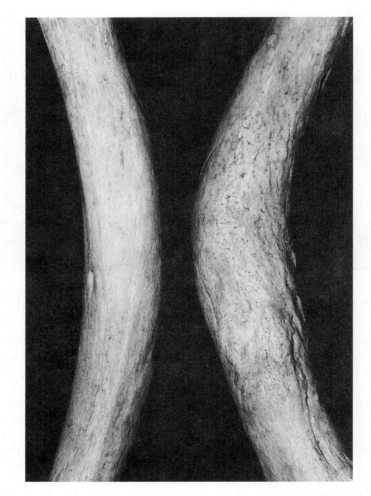

Figure 10–27   Left and right clavicles (*left* and *right*) of same individual; the right clavicle shows evidence of healing [with sclerosis and secondary periostitis (from localized osteomyelitis?)] of compound fracture (prehistoric, Pennsylvania).

closed fracture (Merbs, 1989).] Sometimes one finds evidence of *multiple fractures*. If healed fractures are not identifiable visually (e.g., because of obvious displacement or angulation or by the presence of an external callus), they can often be diagnosed radiographically on the basis of localized cortical bone thickening.

Wounds caused by projectiles and other weapons as well as by surgical procedures (e.g., trephination) can also be classified as fractures (Ortner and Putschar, 1981). The extent to which such wounds may have led to the death of the individual can be inferred from the degree of healing of the wound. In the case of *trephination*—the excising of pieces of bone from the flat cranial bones (Figure 10–28)—the archaeological record is replete with examples of individuals who survived multiple surgical events, often with much in-filling of the trephined region(s).

The surgical extraction or blow-induced exfoliation (shedding) of teeth constitutes trauma to the jaws and, as pointed out by Merbs (1989), produces a "compound fracture" in the sense that the wound is exposed to the outside.

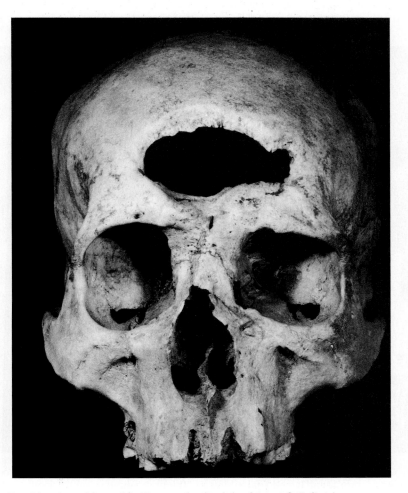

Figure 10–28 Trephination of frontal (with some healing) (prehistoric?, Bolivia).

The term "trauma" has been extended to embrace the results of surgery-like or other body-altering behaviors (e.g., Merbs, 1989). Dental examples—a subset of "cultural modification of the living" (Buikstra and Ubelaker, 1994), sometimes referred to as *dental modification* or *mutilation*—abound, from the archaeological record and recent history. The teeth most frequently affected are upper incisors and canines, and common expressions are seen, for example, in the filing of teeth into different shapes, the incising of patterns on the buccal (labial) surfaces, and the encrusting of teeth with semiprecious stones, gold, and/or silver (the latter accomplished by drilling the buccal surfaces to receive the implants) (Carter et al., 1987; Merbs, 1989; Romero, 1970). Nondental examples of body-altering trauma reflected in skeletal remains include the banned practice of *foot binding* in China and numerous examples of artificial *cranial deformation*. Some types of cranial deformation—such as head binding in pre-Columbian Andean and Mesoamerican populations (Romero, 1970)—were purposeful in that specific devices were employed on the very young to determine a specific head shape [often somewhat conical (e.g., Figure 10–29)]. Occipital flattening, which is quite commonly observed in archaeological populations of the North American Southwest [and has also been recorded for the

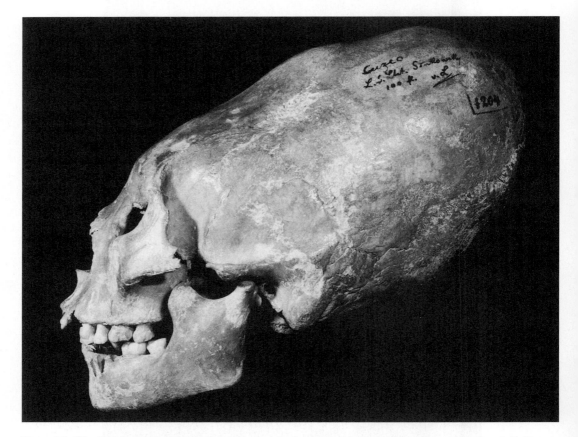

Figure 10–29   One form of cranial deformation; note, for example, elongation of frontal and occipital bones with displacement of coronal and lambdoidal sutures and the landmarks bregma and lambda; also note three-dimensional calculus on M^1 and M^2 (prehistoric, Peru).

Neolithic of Cyprus (Angel, 1953; Schwartz, 1974)], may, however, be considered secondary if it results from the primary activity of binding an infant to a swaddling board.

Infectious and Degenerative Diseases of the Jaws and Teeth

Since developmental (e.g., hypoplasia) and genetic (e.g., hypo- and hyperdontia) aspects of tooth abnormalities were dealt with in Chapter 7 and following the format used for discussing the nondental skeleton, only infectious and degenerative diseases of the jaws and teeth are dealt with here.

Lukacs (1989, p. 264) cites the following as representative of each category of disease: (1) infectious—dental caries, dental caries–induced pulp chamber exposure, dental abscess, abscess or caries-induced antemortem tooth loss, periodontal disease—and (2) degenerative—attrition-induced antemortem tooth loss, attrition-induced pulp chamber exposure, periodontal disease, calculus (tartar) accumulation [calculus/tartar accumulation results from the mineralization of plaque (Hillson, 1986)]. The listing of periodontal disease as both an infectious and a degenerative disease may seem confusing in that, in each case, the disease is often referred to broadly as "periodontitis." As we shall see, the study and identification of periodontal disease is not problem-free (Clarke and Hirsch, 1991).

Unlike the antiseptic, enclosed environment around bone, teeth are exposed upon eruption to a host of microorganisms (e.g., bacteria, viruses, yeasts, and even protozoa) inhabiting the moist and warm oral cavity, a corridor to the outside environment. Most commonly, it is a strain of bacterium that causes infections of teeth and associated tissues. Bacteria, plus a matrix they produce, as well as salivary proteins make up **plaque**. Plaque adherent to smooth coronal and root surfaces, coronal crevices and grooves, and even the periodontal/alveolar region is

populated, respectively, by different bacteria (Hillson, 1986). Except for a brief period after extensive cleaning, an individual's dental surfaces are never free of plaque. (For archaeological or forensic analyses, plaque should not be removed from teeth to make them look "pretty.")

Under normal circumstances and when the individual is not consuming food (especially sugar, and sucrose in particular), the pH of saliva is neutral. As pH levels rise, mineral crystals form in the plaque matrix, producing **dental calculus** or **tartar** (Figure 10–29, see also Figures 9–6 and 9–17). First as isolated specks and then as three-dimensional flakes, these deposits can become continuous bands around the perimeters of the crowns near the neck. If not cleaned, these deposits can expand down and up the crown as well as outward. Various ways of scoring degree and amount of calculus buildup on individual teeth have been developed by osteologists who categorize such buildup as, for example, isolated patches, continuous, three-dimensional, slight, medium, heavy. Buikstra and Ubelaker (1994) provide an example of a widely used standardized form (as well as others for recording tooth presence/absence, location and degree of carious lesions, wear, etc.).

As soon as carbohydrates and especially sucrose are ingested, plaque bacteria begin to ferment them, which produces acids that lower the pH of the saliva. More injurious for teeth, this lowers the pH of the plaque matrix adherent to part of or an entire tooth or to a number of teeth. Over time, an acidic environment can demineralize not only tooth enamel but also cementum and dentine and lead to **dental caries** or **carious lesions** (e.g., Hillson, 1986; Lukacs, 1989).

A carious lesion can develop anywhere on a crown (*coronal caries*) or exposed root surface (*root caries*). The osteologist should note the locations of carious lesions: for example, occlusal, interstitial (interproximal), buccal pit (foramen cecum), cervical, root (see Figure 10–30). (Certain analyses might demand

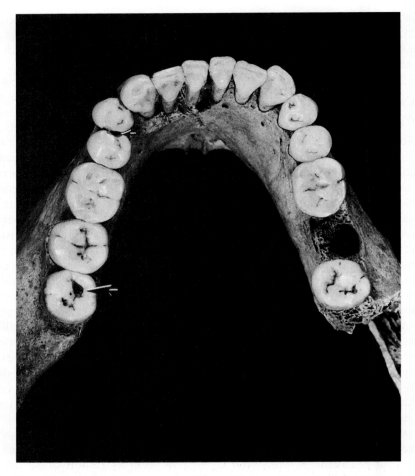

Figure 10–30   Examples of occlusal (*large arrow*) and interstitial (*small arrow*) caries; note the hypoconulid on all three molars (prehistoric, Pennsylvania).

greater specificity about location on crown and root.) The onset of enamel caries is noted in the presence of a spot of opacity (often white but sometimes dark, which characterizes more mature carious lesions); root caries is usually indicated by somewhat yellow or light brown spots (Hillson, 1986). As demineralization proceeds, a cavity forms. A carious lesion is typically shallow on roots; on crowns, depth and breadth of the cavity are more variable but often narrow and deep if in fissures. (Care should be taken not to identify

as a carious lesion the concavity that normally develops in the darker dentine when it is exposed through the lighter enamel.)

In root lesions, the cavity becomes narrower as it penetrates the dentine. In coronal caries, the lesion spreads out along the enamel–dentine junction, thinning the enamel and causing it to collapse, thereby enlarging the cavity. If infection proceeds into the dentine, acidogenic (acid-producing) bacteria first demineralize the dentine, leaving a matrix that will subsequently be attacked by proteolytic (protein-destroying) bacteria. Secondary dentine may be deposited in response to this assault. If the process of dentine caries formation stops, the secondary dentine will appear darker than the primary dentine around it. If the infection spreads into the pulp cavity (or infection is introduced into the pulp cavity of a severely worn tooth), it can travel through the root tip and into the surrounding bone.

As in bone, bacterial infection in teeth leads to necrosis of living tissue (e.g., dentine tubules, nerves, blood supply) and often to the production of pus. If the infection spreads through the root apex, the surrounding alveolar bone can become necrotic and a balloon-like, pus-filled cavity—a *periapical* or *apical abscess*—will form; at this stage, it can only be recognized radiographically. If infection is unimpeded, the periapical abscess will enlarge (filling with pus), bone necrosis will continue, and a drainage hole may erupt through the alveolar surface (usually on the buccal side); in dried specimens, the root tip or tips will be visible through this hole, which can be quite large (Figure 10–31). In contrast to the jagged edges of broken bone or the developmental thinning of bone that exposes underlying roots [either as a *dehiscence* (see Figure 9–16), which is a deficiency that begins at the alveolar margin and can course much of the length of a root, or a *fenestration* (Figure 9–18), which is a developmental lacuna in the bone above the alveolar margin that exposes a portion of the root], the drainage hole of an abscess is typically ovoid to circular in shape, with thin but crisply defined edges. The balloonlike space of the abscess is three-dimensionally symmetrical (typically circular). Continued enlargement of the abscess can progress to the alveolar margin, leaving the tooth sitting precariously in soft tissue. Since these infections also affect the periodontal ligaments, antemortem exfoliation or evulsion of the involved tooth or teeth may occur. In the upper jaw, infection can spread into the maxillary sinuses and nasal cavity. (Sometimes teeth that in life would have had little to tether them in place will fall out postmortem and should be looked or screened for when excavating the skull.)

Infection and inflammation not related to coronal caries can also affect the periodontal ligaments and surrounding alveolar bone and eventually lead to tooth loss. For example, if severe attrition exposes pulp cavities, they can become inflamed. Such inflammation can spread through root canals and affect periodontal ligaments, not just apically but anywhere along the length of the root. Eventually, inflammation turns to infection, resulting in lesion and abscess formation within the periodontium, away from the alveolar margin. As with more stereotypical periapical abscess, these nonapical periodontal abscesses (of pulpal origin) can expand three-dimensionally, destroying bone toward and away from the alveolar margin as well as toward the exterior. Because, like periapical abscesses, these periodontal abscesses are tooth-, rather than region- specific, the alveolar bone on either side of the affected tooth/alveolar bone will present normal (nonpathological) morphology —unless adjacent teeth had suffered pulpal inflammation and subsequent periodontal abscessing. Still, and even in the case of multiple adjacent teeth being lost due to periodontal abscesses, the restrictiveness rather than the pervasiveness of alveolar marginal destruction will be discernible.

Along with caries, the most frequently diagnosed pathological condition involving the jaws and teeth is **periodontal disease** or **periodontitis**. Technically, periodontitis is an inflammatory disease ("itis"). Osteologists have generally thought of it as being characterized by destruction of the alveolar margin, sometimes leading to destruction of tooth attachment and then tooth loss. However, as summarized by Clarke and Hirsch (1991, p. 241), the effects of periodontitis are less extensive: "In periodontitis, the crestal [alveolar] margin of bone undergoes loss of the surface cortical bone, exposing the porous cancellous structure of the supporting bone, usually with an accompanying change of the contour of the crest."

Historically, periodontitis has been conceived as the end of a sequence that begins with the accumulation of plaque, which then causes inflammation of the gums (gingivitis), which in turn leads to inflammation of periodontal tissues and destruction of the alveolar margin. This, however, may be the less frequent sequence of events. Rather, periodontitis often appears to be its own malady (Clarke and Hirsch, 1991). Osteologically, periodontitis has typically been diagnosed on the basis of increased distance (greater than

2 mm) between the alveolar margin and the cervix or enamel–cementum juncture of a tooth. This criterion is evident in Lukacs' (1989, p. 271) scoring of degrees of alveolar resorption: "(0) absent—no resorption; (1) slight—less than one-half of the root exposed; (2) moderate—more than one-half [of] the root exposed; (3) severe—evulsion of the tooth, remnants of the alveolus discernible; and (4) complete—tooth avulsed, alveoli completely obliterated." Factors other than periodontitis can, however, also lead to root exposure and eventual tooth loss.

Clarke and Hirsch (1991) reviewed the field and literature and, in addition to providing new data, pointed out that continued

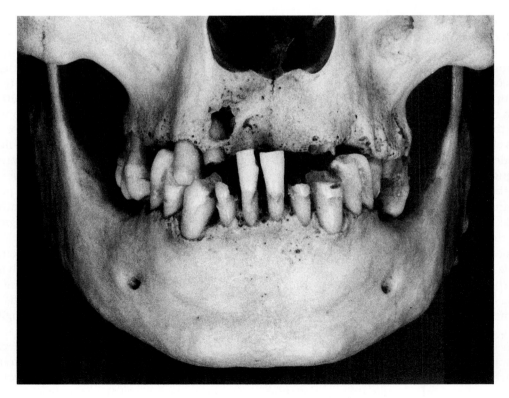

Figure 10–31   Loss of upper anterior teeth probably due to periapical inflammation (note severe attrition of various lower anterior teeth); also note, for example, periapical abscess exposing root of right upper canine, fenestration in left maxilla (exposing root of second premolar), and continued growth of various upper and lower teeth in the absence of occlusal counterparts (provenience unknown, University of Pittsburgh Dental School collection).

facial growth (particularly the lower face) throughout adult life as well as progressive dental attrition can lead to a distancing between the enamel–cementum juncture of a tooth and the alveolar margin. That is, in the absence of other factors, continued growth of the lower face, dental attrition, or especially both together should lead to the dissociation of upper and lower teeth that would otherwise occlude with one another. The reason, however, that an individual's upper and lower teeth remain in occlusion is that they continue to erupt. And it is the continual, albeit incremental, eruption of teeth throughout an individual's life—not periodontitis—that can (and often does) lead to the distancing between a tooth's enamel–cementum juncture and the associated alveolar margin (Figure 10–31). Thus, if one uses, for example, the inferior dental canal (through which the mandibular nerve courses within the mandible) as a stable landmark, one will find that the distance between it and the alveolar margin changes little over the life of an individual in whom an increasing amount of tooth root has become exposed. Critical to a diagnosis of periodontitis, therefore, is not a measurement of amount of root exposure but an inspection

of the alveolar margin in order to determine if crestal bone is still healthy and crisply defined or if it is indeed pathological (e.g., being resorbed and/or lacking contour) (Figure 10–31, see also Figures 7–16 and 9–6).

It might make sense to think that non-traumatic tooth loss, if not due to unabated abscessing, would have to result from an infectious disease, such as periodontitis. But, as Clarke and Hirsch (1991) point out, continued tooth eruption can lead to tooth loss simply by diminishing the amount of root available within the alveolus to be adequately anchored by the periodontal ligaments. Although a tooth may generate additional root cementum, the rate at which it is deposited (and the rate at which compensatory root lengthening would occur) is not equal to the rate of compensatory tooth eruption. The cautionary note here is that one must be careful "to separate physiological attachment loss resulting from continuous tooth eruption from pathological bone loss" (Clarke and Hirsch, p. 247). Indeed, Clarke and Hirsch find that, in contrast to pulpal infections (leading to alveolar bone destruction), the role (if any) of periodontitis in causing tooth loss is and has in the past been minimal.

Osteometry

O steometry is the measurement of bones. Historically, such measurement has often been focused on the skull, the metric analysis of which is specifically referred to as *craniometry*. Although compiling long lists of measurements for its own sake is of little value, the quantification (or at least the attempted quantification) of morphological features is necessary in order to make accurate comparisons between and within skeletal populations. Osteometric measurements and the use of statistical techniques (such as multivariate analysis) allow the researcher to make such comparisons. However, in making these skeletal comparisons, one should never lose sight of the fact that the bone (and parts thereof) being measured was once a living and dynamic tissue and, thus, the size and development of particular features may have been modified by surrounding musculature and other extraosseous tissues.

Skeletal measurements include linear dimensions between two osteometric landmarks (e.g., length, breadth, and height), circumferential dimensions around a bone or structure, arc dimensions (partial circumference) from one point to another on the margin of a rounded structure, and angles between two planes of bone. All measurements should be taken and recorded in millimeters or centimeters. Measurements that can be taken on either the right or left side (e.g., orbital height, mastoid length, minimum breadth of ascending ramus of mandible) should, as a rule, be taken on the left side so that measurements from one study to the next will be comparable. The instruments used to take these measurements include the spreading caliper, sliding caliper, coordinate caliper, head spanner, tape, and osteometric board. The measurement of angles often requires more sophisticated and specialized instruments.

Skeletal indices are ratios that attempt to describe mathematically the shape or configuration of bones. Specifically, an index considers two traits simultaneously and is defined as a ratio of one measurement to another expressed as a percentage. Indices provide a quick and simple technique for the comparison and delineation of differences between individuals with regard, for example, to race and sex.

Cranial and Mandibular Measurements

Braincase

1. *Maximum cranial length* (*g–op, GOL*): Glabella to opisthocranion (spreading caliper).
2. *Maximum cranial breadth* (*eu–eu, XCB*): Right euryon to left euryon (spreading caliper).
3. *Maximum cranial height* (*ba–b, BBH*): Basion to bregma (spreading caliper).
4. *Porion–bregma height* (*po–b*): Porion to bregma (head spanner).
5. *Basion–porion height* (*ba–po*): Basion to porion (coordinate caliper).
6. *Auricular height* (*po–ap*): Porion to apex (head spanner).
7. *Minimum frontal breadth* (*ft–ft*): Right frontotemporale to left frontotemporale (spreading or sliding caliper).
8. *Biasterionic breadth* (*as–as*): Right asterion to left asterion (sliding caliper).
9. *Foramen magnum* (*foraminal*) *length* (*ba–o, FOL*): Basion to opisthion (sliding caliper).
10. *Foramen magnum* (*foraminal*) *breadth* (*FOB*): Maximum (sliding caliper).
11. *Frontal* (*nasion–bregma*) *chord* (*n–b, FRC*): Nasion to bregma (sliding caliper).
12. *Parietal* (*bregma–lambda*) *chord* (*b–l, PAC*): Bregma to lambda (sliding caliper).
13. *Occipital* (*lambda–opisthion*) *chord* (*l–b, OCC*): Lambda to opisthion (sliding caliper).
14. *Frontal arc* (*n–b*): Nasion to bregma (tape).
15. *Parietal arc* (*b–l*): Bregma to lambda (tape).
16. *Occipital arc* (*l–o*): Lambda to opisthion (tape).
17. *Transverse biporial arc* (*po–po*): Right porion to left porion (tape).

Facial Skeleton

18. *Basion–nasion length* (*b–n, BNL*): Basion to nasion (spreading or sliding caliper).
19. *Basion–alveolare length* (*b–ids*): Basion to alveolare (sliding caliper).
20. *Total facial height* (*n–gn*): Nasion to gnathion (spreading or sliding caliper).
21. *Upper facial height* (*n–ids*): Nasion to alveolare (spreading or sliding caliper).
22. *Bizygomatic breadth or facial width* (*zy–zy, ZYB*): Right zygion to left zygion (spreading or sliding caliper).
23. *Bimaxillary breadth* (*zm:a–zm:a, ZMB*): Right zygomaxillare to left zygomaxillare (spreading or coordinate caliper).

Nasal Region

24. *Nasal height* (*n–ns, NLH*): Nasion to nasospinale (sliding caliper).
25. *Nasal breadth* (*al–al, NLB*): Right alare to left alare (sliding caliper).
26. *Simotic chord or least nasal breadth* (*WNB*): Shortest distance between right and left nasomaxillary sutures (sliding or coordinate caliper).

Orbit

27. *Orbital height (orb. h.; OBH)*: Maximum vertical (i.e., perpendicular to the horizontal plane of the orbit) distance between superior and inferior orbital margins (sliding caliper).
28. *Orbital breadth (width) (OBB)*: Variously defined as the distance from ectoconchion (*ec*) to either maxillofrontale (*mf*), dacryon (*d*), or lacrimale (*la*) (sliding caliper).
29. *Biorbital breadth (ec–ec, EKB)*: Right ectoconchion to left ectoconchion (sliding or coordinate caliper).
30. *Interorbital, bidacryonic (chord) or maxillofrontal breadth*: Right maxillofrontale to left maxillofrontale (*mf–mf*) or dacryon to dacryon (*d–d*) (sliding or coordinate caliper).
31. *Midorbital breadth (zo–zo)*: Right zygoorbitale to left zygoorbitale (sliding caliper).
32. *Bidacryonic arc (d–d)*: Right dacryon to left dacryon (tape).

Palate

33. *Maxilloalveolar (external palatal) length (pr–alv)*: Prosthion to alveolon (sliding caliper with reflexed arm).
34. *Maxilloalveolar (external palatal) breadth (MAB)*: Right ectomolare to left ectomolare (*ecm–ecm*) or maximum wherever found (sliding caliper).
35. *Palatal (internal) length (ol–sta)*: Orale to staphylion (sliding caliper).
36. *Palatal (internal) breadth*: Right endomolare to left endomolare (*enm–enm*) or maximum wherever found (sliding caliper).

Mandible

37. *Bicondylar (intercondylar) breadth (cdl–cdl)*: Right condylion laterale to left condylion laterale (sliding caliper).
38. *Bigonial breadth (go–go)*: Right gonion to left gonion (sliding caliper).
39. *Foramen mentalia breadth (z–z)*: Right mentale to left mentale (sliding caliper).
40. *Ascending ramus height (go–cdl)*: Gonion to condylion laterale parallel to the vertical axis of the ramus (sliding caliper).
41. *Coronoid process height (CrH)*: Maximum vertical height of coronoid process from planar surface on which body of mandible is placed (sliding caliper).
42. *Maximum ramus breadth*: Greatest distance between a line connecting the posteriormost point on a condyle with the angle and the anteriormost point on the ramus, taken at a right angle to the imaginary line (osteometric board).
43. *Minimum ramus breadth (RB)*: Minimum distance wherever found between anterior and posterior border of ascending ramus (sliding caliper).
44. *Symphyseal height (gn–idi)*: Gnathion (or tangent to lowest point on either side of gnathion) to infradentale (sliding caliper).
45. *Mandibular body (corpus) height*: Maximum height at the position of the second molar taken to the alveolar margin from the surface on which the

mandible is placed; also sometimes taken at the position between the second premolar and the first molar (sliding caliper).

46. *Mandibular body (corpus) breadth*: Maximum thickness at the position of the second molar taken perpendicular to the long axis of the body; also sometimes taken at the position between the second premolar and the first molar (sliding caliper).

47. *Maximum projective mandibular length*: Distance between the posteriormost extension of the mandibular condyles and gnathion; taken when mandibular body is placed on a horizontal surface (osteometric board).

48. *Mandibular body (corpus) length (gn–go)*: Gnathion to gonion (sliding caliper or mandibulometer).

Postcranial Measurements

Length measurements on long bones (see individual bones below) can be of one of two basic kinds: *morphological* and *physiological (functional, oblique, bicondylar, or anatomical)*. Morphological length is the *maximum* length that can be measured from anywhere on one end of the bone to anywhere on the opposite end; morphological length lies parallel to the long axis of the bone. When maximum length is specified below for a particular long bone, morphological length is to be taken. Physiological (functional, oblique, bicondylar, or anatomical) length measures the length of the bone as it would be oriented when articulated and in the anatomical position. When physiological length is specified for a particular bone, it will be identified as such. Transverse measurements (e.g., widths, diameters) are taken perpendicular to the long axis of the bone.

Sternum (All Are Maximums)

1. *Manubrium length (midsagittal)*: Distance from the jugular notch to the facet of articulation with the body of the sternum (sliding or spreading caliper).

2. *Manubrium width*: Distance between midpoints of the right and left facets for the first costal cartilage taken perpendicular to the long axis (sliding or spreading caliper).

3. *Maximum length of sternal body (mesosternal) (midsagittal)*: Distance from facet for articulation with the manubrium to facet for articulation with the xiphoid process (sliding or spreading caliper or osteometric board).

4. *Sternal body (mesosternal) width at first sternebra*: Minimum distance from side to side across third sternebra (lowest point between the facet for the second and third costal cartilage on each side) (sliding caliper).

5. *Sternal body (mesosternal) width at third sternebra*: Minimum distance from side to side across first sternebra (lowest point between the facet for the fourth and fifth costal cartilage on each side) (sliding caliper).

6. *Maximum sternal (manubrium + mesosternum) length (midsagittal)*: Distance from the jugular notch (i.e., lowest midline point superiorly) of the manubrium to the facet for the xiphoid process; Stewart and McCormick (1983) added the lengths taken separately on the manubrium and mesosternum (sliding caliper or osteometric board).

Clavicle

7. *Maximum length* (as if measuring a straight bone): Maximum distance from sternal to scapular articular end (sliding caliper or osteometric board).
8. *Maximum breadth*: Widest distance taken perpendicular to the "midline" (sliding caliper).
9. *Maximum breadth of sternal end*: Widest distance taken perpendicular to the midline of the segment (sliding caliper).
10. *Midshaft circumference*: Taken midway between the articular ends (tape).

Scapula

11. *Scapular (total) height (maximum length)*: Maximum distance from the superior to the inferior angle (sliding caliper or osteometric board).
12. *Scapular (maximum) breadth*: Maximum distance (following the long axis of the spine) from the middle of the posterior (dosal) margin of the glenoid cavity to the vertebral border (sliding caliper).
13. *Length of scapular spine*: Maximum distance from the lateralmost extent of the acromion to the vertebral border; this point on the vertebral border will be the same as that determined for scapular (maximum) breadth (sliding caliper).
14. *Length of supraspinous line*: Distance between the confluence of the spine and vertebral border (i.e., the point used for maximum scapular breadth and scapular spine length) and the most superior extent of the superior angle (sliding caliper).
15. *Length of infraspinous line*: Distance between the confluence of the spine and vertebral border (see measurements 11 to 13 above) and most inferior extent of the inferior angle (sliding caliper).
16. *Length of glenoid cavity*: Maximum distance across glenoid cavity perpendicular to the anteroposterior axis (sliding caliper).
17. *Breadth of glenoid cavity*: Maximum distance across glenoid cavity measured at a right angle to the axis of the length of the glenoid cavity (sliding caliper).

Humerus

18. *Maximum length*: (osteometric board).
19. *(Maximum) Vertical diameter of the head*: Taken parallel to the coronal plane on the margin of the articular surface of the head to obtain the maximum distance between one point and a point on the opposite side (sliding caliper).
20. *(Maximum) Transverse diameter of the head*: Maximum anteroposterior distance measured between one point on the margin of the articular surface and a point on the opposite side (sliding caliper).
21. *Maximum diameter of the head*: Taken anywhere along the margin of the articular surface to obtain the maximum distance from one point to a point on the opposite side (sliding caliper).

22. *Minimum circumference of shaft*: Taken inferior to the deltoid tuberosity (tape).
23. *Midshaft circumference*: Taken on the shaft perpendicular to the midline axis at the true midpoint of maximum length (tape).
24. *Maximum midshaft diameter*: Maximum distance measured perpendicular to the midline axis at the true midpoint of maximum length (sliding caliper).
25. *Minimum midshaft diameter*: Taken as above but for minimum measurement (sliding caliper).
26. *Biepicondylar (bicondylar, distal epiphyseal) width (breadth)*: Taken from the medial to the lateral epicondyle (sliding caliper).
27. *(Distal) Articular width*: Distance from the most medial extent of the trochlen to the most lateral extent of the capitulum (sliding caliper).
28. *Maximum diameter (width) of capitulum*: (sliding caliper).

Radius

29. *Maximum length*: (osteometric board).
30. *Minumum shaft circumference*: Taken perpendicular to the midline axis wherever necessary to obtain minumum (tape).
31. *Midshaft circumference*: Taken perpendicular to the midline axis at the true midpoint of maximum length (tape).
32. *Minimum shaft diameter*: Taken perpendicular to the midline axis at the true midpoint of maximum length (sliding caliper).
33. *Head circumference*: (tape).
34. *Maximum (mediolateral) distal (epiphyseal) breadth (width)*: Maximum distance from the most medial extension (beyond the ulnar notch) to the lateral side (sliding caliper or osteometric board).

Ulna

35. *Maximum length*: (osteometric board).
36. *Physiological length*: (osteometric board).
37. *Minimum shaft circumference*: Taken perpendicular to the midline axis wherever necessary to obtain minimum (tape).
38. *Midshaft circumference*: Taken perpendicular to the midline axis at the true midpoint of maximum length (tape).
39. *Transverse diameter (width) of shaft*: Maximum mediolateral width taken at the point of greatest development of the interosseous crest (sliding caliper).

Metacarpals and Phalanges

40. *Physiological length*: In general, measured from the midpoint of the proximal epiphysis to the most distal point of the distal end (head); the proximal point specifically with regard to metacarpal II is the midpoint of the notch, to metacarpal III is the midpoint of the longitudinal ridge, and to metacarpal V is the deepest point in the concavity (sliding caliper).
41. *Transverse diameter*: Width measured at the proximal end (sliding caliper).

Vertebrae

42. *Anterior vertebral body* (*centrum*) *height*: Taken in the midline [median (mid-sagittal) plane], from the upper (cranial or superior) to the lower (caudal or inferior) margin of the anterior (ventral) side of the centrum (sliding caliper).
43. *Posterior vertebral body* (*centrum*) *height*: Same as above but taken on the posterior (dorsal) side of the centrum (sliding or spreading caliper).
44. *Maximum breadth* (*width*): Maximum distance between apices of transverse processes (sliding caliper).

Sacrum

45. *Sacral length* (*maximum anterior height*): Vertical median (midsagittal) distance from the anterior (ventral) rim of the promontory to most caudal (inferior) extent of the last vertebra; only five-segmented sacra should be compared (sliding caliper).
46. *Sacral breadth* (*maximum anterior breadth*): Greatest distance from the anterolateral border of right to left alae (lateral masses) (sliding caliper).
47. *Midventral* (*curved*) *length*: Distance in the median (midsagittal) plane along the anterior (ventral) surface from the rim of the promontory to most caudal (inferior) extent of the last vertebra; only five-segmented sacra should be compared (tape).
48. *Maximum depth of curvature*: Greatest distance from the cord representing sacral length to the anterior (ventral) surface (coordinate caliper).
49. *Transverse diameter* (*external*): Maximum bilateral distance perpendicular to the median (midsagittal) plane of the body of the first sacral vertebra taken to the outside of the epiphyseal ring (sliding caliper).
50. *Transverse diameter* (*internal*): Maximum bilateral distance perpendicular to the median (midsagittal) plane of the body of the first sacral vertebra taken to the inside of the epiphyseal ring (sliding caliper).
51. *Anteroposterior diameter* (*external*): Distance in the median (midsagittal) plane between the anterior (ventral) and posterior (dorsal) outer margins of the epiphyseal ring (sliding caliper).
52. *Anteroposterior diameter* (*internal*): Distance in the median (midsagittal) plane between the anterior (ventral) and posterior (dorsal) inner margins of the epiphyseal ring (sliding caliper).
53. *Length of ala* (*lateral part, lateral mass, transverse process element*): Maximum distance on one side from the point that yields the maximum transverse diameter of the first sacral vertebra to the lateral margin of the ala (sliding caliper).

Os Coxa and Pelvis

54. *Maximum* (*os coxa*) *height*: Greatest distance obtainable between the ischium and the iliac crest of an os coxa (osteometric board).
55. *Anteroposterior height of* (*articulated*) *pelvis* (*sagittal, anteroposterior, or true conjugate diameter*): Average of measurements taken from the midpoint of the promontory (sacrum) to the right and left pubic crests (i.e., anterior border between the pubic tubercle and symphysis) (sliding caliper).

56. *Acetabular diameter*: Variably measured parallel to the superior pubic ramus, the iliac pillar, and the ascending ramus of the ischium (sliding caliper).
57. *Bi-iliac (articulated pelvic) breadth*: Greatest distance obtainable between right and left iliac crests (sliding caliper).
58. *Transverse breadth (diameter) of (articulated) pelvis*: Greatest distance obtainable between right and left arcuate lines (sliding caliper).
59. *Iliac breadth*: Taken from the anterior to the posterior superior iliac spine (sliding caliper).
60. *(Maximum) Iliac height*: Greatest distance obtainable between the point of confluence in the acetabulum of the ilium, pubis, and ischium and the iliac crest (sliding caliper).
61. *(Maximum) Ischial length*: Greatest distance obtainable between the point of confluence in the acetabulum of the ilium, pubis, and ischium and the ischial tuberosity (sliding caliper).
62. *(Maximum) Pubic length*: Greatest distance obtainable between the point of confluence in the acetabulum of the ilium, pubis, and ischium and the anterior (ventral) face of the superior pubic ramus (sliding caliper).
63. *(Greater) Sciatic notch position*: Taken from the point at which the line of greatest sciatic notch width and greatest depth measured from this line meet to the ischial spine (sliding caliper).
64. *(Greater) Sciatic notch width or height*: Variably measured from the pyramidal spine to the base or tip of the ischial spine (sliding caliper).
65. *Acetabulosciatic breadth*: The distance from a point on the anterior border of the greater sciatic notch that is midway between the apex of the notch and the ischial spine to the acetabular border, all the while trying to stay at right angles to both borders.
66. *Acetabulum–innominate line length*: The distance from the lateralmost point on the pubic portion of the acetabular border to the innominate line, taken at a right angle to the plane of the obturator foramen.
67. *Anterior superior iliac spine "measure"*: The shortest distance from the anterior iliac spine to the edge of the sciatic notch less the shortest distance from the anterior iliac spine to the auricular surface.

Femur

68. *Maximum length*: (osteometric board).
69. *Physiological (trochanteric oblique) length*: (osteometric board).
70. *Trochanteric oblique length*: Taken in the same position as physiological length but from the level of the condyles to the most proximal extent of the greater trochanter (osteometric board).
71. *Maximum vertical diameter of head*: Greatest distance obtainable from a point on the inferior edge of the margin of the articular surface of the head to a point opposite it on the superior edge of the margin (sliding caliper).
72. *Maximum subtrochanteric anteroposterior (sagittal) diameter*: Taken in the median (sagittal) plane at a right angle to the long axis of the femur immediately below the lesser trochanter (sliding caliper).

73. *Maximum subtrochanteric transverse (mediolateral) diameter*: Taken in the same plane as above but perpendicular to the maximum subtrochanteric anteroposterior diameter above (sliding caliper).

74. *Midshaft circumference*: Taken at the true midpoint of morphological length, perpendicular to the long axis of the bone, following the topography of the shaft (tape).

75. *Maximum anteroposterior midshaft diameter*: Taken at the true midpoint of morphological length, perpendicular to the long axis of the bone (sliding caliper).

76. *Maximum transverse (mediolateral) midshaft diameter*: Taken at the true midpoint of morphological length, perpendicular to the long axis of the bone (sliding caliper).

77. *Maximum bicondylar (epicondylar, distal epiphyseal) breadth (width)*: Greatest distance obtainable between medial and lateral epicondyles (sliding caliper or osteometric board).

78. *Popliteal length*: Taken on the posterior (dorsal) surface from the midpoint of the intercondylar margin of the politeal surface to the point of confluence of the medial and lateral supracondylar lines (sliding caliper).

Tibia

79. *Maximum length*: (osteometric board).

80. *Length*: Taken from the highest elevation of the lateral condyle to the distal-most extension of the medial malleolus (osteometric board).

81. *Physiological length*: Taken from the distal articular surface (just lateral to the malleolus) to the lowest point on the medial condylar surface (large sliding caliper or osteometric board).

82. *Maximum proximal (bicondylar, epiphyseal) breadth (width)*: Greatest mediolateral measurement across the condyles obtainable (sliding caliper).

83. *Midshaft circumference*: Taken at the true midpoint of maximum length, following the topography of the bone (tape).

84. *Maximum midshaft (anteroposterior) diameter*: Taken at the true midpoint of maximum length (sliding caliper).

85. *Circumference at nutrient foramen*: Maximum circumference obtained following the topography of the bone (tape).

86. *Anteroposterior (sagittal cnemic) diameter at the nutrient foramen*: Maximum anteroposterior measurement obtainable at the level of the nutrient foramen, perpendicular to the long axis of the bone (sliding caliper).

87. *Mediolateral (transverse cnemic) diameter at the nutrient foramen*: Maximum mediolateral measurement obtainable at the level of the nutrient foramen, perpendicular to the long axis of the bone and in the same plane as the previous measurement (sliding caliper).

88. *Minimum shaft circumference*: Usually found toward the distal portion of the shaft and taken following the topography of the bone (tape).

89. *Distal (epiphyseal) breadth (width)*: Maximum distance obtainable from the fibular notch to a point on the medial surface of the malleolus (sliding caliper).

Fibula

90. *Maximum length*: (osteometric board).
91. *Midshaft circumference*: Taken at the true midpoint of maximum length, following the topography of the bone (tape).
92. *Anteroposterior diameter*: Taken at the true midpoint of maximum length and perpendicular to the long axis of the bone, the maximum anteroposterior measurement obtainable (sliding caliper).
93. *Transverse diameter*: Taken at the true midpoint of maximum length, perpendicular to the long axis of the bone and in the same plane as the previous measurement, the maximum anteroposterior measurement obtainable (sliding caliper).
94. *Maximum distal (epiphyseal) breadth (width)*: (sliding caliper).

Calcaneus

95. *Maximum length*: Maximum measurement obtainable between the posterior surface of the tuberosity and the anterosuperior margin of the cuboidal facet (sliding caliper or osteometric board).
96. *Minimum (mediolateral) width*: Taken perpendicular to the long axis of the bone and found between the tuberosity and the posterior facet for the talus (sliding caliper).
97. *Body height*: Maximum measurement obtainable between the inferior (plantar) surface of the tuberosity and the superiormost elevation of the posterior facet for the talus (osteometric board).
98. *Height of cuboidal facet*: Maximum vertical measurement from the inferiormost to the superiormost margins of the articular surface (sliding caliper).
99. *Length posterior to the posterior facet for the talus*: With the bone oriented as for maximum length, the maximum measurement obtainable from the highest elevation of the posterior facet for the talus to the posterior surface of the tuberosity (sliding caliper).
100. *Load arm length*: With the bone oriented as for maximum length, the maximum measurement obtainable from the posterior margin of the posterior facet for the talus to the anterosuperior margin of the cuboidal facet (sliding caliper).
101. *Load arm width*: The maximum measurement obtainable between the lateral margin of the posterior facet for the talus and the medial extension of the sustentaculum tali (osteometric board).

Talus

102. *Maximum length*: The maximum measurement obtainable from the sulcus for the hallucis longus muscle posteriorly to the anterior surface of the head (i.e., the facet for the navicular) (sliding caliper).
103. *Width*: The greatest distance between the lateralmost projection of the surface that articulates with the (lateral) malleolus of the fibula to a point opposite on the surface that articulates with the tibia; the latter point is

often found just anterior to the midpoint of the articular surface for the tibia (sliding caliper).
104. *Body height*: With the inferior surface of the bone lying on a flat surface (e.g., the vertical, permanent end of an osteometric board), the maximum measurement obtainable between the inferior and the highest elevation on the superior surface; the latter is often the medial rim of the facet for the tibia (osteometric board).
105. *Maximum length of the trochlea for the tibia*: Taken in the anteroposterior plane (sliding caliper).
106. *Maximum width of the trochlea for the tibia*: Taken at a right angle to the previous measurement (sliding caliper).

Metatarsals and Phalanges

107. *Maximum length*: (sliding caliper or osteometric board).
108. *Maximum width of proximal end*: (sliding caliper).
109. *Minimum dorsal–plantar width*: (sliding caliper).
110. *Minimum mediolateral width*: (sliding caliper).

Postcranial Indices

1. Scapular index = max. breadth × 100/max. length
(Reflects relative squatness or narrowness of scapula; the higher the index, the broader the bone.)
2. Claviculohumeral index = max. clavicular length × 100/max. humeral length
(Reflects breadth superiorly of the thorax relative to the length of the upper arm; the higher the index, the broader the thorax.)
3. Clavicular robustness (robusticity, circumference: length) index = midclavicular circumference × 100/max. clavicular length
[Reflects relative size of the shaft; the higher the index, the more robust the bone; (poor) indicator of sex.]
4. Robusticity index (humerus) = min. shaft circumference × 100/max. length
(Reflects the relative size of the humeral shaft; the higher the index, the more robust the bone).
5. Diaphyseal (shaft) index (humerus) = min. shaft diameter × 100/max. length
(Reflects the relative size of the humeral shaft; the higher the index, the more robust the bone.)
6. Brachial (radiohumeral) index = max. radius length × 100/max. humerus length
(Reflects the length of the forearm relative to the upper arm.)
7. Caliber index (ulna) = min. shaft circumference × 100/physiological length
(Reflects the relative size of the shaft; the higher the index, the more robust the bone.)
8. Platymeric index (femur) = subtrochanteric sagittal diameter × 100/ subtrochanteric transverse diameter.
(Reflects the degree to which the most proximal part of the femoral shaft is flattened or compressed anteroposteriorly.)

Platymeria (platymeric, flattened) = <85.0
Eurymeria (eurymeric, moderate) = 85.0–99.9
Stenomeria (stenomeric, rounded) = >99.9

9. Robusticity index (femur) = (mediolateral + anteroposterior diameters) × 100/bicondylar length
(Reflects the relative size of the shaft; the higher the index, the more robust the bone.)

10. Pilastric index (femur) = anteroposterior diameter × 100/mediolateral diameter
(Reflects relative development of the linea aspera; useful in determining sex.)

11. Intermembral index = (max. humeral + max. radial lengths) × 100/(max. femoral + max. tibial lengths)
(Reflects the length of the upper limb relative to the lower limb; the higher the index, the longer the upper limb.)

12. Crural index = max. tibial length × 100/max. femoral length
(Reflects the length of the lower leg relative to the thigh; the higher the index, the longer the lower leg.)

13. Humerofemoral index = max. humeral length × 100/max. femoral length
(Reflects the length of the upper arm relative to the thigh; the higher the index, the longer the length of the upper arm.)

14. Platycnemic index (tibia) = mediolateral diameter × 100/anteroposterior diameter
(Reflects the degree to which the most proximal part of the tibial shaft is flattened in an anteroposterior direction.)

Hyperplatycnemia (hyperplatycnemic, extremely flat) = <55.0
Platycnemia (platycnemic, very flat) = 55.0–62.9
Mesocnemia (mesocnemic, moderately flat) = 63.0–69.9
Eurycnemia (eurycnemic, broad, wide) = >69.9

15. Vertebral index = posterior body height × 100/anterior body height
(Reflects the relative uniform thickness of a vertebra; the higher the index, the more compressed the body is anteriorly.)

16. Sacral (breadth) index = sacral breadth × 100/sacral length
[Reflects the breadth (maximum anterior breadth) of the sacrum relative to its length (maximum anterior height); the higher the index, the broader the sacrum.]

	Female (Average)	Male (Average)
Egyptian	99.1	94.3
Andaman	103.4	94.8
African	103.6	91.4
Japanese	107.1	101.5
Australian	110.0	100.2
European	112.4	102.9

17. Ischiopubic index = pubic length × 100/ischial length
(Reflects the length of the pubis relative to the ischium; used in determining sex, with female indices being on average 15% higher than those of males.)

White Americans		African-Americans	
Male	= <90.0	Male	= <84.0
Indeterminate	= 90.0–95.0	Indeterminate	= 84.0–88.0
Female	= >95.0	Female	= >88.0

18. Pelvic brim index = anteroposterior height × 100/transverse diameter (Reflects the relative size of the pelvic cavity.) Classification according to Turner (1886):

Platypellic (transversely flattened pelvic cavity)	= <90.0
Mesatipellic (intermediate)	= 90.0–94.9
Dolichopellic (long anteroposteriorly)	= >94.9

According to Clyne (1963, cited in Clemente, 1984):

	Average
Platypelloid (platypellic)	59.2
Android (brachypellic)	78.1
Gynecoid (mesatipellic)	78.9
Anthropoid (dolichopellic)	90.8

19. Acetabulum-pubic index = acetabular diameter × 100/pubic length (Used in determining sex.)

White Americans	
Males	= >71
Females	= <70

20. Sciatic notch-acetabular index = sciatic notch width × 100/acetabular diameter (Used in determining sex.)

White Americans		African-Americans		Native Americans	
Males	= <87.0	Males	= <87.6	Males	= <87.0
Females	= >87.0	Females	= >87.6	Females	= >87.0

21. Calcaneal load index = load arm width × 100/load arm length (Reflects the relative bulk of the anterior segment of the calcaneus.)
22. Calcaneal length index = load arm length × 100/max. length (Reflects the length of the anterior element relative to the total length of the bone; used most often in studies on locomotory differences among different species.)

Comparative Osteology

Human, Deer, Bear, Pig

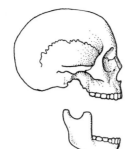
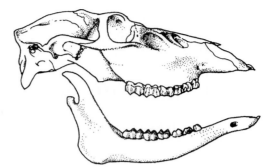
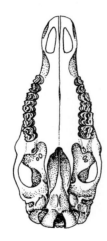

Figure B–1   Comparison of skulls of (*from left to right*) a human, a caribou [representative of a cervid (deer), which is a ruminating, "cloven-hooved" artiodactyl], a bear (representative of a carnivore), and a pig (a suid, which is a nonruminating, non-"cloven-hooved" artiodactyl): (*top row*) lateral view, (*bottom row*) occlusal view.

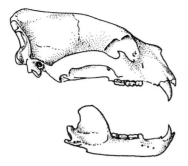
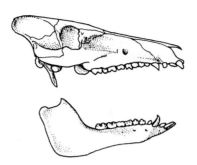

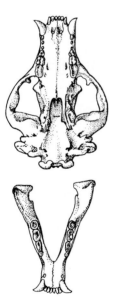
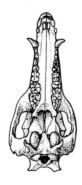
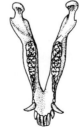

Figure B–1   (*continued*)

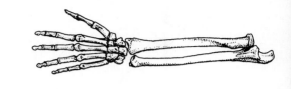

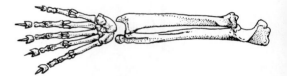

Figure B–2   The elements of the right forelimb of (*from top to bottom*) a human, a caribou, a bear, and a pig.

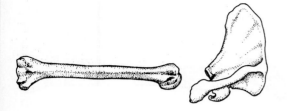

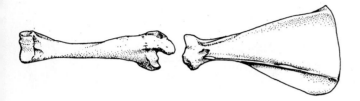

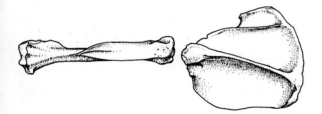

Figure B–2   (*continued*)

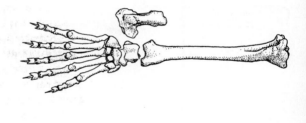

Figure B–3   The sacrum, right os coxa, and elements of the right hindlimb of (*from top to bottom*) a human, a caribou, a bear, and a pig.

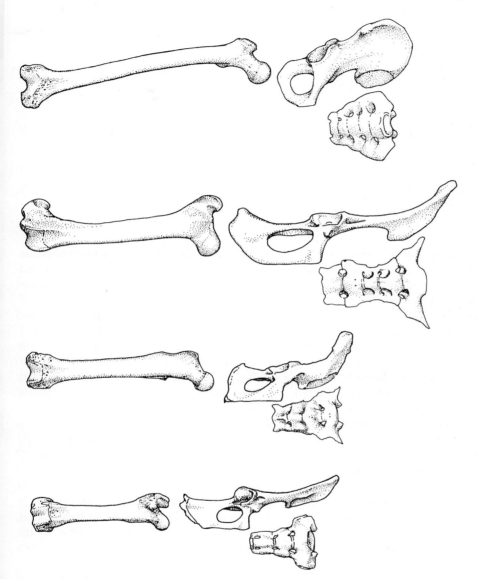

Figure B–3   (*continued*)

Glossary

Abduction: A laterally directed movement in the coronal plane away from the median (sagittal) plane; opposite of **adduction**.

Acanthion: The point at the base of the anterior nasal spine.

Acidogenic: Acid-producing.

Acromion: The most lateral and superior point on the acromial process.

Adduction: A medially directed movement in the coronal plane toward the median sagittal plane; opposite of **abduction**.

Alare (*al*): The lateralmost point on the margin of the nasal aperture.

Alveolare (abbr. *ids*; also *infradentale superius* or *hypoprosthion*): The inferiormost point of the alveolar margin between the upper central incisors; often confused with **prosthion**. Compare with **infradentale** and **prosthion**.

Alveolon (*alv*): The point on the hard palate at which a line drawn through the distal margin of the maxillary alveolar processes intersects the median (sagittal) plane.

Alveolus (pl. *alveoli*): A small saclike dilatation or socket in which the root of a tooth is nestled.

Antebrachium (pl. *antebrachia*): The forearm; the distal half of the upper limb, which lies between the elbow and the wrist and encompasses the radius and the ulna.

Anterior (also *frontal*, *ventral*): Toward the front; opposite of **posterior**.

Anteroposterior (AP) plane: see **median plane**.

Aperture: In general, a medium-to-large opening (e.g., the nasal aperture or a septal aperture in the humerus).

Apex (also *tip*; *ap*): 1. The top of a pointed structure (e.g., patella, fibular head, petrous portion of temporal bone). 2. The highest point on the skull, which also lies in the same vertical plane as **porion**; the skull must be in the Frankfort horizontal.

Apical: Pertaining to the apex or highest part of a structure; opposite of **basilar**.

Appendicular skeleton: Consists of the 63 paired bones of the upper and lower limbs (appendages) and the pectoral and pelvic girdles; does not include the sacrum.

Asterion (*as*): The juncture of temporal, parietal, and occipital bones (also of the lambdoid, temporoparietal, and occipitomastoid sutures).

Axial skeleton: The skeleton of the axis of the body, which consists of 80 bones and includes the skull, hyoid, sternum, ribs, vertebrae, sacrum, and coccyx.

Basilar: Pertaining to the base or lowest part of a structure; opposite of **apical**.

Basion (*ba*): 1. (General) A point at the midline of the anterior border of the foramen magnum. 2. A point at the position pointed to by the apex of the triangular surface at the base of either condyle (i.e., the average position from the crests bordering this area), about halfway between the inner border directly facing the posterior border (**opisthion**) and the lowermost point on the border (i.e., between **endobasion** and **hypobasion**); it is the same as **endobasion** if the anterior border of the foramen magnum is thin and sharp (Howells, 1973).

Brachium (pl. *brachia*): The arm or upper arm; the proximal half of the upper limb, which lies between the shoulder and the elbow and encompasses the humerus.

Braincase (also the *cranium*; not including the facial skeleton): The bony cavity that houses the brain; formed by the occipital, parietal, temporal, frontal, ethmoid, and sphenoid bones.

Bregma (*b*): The juncture of the coronal and sagittal sutures in the median (sagittal) plane; should an ossicle be present, the landmark can be located by drawing in pencil a continuation of the sutures until these lines intersect.

Calvaria (pl. *calvariae*; also *calotte, concha of cranium, skullcap*, and, incorrectly, *calvarium*): The superior portion of the **braincase** or the roof of the cranial vault; formed by portions of the frontal, parietal, and occipital bones.

Canal (also *duct*): A narrow tubular channel (e.g., the hypoglossal canal of the occipital bone, the semicircular canals of the bony labyrinth of the ear, or the haversian canals of compact bone).

Caput (pl. *capita*; also *skull*): The head; includes the **cranium** and the **mandible**.

Carpus (pl. *carpi*): The wrist; the region between the forearm and the hand containing eight irregular carpal bones (the scaphoid, lunate, triquetrum, pisiform, hamate, capitate, trapezoid, and trapezium).

Cavity (also *fossa* and *sinus*): A hollow space or depression [e.g., the glenoid cavity (also fossa) of the scapula, the cranial cavity (composed of numerous bones), and the pulp cavity of teeth].

Cervix (pl. *cervices*): The neck; the region between the **thorax** and the **skull** formed by the cervical vertebrae and the hyoid bone.

Circumduction: A circular movement created by the sequential combination of **abduction**, **flexion**, **adduction**, and **extension**.

Condyle: A rounded, knucklelike projection often associated with articular eminences (e.g., the lateral and medial condyles of the femur, the left and right condyles of the mandible, or the occipital condyles).

Condylion laterale (*cdl*): The most lateral point on the mandibular condyle.

Contralateral: Refers to a structure, feature (even motion), etc., on the side of the body opposite that on which a structure, etc., of interest occurs; opposite of **ipsilateral**.

Cornu (pl. *cornua*): A hornlike projection (e.g., the complementary, articulating cornua of the sacrum and coccyx or the greater and lesser horns of the hyoid bone).

Coronal plane: A vertical plane that passes through the body (or structure, e.g., skull) from side to side (i.e., parallel to the coronal suture) and divides the body (or structure) into anterior and posterior portions; it lies perpendicular to the **median plane**.

Coronale (obsolete): The most lateral point on the coronal suture.

Cranium (pl. *crania*): All of the bones of the head with the exception of the mandible.

Crest (also *crista, line,* or *ridge*): A raised, linear structure that surmounts the surface of a bone or forms its border and which is typically more prominent than a ridge (e.g., the iliac crest of the ilium, the supinator crest of the ulna, or the intertrochanteric crest of the femur).

Crista (pl. *cristae*): Synonymous with "crest" but sometimes used interchangeably with "ridge" (e.g., the crista galli of the ethmoid bone).

Crus (pl. *crura*): The leg or lower leg; the region, between the knee and the foot, of the two long bones, the tibia and the fibula.

Cubitus (pl. *cubiti*): The elbow; the complex synovial joint formed by the humerus, ulna, and radius.

Dacryon (*d*; may also be abbreviated as *dk*): 1. (Traditional; considered obsolete by Vallois, 1965) The juncture of the frontolacrimal, frontomaxillary, and lacrimomaxillary sutures; lies between **lacrimale** and **maxillofrontale**. 2. "The apex of the lacrimal fossa, as it impinges on the frontal bone" (Howells, 1973, p. 167); considered to be more easily located and useful in anteroposterior measurements.

Dehiscence: With regard to jaws, a deficiency in bone formation that begins at the alveolar margin and can expose much of the length of a root.

Depression: An inferiorly or inwardly directed movement; opposite of **elevation**.

Digit: A finger (manual digit) or toe (pedal digit) consisting either of two (first digit) or of three (digits II–V) articulating phalanges.

Distal: (Limbs) Away from the trunk or from the articulation of the limb with the axial skeleton; opposite of **proximal**.

Dorsal (also *posterior*): Pertaining to (1) the back (i.e., posterior surface) of the hand or (2) the superior surface of the foot; opposite, respectively, of **palmar** and **plantar**.

Dorsiflexion: A bending of the foot that decreases the angle between it and the anterior surface of the leg (e.g., as in squatting); opposite of **plantar flexion**.

Duct (also *canal*): A tubelike passage [e.g., the frontonasal duct and the acoustic duct (the external auditory meatus)].

Ectoconchion (*ec*; also abbreviated *ek*): The anteriormost point on the lateral rim of the orbit that contributes to the determination of maximum orbital breadth (as measured from **dacryon** or **maxillofrontale**); this point is also used in the measurement of facial flatness.

Ectomolare (*ecm*): The most lateral extent of the alveolar process in the middle of the second maxillary molar on its buccal side.

Elevation: A superiorly or outwardly directed movement; opposite of **depression**.

Eminence: A variably swollen, projecting area of bone (e.g., the parietal eminence, the canine eminence of the maxilla, or the iliopubic eminence).

Endobasion (*endoba*): The most posterior point on the midline of the anterior margin of the foramen magnum; it is usually behind and below **basion** but coincident with it when this margin is thin and sharp.

Endomolare (*enm*; may also be abbreviated as *endo*): The most medial extent of the alveolar process in the middle of the second maxillary molar on its lingual side.

Euryon (*eu*): The point on a parietal that contributes to determining maximum (biparietal) cranial breadth.

Eversion: A movement of the foot that causes its plantar surface to face laterally; opposite of **inversion**.

Extension: A movement in the sagittal plane around a transverse axis that separates two structures (e.g., straightening the leg or arm); opposite of **flexion**.

External (also *exterior*; sometimes incorrectly referred to as *lateral*): Toward the outside; opposite of **internal**.

Facial skeleton: The skeleton of the face, which is composed of two single bones (the vomer and the mandible) and six paired bones (the zygomatics, maxillae, nasals, lacrimals, palatines, and inferior nasal conchae).

Fenestra (pl. *fenestrae*): A windowlike opening; not present in the human skeleton, but most mammals develop anterior palatine fenestrae.

Fenestration: With regard to jaws, a developmental lacuna in the bone above the alveolar margin that exposes a portion of a root.

Fissure: A cleft or slit (e.g., the superior and inferior orbital fissures, the petrotympanic fissure, or the pterygoid fissure).

Flexion: A bending movement in the sagittal plane and around a transverse axis that draws two structures toward each other; opposite of **extension**.

Foramen (pl. *foramina*): A circular to ovoid hole or opening that perforates bone (e.g., the infraorbital foramen of the maxilla, the foramen ovale of the sphenoid, or the foramen magnum of the occipital bone).

Fossa (pl. *fossae*; also *cavity* and *sinus*): A general term for a hollowed-out area (e.g., the mandibular fossa of the temporal bone, the acetabular fossa, or the coronoid fossa of the humerus).

Fovea (pl. *foveae*): A small pit (e.g., the fovea capitis of the femur, the pterygoid fovea of the mandible, or the dental fovea of the atlas).

Frankfort horizontal (FH) plane (also *plane of Virchow*): A horizontal plane on which the anthropometric landmarks porion (i.e., the midpoint of the superior margin of the external auditory meatus) and orbitale (i.e., the inferiormost point on the inferior margin of the orbit) are positioned in order to provide a standard orientation of the skull (e.g., for measuring, describing, illustrating).

Frontal plane: see **coronal plane**.

Frontomalare anterior (*fm:a*): The most anteriorly prominent point on the zygomaticofrontal suture.

Frontotemporale (*ft*): The point at which the temporal ridge is most inwardly or medially curved; also, the point which contributes to the determination of minimum frontal breadth.

Genu (pl. *genua*): The knee; the complex synovial joint formed by the distal femur, patella, or proximal tibia.

Gibbus: Sharply angular kyphosis.

Glabella (*g*; also *metopion*): The anteriormost region (not a point) of the frontal bone above the frontonasal suture and between the superciliary arches. This region generally protrudes anteriorly, but in certain skulls this region may be flat ("hyperfeminine" skulls, infants, and juveniles) or even depressed. When it is not protrusive,

glabella may be identified by a change in the direction of the frontal bone.

Gnathion (*gn*; also *menton*): The "lowest" (anterior- and inferiormost) point on the chin.

Gonion (*go*; may also be abbreviated as *g* or *gn*): The intersection of a line tangent to the posterior margin of the ascending ramus and a line tangent to the inferior margin of the body of the mandible; also, the most marked point of transition in the upward curvature of the posteroinferior margin of the mandible.

Groove: A trench or channel (e.g., the lacrimal groove, the groove on the clavicle for the subclavius muscle, or the intertubercular groove of the humerus).

Hallux (pl. *halluces*): The major or great toe; the first digit of the foot.

Hamulus (pl. *hamuli*): A hooklike projection (e.g., the hamulus of the hamate or the lacrimal hamulus).

Hematogenous: Spread through the bloodstream.

Hiatus (pl. *hiatus, hiatuses*): A gap or gashlike opening [e.g., the sacral hiatus, the maxillary hiatus, or the hiatus of the canal for the greater petrosal nerve (temporal bone)].

Horizontal plane: see **transverse plane**.

Iliospinale: The peak of the anterior superior iliac spine.

Incision (*inc*): The point of contact [identified on the buccal (labial) side] between the mesial occlusal edges of the upper central incisors.

Incisura (pl. *incisurae*; also *notch*): An indentation or notch on the edge of a bone [e.g., the costal incisurae of the sternum (costal notches), the incisura of the mandible (the mandibular notch), and the parietal incisura (notch) of the temporal bone].

Inferior (used in reference to elements of the axial skeleton; in a biped, also *caudal*): In a biped, toward the feet (opposite of **superior**); in a quadruped, synonymous with *ventral*.

Infradentale (*idi*; also *infradentale inferius* and *symphysion*): The superiormost point of the alveolar margin (identified on the anterior surface) between the mandibular central incisors.

Infundibulum (pl. *infundibula*): A funnel-shaped passage. The sole example is the infundibulum of the ethmoid bone.

Inion (*i*): A point at the midline of the superior nuchal lines, which often coincides but should not be identified with the base of the external occipital protuberance. Measurement should not be taken from the protuberance.

Internal (also *interior*; sometimes incorrectly referred to as *medial*): Toward the inside; opposite of **external**.

Inversion: A movement of the foot that causes its plantar surface to face medially; opposite of **eversion**.

Ipsilateral: Referring to a structure, feature (even motion), etc., occurring on the same side of the body as the structure, etc., of interest; opposite of **contralateral**.

Jugale (*ju*): The juncture on the posterior margin of the zygomatic bone of the horizontal and vertical portions (i.e., temporal and frontal processes, respectively) of this bone.

Krotaphion (*k*): The juncture of the spheno-parietal, sphenotemporal, and squamosal sutures; the location of this landmark must be estimated on those individuals in whom there is frontotemporal articulation.

Kyphosis: Anterior deflection of the vertebral column (producing a concomitant outward curvature); severe angulation produces **gibbus**.

Lacrimale (*la*): The juncture of the posterior lacrimal crest and the frontolacrimal suture; lies posterior to **dacryon**.

Lambda [*l*; sometimes abbreviated as *la* (but see lacrimale)]: The juncture in the median (sagittal) plane of the lambdoid and sagittal sutures; should an ossicle be present, the landmark is located by drawing in pencil a continuation of the sutures until these lines intersect.

Lateral: Away from the median plane; opposite of **medial**.

Lateral rotation: Movement of a structure around its longitudinal axis, causing its anterior surface to face laterally; opposite of **medial rotation**.

Line: In general, the least marked of the linear elevations (e.g., the temporal line, the pectineal line of the pubis, or the soleal line of the tibia).

Lipping: The effect of osteophytosis on superior and inferior margins of vertebral bodies.

Lumbus (pl. *lumbi*): The lower back or loin; the region of the back, between the **thorax** and the **pelvis**, consisting of the five lumbar vertebrae.

Lytic: Destructive.

Malleolare: The distalmost extension of the medial malleolus.

Malleolus (pl. *malleoli*): A rounded projection. The only examples are the medial malleolus of the tibia and the lateral malleolus of the fibula.

Mandible: The lower jaw.

Manus (pl. *manus*): The hand (including the **carpus**, **metacarpus**, and manual **digits**), which is formed by 54 bones (including the scaphoid, lunate, triquetrum, pisiform, hamate, capitate, trapezoid, trapezium, metacarpals, and phalanges).

Mastoidale (*ms*): The inferiormost point on the mastoid process when the skull is in the Frankfort horizontal.

Maxillofrontale (*mf*): The juncture of the anterior lacrimal crest and the fronto-maxillary suture; lies anterior to **dacryon**.

Meatus (pl. *meati, meatuses*): A passage or opening, especially the external opening of a duct or canal (e.g., the external and internal acoustic meatus or the three nasal cavity meatuses).

Medial: Toward the midline or median plane; opposite of **lateral**.

Medial rotation: Movement of a structure around its longitudinal axis, causing its anterior surface to face medially; opposite of **lateral rotation**.

Median (sagittal) plane: A vertical plane that passes through the midline of the body (or the skull coincident with the sagittal suture) from front to back, dividing it into symmetrical left and right halves; it lies perpendicular to the **coronal plane**.

Mentale (*z*): The anteriormost point on the margin of the mental foramen (mandible).

Menton: see **gnathion**.

Mesosternale: A point in the midline of the anterior surface of the sternum taken at the level of the facet for the fourth costal cartilage (i.e., the fourth chondrosternal joint).

Metacarpus (pl. *metacarpi*): The palm of the hand; the region of the hand between the **carpus** and the **digits** represented by the metacarpals.

Metatarsus (pl. *metatarsi*): The portion of the foot between the **tarsus** and the pedal **digits** represented by the metatarsals.

Metopion: see **glabella**.

Midsagittal plane: see **median plane**.

Morphogenetic: Shape-producing.

Nariale: The most inferior extent of the anterior margin of the nasal aperture on one side or the other of the anterior nasal spine.

Nasion (*n*): The juncture in the median (sagittal) plane of the frontonasal and (inter)nasal sutures.

Nasospinale (*ns*): A midline point on a line drawn through the inferiormost margins of the nasal aperture; it may coincide but is not synonymous with **acanthion**.

Notch (also *incisura*): A cleft or indentation, particularly on the edge of a bone (e.g., the mastoid notch of the temporal bone, the trochlear notch of the ulna, or the greater sciatic notch of the ilium).

Obelion (obsolete): The point of intersection between a line drawn through the two parietal foramina and the sagittal suture.

Oblique plane: A plane not parallel to the **coronal**, **median**, or **transverse planes**.

Occiput: The back of the head.

Odontogenetic (also **odontogenic**): Tooth-producing.

Ophryon (obsolete): A midline point on a line drawn across the forehead and which lies on a plane that passes through right and left **frontotemporale**.

Opisthion (*o*): A point in the midline of the posterior margin of the foramen magnum.

Opisthocranion (*op*): The most posterior point on the midline of the **skull** when it is oriented in the Frankfort horizontal.

Opposition: A movement of the thumb across the palm such that its "pad" contacts the "pad" of another digit; this movement involves **abduction** with **flexion** and **medial rotation** at the carpometacarpal joint of the thumb.

Orale (*ol*): The point of intersection of a line drawn through the lingualmost surfaces of the upper central incisors and the midline of the premaxillary region of the hard palate.

Orbit: The bony eye socket, the walls of which are formed by portions of the frontal, ethmoid, sphenoid, lacrimal, maxillary, and zygomatic bones.

Orbitale (*or*; may also be abbreviated as *orb*): The most inferior point on the orbital rim; used with **porion** to orient the skull in the Frankfort horizontal.

Osteogenetic (also **osteogenic**): Bone-producing.

Osteophyte: A bony strut or spicule that results from disruption of the bone–cartilage interphase (contrast with *syndesmophyte*) but which is not diagnostic of only one pathology.

Osteophytosis: The process of producing osteophytes.

Ostium (pl. *ostia*): An opening into a tubular structure or between two distinct cavities (e.g., the sphenoidal ostium and the tympanic ostium of the auditory tube).

Palmar (also *volar*): The anterior surface or palm of the hand; opposite of **dorsal**.

Paramedian (parasagittal) plane: A vertical plane parallel to the **median plane**.

Pectoral girdle: The incomplete ring of bone, composed of the scapulae and the clavicles, that provides for the attachment of the upper limbs to the **thorax**.

Pelvis (pl. *pelves*; also *pelvic girdle*): The bowl-shaped ring of bone formed by the two os coxae laterally and the sacrum and coccyx posteriorly that supports the trunk of the body and transmits the weight of the upper body to the legs.

Periodontosis: Recession of the gums leading to exposure of the root.

Pes (pl. *pedes*): The foot, which consists of 52 bones, including the talus, calcaneus, navicular, cuboid, cuneiforms, metatarsals, and phalanges.

Plantar (also *volar*): The inferior surface or sole of the foot; opposite of **dorsal**.

Plantar flexion: A bending of the foot in the direction of its plantar surface such that the top of the foot lies in the same plane as the anterior surface of the leg (e.g., standing on tiptoe); opposite of **dorsiflexion**.

Pogonion (*pg*): The anteriormost midline point of the chin; it lies anterior to and above **gnathion**.

Pollex (pl. *pollices*): The thumb or first **digit** of the hand.

Porion (*po*; also abbreviated as *p*): The most lateral point on the superior margin of the external acoustic (auditory) meatus; it is located vertically over the center and thus at the middle of the superior margin of the meatus.

Posterior (also *dorsal*): Toward the back; opposite of **anterior**.

Process: A general term for a projection (e.g., the anterior clinoid process of the sphenoid, the frontal process of the maxilla, or the coracoid process of the scapula).

Prominence: Another term for a projection (e.g., the styloid prominence at the base of the styloid process of the temporal bone or the mental prominence of the mandible).

Promontory: Usually refers to a smooth elevation (e.g., the promontory of the sacrum).

Pronation: A medial rotation of the forearm, which causes the palm of the hand to face

posteriorly and the radius to cross over the ulna: opposite of **supination**.

Prone: Facing downward; opposite of **supine**.

Prosthion (*pr*; also *exoprosthion*): The anteriormost midline point on the premaxillary alveolar process in the median (sagittal) plane; it is located slightly above **alveolare**, with which it is often confused.

Proteolytic: Protein-destroying.

Protraction (also *protrusion*): An anteriorly directed movement, often used to describe the forward movement of the mandible at the temporomandibular joint; opposite of **retraction**.

Protuberance: A small to medium elevation that may be rounded and smooth (e.g., the internal occipital protuberance and the "hyperfeminine" expression of the external occipital protuberance) or distended and peaked (e.g., the "hypermasculine" expression of the external occipital protuberance).

Proximal: (Limbs) Toward the trunk or toward the articulation of the limb with the axial skeleton; opposite of **distal**.

Pterion (*pt*): A region (not a point) delineated by the frontal bone anteriorly, the squamous portion of the temporal bone posteriorly, and the area along the sphenoparietal suture; the landmark would not be recognized in those individuals with frontotemporal articulation (see **sphenion**).

Pubes: A point in the midline of the superior border of the pubic symphysis.

Pyogenic: Pus-forming.

Radiale: A point on the proximal and posterior margin of the radial head.

Retraction: A posteriorly directed movement, usually used to describe the backward movement of the mandible at the temporomandibular joint; opposite of **protraction**.

Rhinion (*rhi*): The inferiormost point on the (inter)nasal suture.

Ridge: A flaring linear elevation, intermediate in development between a line and a crest, which commonly results from the confluence of two adjacent surfaces (e.g., the lateral and medial supracondylar ridges of the humerus, the longitudinal ridge of the ischium, or the vertical ridge of the patella).

Sagittal plane: see **paramedian plane**.

Sinus (pl. *sinuses*; also *cavity*): A cavity or (less frequently) a channel [e.g., the (venous) sigmoid sinus of the temporal bone, the maxillary sinus, and the frontal sinus].

Skull (also *caput* or, incorrectly, *cranium*): The skeleton of the head, which includes the bones of the **braincase** and those of the **facial skeleton**.

Sphenion (*sphn*): The juncture of the frontal bone with the parietal bone, the greater wing of the sphenoid, and the squamous portion of the temporal bone.

Spine (also *spinous process*): A narrowly elongated projection that is typically broad at its base and blunter than a stylus (e.g., the anterior superior iliac spine, the ischial spine, or the anterior nasal spine of the maxilla).

Staphylion (*sta*): The midline point of a line drawn through the anteriormost invaginations of the posterior margins of the right and left palatine bones; it typically falls anterior to the posterior nasal spine.

Stephanion (*st*): The point at which the inferior temporal line crosses the coronal suture.

Stylion: The distalmost extension of the styloid process of the radius.

Stylus (also *styloid process*): A long, pointed projection (e.g., the styloid process of the temporal bone, the styloid process of the ulna, or the styloid process of the fibula).

Subspinale (*ss*): The deepest midline point in the concavity between the anterior nasal spine and the alveolar margin; it is located on the crest of the midline suture.

Sulcus (pl. *sulci*; also *groove* or *sinus*): A trench or channel [e.g., the preauricular sulcus of the ilium (variably present), the calcaneal sulcus, or the costal sulcus (groove) of the ribs].

Superior (used in reference to the axial skeleton; in a biped, also *cephalic*, *cranial*, or *rostral*): In a biped, toward the head (opposite of **inferior**); in a quadruped, synonymous with **dorsal**.

Supination: A lateral rotation of the forearm that causes the palm of the hand to face anteriorly and the ulna and radius to lie parallel to one another; opposite of **pronation**.

Supine: Facing upward with the back down; opposite of **prone**.

Supradentale: see **prosthion**.

Suprasternale: The most inferior point in the midline of the jugular notch of the sternum.

Symphysion: see **infradentale**.

Syndesmophyte: A bony strut of spicule that derives from ossification of soft tissue (contrast with osteophyte).

Tarsus (pl. *tarsi*): The ankle; the articular region between the foot and the leg formed by seven irregular bones (i.e., the talus, calcaneus, navicular, cuboid, and three cuneiforms).

Thorax (pl. *thoraces*): The chest or upper trunk; the region between the **cervix** and the abdomen that is formed by the thoracic vertebrae, ribs, and sternum.

Tibiale: The most proximal and medial point on the edge of the tibia's medial condyle.

Torus (pl. *tori*): A swollen or bulging projection in the shape of a bar or strut (e.g., a supraorbital torus of the frontal bone, an occipital torus, a mandibular torus, or a palatine torus).

Transcription: The reading of a DNA sequence into its complementary RNA sequence (ultimately leading to RNA translation).

Transcription Factor: A protein that binds to DNA and plays a role in the regulation of gene expression by promoting the transcription or reading of DNA into its complementary messenger RNA sequence (ultimately leading to DNA translation).

Translation: The process of producing proteins from groups of three nucleotide bases (a codon) represented in the messenger RNA that is produced during DNA transcription.

Transverse plane: A horizontal plane that passes through the body (or structure) at right angles to both the **median** and **coronal planes**, dividing it into superior and inferior portions and creating a cross section of the body (or structure).

Trochanter: An expansive, roughened area of bone that is much larger than a *tuberosity*. The sole examples are the greater and lesser trochanters of the femur.

Trophic level: An animal or plant's position in the food chain.

Tubercle: A small, variably rounded, roughened elevation that is smaller than a *tuberosity* (e.g., the infraglenoid tubercle of the scapula, the dorsal tubercle of the radius, or the pubic tubercle of the pubis).

Tuberosity: A medium-sized, variably rounded, and roughened elevation that is often larger than a *tubercle* (e.g., the tibial tuberosity, the gluteal tuberosity of the femur, or the deltoid tuberosity of the humerus).

Vertebral column: The column of bony segments which encapsulates the spinal cord; it is normally composed of 33 vertebrae [seven cervical, 12 thoracic, five lumbar, five sacral (fused into one bone, the sacrum), and four coccygeal (fused into one bone, the coccyx)].

Vertex (*v*): The highest midline point on the skull when it is in the Frankfort horizontal; it may coincide with **apex**.

Zygion (*zy*): The most lateral point on the zygomatic arch.

Zygomaxillare (inferior) (*zm*): The most inferior point on the zygomaticomaxillary suture.

Zygomaxillare anterior (*zm:a*): The point on the facial (anterior) surface where the zygomaticomaxillary suture intersects the attachment scar of the masseter muscle; considered more appropriate than **zygomaxillare (inferior)** for measuring flatness/protrusion of the face.

Zygoorbitale (*zo*): The juncture of the zygomaticomaxillary suture and the inferior orbital rim.

References

Acsádi, Gy., and Nemeskéri, J. 1970. *History of Human Life Span and Mortality*. Budapest: Akademiai Kiado.

Ahlqvist, J., and Damsten, O. 1969. Modification of Kerley's method for the microscopic determination of age in human bone. *J Forensic Sci* 14:205–212.

Andersen, H., and Matthiessen, M. 1967. Histochemistry of the early development of the human central face and nasal cavity with special reference to the movements and fusion of the palatine processes. *Acta Anat 68*:483–508.

Angel, J. L. 1953. The human remains from Khirokitia. In: Dikaios, P. (ed.): *Khirokitia, Final Report*. Nicosia: Cyprus Department of Antiquities, pp. 416–430.

Angel, J. L. 1966. Porotic hyperostosis, anemias, malarias, and marshes in the prehistoric eastern Mediterranean. *Science 153*:760–763.

Angel, J. L., Suchey, J. M., İşcan, M. Y., and Zimmerman, M. R. 1986. Age at death from the skeleton and viscera. In: Zimmerman, M. R., and Angel, J. L. (eds.): *Dating and Age Determination in Biological Materials*. London: Croom Helm, pp. 179–220.

Aufderheide, A. C., and Rodríguez-Martín, C. 1998. *The Cambridge Encyclopedia of Human Paleopathology*. Cambridge: Cambridge University Press.

Baker, B., and Armelagos, G. 1988. The origin and antiquity of syphilis. *Curr Anthropol 29*:703–737.

Bedford, M. E., Russell, K. F., Lovejoy, C. O., Meindl, R. S., Simpson, S. W., and Stuart-Macadam, P. L. 1993. Test of the multifactorial aging method using skeletons with known ages-at-death from the Grant Collection. *Am J Phys Anthropol 91*:287–297.

Berkovitz, B. K. B., and Thomson, P. 1973. Observations on the aetiology of supernumerary upper incisors in the albino ferret (*Mustela putorious*). *Arch Oral Biol 18*:457–463.

Berry, R. J. 1968. The biology of non-metrical variation in mice and men. In: Brothwell, D. R. (ed.): *The Skeletal Biology of Earlier Human Populations*. Oxford: Pergamon, pp. 103–133.

Blumberg, J. M., and Kerley, E. R. 1966. Discussion: A critical consideration of roentgenology and microscopy in palaeopathology. In: Jarcho, S. (ed.): *Human Palaeopathology*. New Haven, CT: Yale University Press, pp. 150–170.

Blumenbach, J. F. 1775 and 1795. *On the Natural Varieties of Mankind (De Generis Humani Varietate Nativa)* (reprinted 1969). New York: Berman.

Bouvier, M., and Ubelaker, D. 1977. A comparison of two methods for the microscopic determination of age at death. *Am J Phys Anthropol 46*:391–394.

Boyd, A. 1971. Comparative histology of mammalian teeth. In: Dahlberg, A. A. (ed.): *Dental Morphology and Evolution*. Chicago: University of Chicago Press, pp. 81–94.

Brooks, S. T. 1955. Skeletal age at death: The reliability of cranial and pubic age indicators. *Am J Phys Anthropol 13*:567–597.

Brooks, S. T., and Suchey, J. M. 1990. Skeletal age determination based on the os pubis: A comparison of the Acsádi and Nemeskéri and Suchey-Brooks methods. *Hum Evol 5*:227–238.

Brothwell, D. R. 1972. *Digging Up Bones*, 2nd ed. London: British Museum (Natural History).

Brown, W. A. B., Molleson, T. I., and Chinn, S. 1984. Enlargement of the frontal sinus. *Ann Hum Biol 11*:221–226.

Buikstra, J. E. 1976. The Caribou Eskimo: General and specific disease. *Am J Phys Anthropol 45*:351–368.

Buikstra, J. E., and Ubelaker, D. E. (eds.) 1994. *Standards for Data Collection from Human Skeletal*

Remains, Research Series 44. Fayetteville: Arkansas Archaeological Survey.

Campbell, T. B. 1939. Food, food values and food habits of the Australian Aborigines in relation to their dental conditions: Part V. *Aust J Dent* 43:177–199.

Caplan, A. I. 1988. The cellular and molecular embryology of bone formation. In: Peck, W. A. (ed.): *Bone and Mineral Research*. New York: Elsevier, pp. 117–184.

Carbonell, V. M. 1963. Variations in the frequency of shovel-shaped incisors in different populations. In: Brothwell, D. R. (ed.): *Dental Anthropology*. New York: Macmillan, pp. 211–234.

Carroll, S. B., Grenier, J. K., and Weatherbee, S. D. 2005. *From DNA to Diversity*, 2nd ed. Malden, MA: Blackwell.

Carter, W., Butterworth, B., Carter, J., and Carter, J. 1987. *Ethnodentistry and Dental Folklore*. Kansas City: Dental Folklore Books.

Chambers, J. T. 1988. The regulation of osteoclastic development and function. In: *Cell and Molecular Biology of Vertebrate Hard Tissues*, Ciba Foundation Symposium 136. New York: John Wiley, pp. 92–104.

Cihák, R. 1972. *Ontogenesis of the Skeleton and Intrinsic Muscles of the Human Hand and Foot*. New York: Springer-Verlag.

Clark, W. E. LeGros. 1966. *The Fossil Evidence for Human Evolution*, 2nd ed. Chicago: University of Chicago Press.

Clarke, N. G., and Hirsch, R. S. 1991. Physiological, pulpal, and periodontal factors influencing alveolar bone. In: Kelley, M. A., and Larsen, C. S. (eds.): *Advances in Dental Anthropology*. New York: Wiley-Liss, pp. 241–266.

Clemente, C. D. (ed.) 1984. *Gray's Anatomy*, 30th American ed. Philadelphia: Lea & Febiger.

Condon, K., Charles, D. K., Cheverud, J. M., and Buikstra, J. E. 1986. Cementum annulation and age determination in *Homo sapiens*. II. Estimates and accuracy. *Am J Phys Anthropol* 71:321–330.

Cruwys, E. 1988. Morphological variation and wear in teeth of Canadian and Greenland Inuit. *Polar Rec* 24:293–298.

Dahlberg, A. A. 1949. The dentition of the American Indian. In: Laughlin, W. S. (ed.): *Papers on the Physical Anthropology of the American Indian*. New York: Viking Fund, pp. 138–176.

Davies, D. M. 1972. *The Influence of Teeth, Diet, and Habits on the Human Face*. London: William Heinemann.

Demirjian, A. 1980. Dental development: A measure of physical maturity. In: Johnston, F. E., and Roche, A. F. (eds.): *Human Physical Growth and Maturation: Methodologies and Factors*. New York: Plenum Press, pp. 83–100.

DeVito, C., and Saunders, S. R. 1990. A discriminant function analysis of deciduous teeth to determine sex. *J Forensic Sci* 35:845–858.

Ducy, P., Schinke, T., and Karsenty, G. 2000. The osteoblast: A sophisticated fibroblast under central surveillance. *Science* 289:1501–1504.

Dudar, J. C., Pfeiffer, S., and Saunders, S. R. 1993. Evaluation of morphological and histological adult sekeltal age-at-death estimation techniques using ribs. *J Forensic Sci* 38:677–685.

Eli, I., Sarnat, H., and Talmi, E. 1989. Effect of the birth process on the neonatal line in primary tooth enamel. *Pediatr Dent* 11:220–223.

Falk, D., and Conroy, G. C. 1983. The cranial venous sinus system in *Australopithecus afarensis*. *Nature* 306:779–781.

Fazekas, I. Gy., and Kósa, F. 1978. *Forensic Fetal Osteology*. Budapest: Akademiai Kiado.

Finnegan, M. 1978. Non-metric variation of the infracranial skeleton. *J Anat* 125:23–37.

Flecker, H. 1942. Time of appearance and fusion of ossification centers as observed by roentgenographic methods. *J Roentgenol Radium Ther* 47:97–157.

Frost, H. M. 1987a. Secondary osteon populations: An algorithm for determining mean bone tissue age. *Ybk Phys Anthropol* 30:221–238.

Frost, H. M. 1987b. Secondary osteon population densities: An algorithm for estimating the missing osteons. *Ybk Phys Anthropol* 30:239–254.

Gans, C., and Northcutt, R. G. 1983. Neural crest and the origin of vertebrates: A new head. *Science* 220:268–274.

Garn, S. M., Silverman, F. N., Hertzog, K. P., and Rohmann, C. G. 1968. Lines and bands of increased density: Their implications to growth and development. *Med Radiogr Photogr* 44:58–88.

Gilbert, B. M., and McKern, T. W. 1973. A method of aging the female os pubis. *Am J Phys Anthropol* 38:31–38.

Giles, E. 1970. Discriminant function sexing of the human skeleton. In: Stewart, T. D. (ed.): *Personal Identification in Mass Disasters*. Washington, DC: National Museum of Natural History, Smithsonian Institution, pp. 99–107.

Giles, E., and Elliot, O. 1963. Sex determination by discriminant function. *Am J Phys Anthropol* 21:53–68.

Glasstone, S. 1967. Morphodifferentiation of teeth in embryonic mandibular segments in tissue culture. *J Dent Res* 46:611–614.

Goodman, A. H., and Armelagos, G. J. 1985. Factors affecting the distribution of enamel hypoplasias within the human permanent dentition. *Am J Phys Anthropol* 68:479–493.

Goodman, A. H., and Rose, J. C. 1991. Dental enamel hypoplasias as indicators of nutritional status. In: Kelley, M. A., and Larlsen, C. S. (eds.): *Advances in Dental Anthropology.* New York: Wiley-Liss, pp. 279–293.

Grüneberg, H. 1963. *The Pathology of Development.* New York: Wiley.

Gustafson, G. 1950. Age determination on teeth. *J Am Dent Assoc 41*:45–54.

Hall, B. K. 1978. *Developmental and Cellular Skeletal Biology.* New York: Academic Press.

Hall, B. K. 1988. The embryonic development of bone. *Am Sci 76*:174–181.

Hanihara, K. 1959. Sex diagnosis of Japanese skulls and scapulae by means of discriminant function [in Japanese with English summary]. *Zinruigaku Zassi 67*:191–197.

Harris, H. A. 1926. The growth of the long bones in childhood, with special reference to certain bony striations of the metaphysis and the role of vitamins. *Arch Intern Med 38*:785–806.

Harris, H. A. 1931. Lines of arrested growth in the long bones in childhood: Correlation of histological and radiographic appearances. *Br J Radiol 4*:561–588.

Hershkovitz, P. 1977. *Living New World Monkeys (Platyrrhini) with an Introduction to the Primates,* vol. 1. Chicago: University of Chicago Press.

Hillson, S. 1986. *Teeth.* Cambridge: Cambridge University Press.

Howells, W. W. 1973. *Evolution of the Genus* Homo. Reading, MA: Addison-Wesley.

Howells, W. W. 1989. *Skull shapes and the map. Craniometric analyses in the dispersion of modern* Homo. Papers of the Peabody Museum of Archaeology and Ethnology, vol. 79. Cambridge: Peabody Museum of Archaeology and Ethnology.

Howells, W. W. 1995. *Who's who in skulls. Ethnic identification of crania from measurements.* Papers of the Peabody Museum of Archaeology and Ethnology, vol. 82. Cambridge: Peabody Museum of Archaeology and Ethnology.

Hunt, E. H., Jr., and Hatch, J. W. 1981. The estimation of age at death and ages of formation of transverse lines from measurements of human long bones. *Am J Phys Anthropol 54*:461–469.

Igarashi, Y., Uesu, K., Wakebe, T., and Kanazawa, E. 2005. New method for estimation of adult skeletal age at death from the morphology of the auricular surface of the ilium. *Am J Phys Anthropol 128*:324–339.

İşcan, M. Y., and Loth, S. R. 1986. Estimation of age and determination of sex from the sternal rib. In: Reichs, K. J. (ed.): *Forensic Osteology.* Springfield, IL: Charles C. Thomas, pp. 68–89.

İşcan, M. Y., and Loth, S. R. 1989. Osteological manifestations of age in the adult. In: İşcan, M. Y., and Kennedy, K. A. R. (eds.): *Reconstruction of Life from the Skeleton.* New York: Alan R. Liss, pp. 23–40.

İşcan, M. Y., Loth, S. R., and Wright, R. K. 1984a. Age estimation from the rib by phase analysis: White males. *J Forensic Sci 29*:1094–1104.

İşcan, M. Y., Loth, S. R., and Wright, R. K. 1984b. Metamorphosis at the sternal rib end: A new method to estimate age at death in white males. *Am J Phys Anthropol 65*:147–156.

İşcan, M. Y., Loth, S. R., and Wright, R. K. 1985. Age estimation from the rib by phase analysis: White females. *J Forensic Sci 30*:853–863.

İşcan, M. Y., Loth, S. R., and Wright, R. K. 1987. Racial variation in the sternal extremity of the rib and its effect on age determination. *J Forensic Sci 32*:452–466.

Jackes, M. 2000. Building the bases for paleodemographic analysis: Adult age determination. In: Katzenberg, M. A., and Saunders, S. R. (eds.): *Biological Anthropology of the Human Skeleton.* New York: Wiley-Liss, pp. 417–466.

Johnston, F. E. 1961. Sequence of epiphyseal union in a prehistoric Kentucky population from Indian Knoll. *Hum Biol 33*:66–81.

Kangas, A. T., Evans, A. R., Thesleff, I., and Jernvall, J. 2004. Nonindependence of mammalian dental characters. *Nature 432*:211–214.

Katayama, K. 1988. Geographic distribution of auditory exostoses in South Pacific human populations. *Man Cult Oceania 4*:63–74.

Katzenberg, M. A. 1989. Stable isotope analysis of archaeological faunal remains from southern Ontario. *J Archaeol Sci 16*:319–329.

Katzenberg, M. A. 1992. Advances in stable isotope analysis of prehistoric bones. In: Saunders, S. R., and Katzenberg, M. A. (eds.): *Skeletal Biology of Past Peoples: Research Methods.* New York: Wiley-Liss, pp. 105–119.

Katzenberg, M. A. 2000. Stable isotope analysis: A tool for studying past diet, demography, and life history. In: Katzenberg, M. A., and Saunders, S. R. (eds.): *Biological Anthropology of the Human Skeleton.* New York: Wiley-Liss, pp. 305–327.

Katzenberg, M. A., and Harrison, R. G. 1987. What's in a bone? Recent advances in archaeological bone chemistry. *J Archaeol Res 5*:265–293.

Kay, R. F. 1977. The evolution of molar occlusion in the Cercopithecidae and early catarrhines. *Am J Phys Anthropol 46*:327–352.

Kelley, M. A. 1989. Infectious disease. In: İşcan, M. Y., and Kennedy, K. A. R. (eds.): *Reconstruction of Life from the Skeleton.* New York: Alan R. Liss, pp. 191–199.

Kemkes-Grottenthaler, A. 1996. Critical evaluation of osteomorphognostic methods to estimate adult age at death: A test of the "complex method." *Homo 46*:280–292.

Kennedy, K. A. R. 1989. Skeletal markers of occupational stress. In: İşcan, M. Y., and Kennedy, K. A.

R. (eds.): *Reconstruction of Life from the Skeleton*. New York: Alan R. Liss, pp. 129–160.

Kerley, E. R. 1965. The microscopic determination of age in human bone. *Am J Phys Anthropol* 23:149–163.

Kerley, E. R. 1970. Estimation of skeletal age: After about age 30. In: Stewart, T. D. (ed.): *Personal Identification in Mass Disasters*. Washington, DC: National Museum of Natural History, Smithsonian Institution, pp. 57–70.

Kerley, E. R., and Ubelaker, D. H. 1978. Revisions in the microscopic method of estimating age at death in human cortical bone. *Am J Phys Anthropol* 49:545–546.

Klepinger, L. L. 1992. Innovative approaches to the study of past human health and subsistence strategies. In: Saunders, S. R., and Katzenberg, M. A. (eds.): *Skeletal Biology of Past Peoples: Research Methods*. New York: Wiley-Liss, pp. 121–130.

Kollar, E. J., and Baird, G. R. 1971. Tissue interactions in developing mouse tooth germs. In: Dahlberg, A. A. (ed.): *Dental Morphology and Evolution*. Chicago: University of Chicago Press, pp. 15–29.

Kollar, E. J., and Fisher, C. 1980. Tooth induction in chick epithelium: Expression of quiescent genes for enamel synthesis. *Science* 207:993–995.

Kovacs, I. 1971. A systematic descripton of dental roots. In: Dahlberg, A. A. (ed.): *Dental Morphology and Evolution*. Chicago: University of Chicago Press, pp. 211–256.

Krogman, W. M. 1962. *The Human Skeleton in Forensic Medicine*. Springfield, IL: Charles C. Thomas.

Krogman, W. M., and İşcan, M. Y. 1986. *The Human Skeleton in Forensic Medicine*, 2nd ed. Springfield, IL: Charles C. Thomas.

Kronfeld, R. 1954. Development and calcification of the human deciduous and permanent dentition. In: *Basic Reading on the Identification of Human Skeletons*. New York: Wenner-Gren Foundation, pp. 3–11.

Lalueza, C., Pérez-Pérez, A., and Turbón, D. 1996. Dietary inferences through buccal microwear analysis of Middle and Upper Pleistocene human fossils. *Am J Phys Anthropol* 100:367–387.

Langdon, J. H. 2005. *The Human Strategy: An Evolutionary Perspective on Human Anatomy*. New York: Oxford University Press.

Larsen, C. S. 1997. *Bioarchaeology: Interpreting Behavior from the Human Skeleton*. Cambridge: Cambridge University Press.

Laughlin, W. S., Harper, A. B., and Thompson, D. D. 1979. New approaches to the pre- and post-contact history of Arctic peoples. *Am J Phys Anthropol* 51:579–588.

Lee-Thorpe, J. A., and van der Merwe, N. J. 1993. Stable carbon isotope studies of Swartkrans fossils. In: Brain, C. K. (ed.): *Swartkrans: A Cave's Chronicle of Early Man*, Transvaal Museum Monograph 8. Pretoria: Transvaal Museum, pp. 251–256.

Lester, C., and Shapiro, H. 1968. Vertebral defects in the lumbar vertebrae of prehistoric American Eskimos. *Am J Phys Anthropol* 28:43–48.

Levine, R. S., Turner, E. P., and Dobbing, J. 1979. Deciduous teeth contain histories of developmental disturbances. *Early Hum Dev* 3:211–220.

Loth, S. R., and İşcan, M. Y. 1987. The effect of racial variation on sex determination from the sternal rib [abstract]. *Am J Phys Anthropol* 72:227.

Lovejoy, C. O. 1985. Dental wear in the Libben population: Its functional pattern and role in the determination of adult skeletal age at death. *Am J Phys Anthropol* 68:47–56.

Lovejoy, C. O., Meindl, R. S., Mensforth, R. P., and Barton, T. J. 1985a. Multifactorial determination of skeletal age at death: A method and blind tests of its accuracy. *Am J Phys Anthropol* 68:1–14.

Lovejoy, C. O., Meindl, R. S., Pryzbeck, T. R., and Mensforth, R. P. 1985b. Chronological metamorphosis of the auricular surface of the ilium: A new method for the determination of adult skeletal age at death. *Am J Phys Anthropol* 68:15–28.

Lukacs, J. R. 1989. Dental paleopathology: Methods for reconstructing dietary patterns. In: İşcan, M. Y., and Kennedy, K. A. R. (eds.): *Reconstruction of Life from the Skeleton*. New York: Alan R. Liss, pp. 261–286.

Lumsden, A. G. S. 1979. Pattern formation in the molar dentition of the mouse. *J Biol Buccale* 7:77–103.

Lumsden, A. G. S. 1980. The developing innervation of the lower jaw and its relation to the formation of tooth germs in mouse embryos. In: Kurtén, B. (ed.): *Teeth: Form, Function, and Evolution*. New York: Columbia University Press, pp. 32–43.

Lumsden, A. G. S. 1988. Spatial organization of the epithelium and the role of neural crest cells in the initiation of the mammalian tooth germ. *Development* 103(Suppl.):155–169.

Lunt, D. A. 1978. Molar attrition in medieval Danes. In: Butler, P. M., and Joysey, K. A. (eds.): *Development, Function, and Evolution of Teeth*. New York: Academic Press, pp. 465–482.

Maresca, B., and Schwartz, J. H. 2006. Sudden origin: A general mechanism of evolution based on stress protein concentration and rapid environmental change. *Anat Rec B New Anat* 289B:38–46.

Maresh, M. M. 1970. Measurements from roentgenograms. In: McCammon, R. W. (ed.): *Human*

Growth and Development. Spring field, IL: C. C. Thomas, pp. 157–200.

Marieb, E. M. 1989. *Human Anatomy and Physiology.* New York: Benjamin/Cummings.

McCollum, M. A., and Sharpe, P. 2002. Developmental genetics and early hominid craniodental evolution. *Bioessays* 23:481–493.

McHenry, H., and Schulz, P. 1976. The association between Harris lines and enamel hypoplasias in prehistoric California Indians. *Am J Phys Anthropol* 44:507–512.

McKern, T. W., and Stewart, T. D. 1957. *Skeletal age changes in young American males.* Technical Report EP-45. Natick, MA: Quartermaster Research and Development Command.

McLean, F. C., and Urist, M. R. 1964. *Bone: An Introduction to the Physiology of Skeletal Tissue,* 2nd ed. Chicago: University of Chicago Press.

Meindl, R. S., and Lovejoy, C. O. 1985. Ectocranial suture closure: A revised method for the determination of skeletal age at death based on the lateral-anterior sutures. *Am J Phys Anthropol* 68:57–66.

Meindl, R. S., Lovejoy, C. O., Mensforth, R. P., and Walker, R. A. 1985. A revised method of age determination using the os pubis, with a review and tests of accuracy of other current methods of pubic symphyseal aging. *Am J Phys Anthropol* 68:29–45.

Meindl, R. S., and Russell, K. F. 1998. Recent advances in method and theory in paleodemography. *Annu Rev Anthropol* 27:375–399.

Melton, D. A. 1991. Pattern formation during animal development. *Science* 252:234–241.

Mensforth, R. P., Lovejoy, C. O., Lallo, J. W., and Armelagos, G. J. 1978. The role of constitutional factors, diet, and infectious disease in the etiology of porotic hyperostosis and periosteal reactions in prehistoric infants and children. *Med Anthropol* 2:1–59.

Merbs, C. F. 1989. Trauma. In: İşcan, M. Y., and Kennedy, K. A. R. (eds.): *Reconstruction of Life from the Skeleton.* New York: Alan R. Liss, pp. 161–189.

Miles, A. E. W. 1963. Dentition in the assessment of individual age. In: Brothwell, D. (ed.): *Dental Anthropology.* New York: Macmillan, pp. 191–209.

Miller, W. A. 1971. Early dental development in mice. In: Dahlberg, A. A. (ed.): *Dental Morphology and Evolution.* Chicago: University of Chicago Press, pp. 31–43.

Molleson, T. 1995. Rates of ageing in the eighteenth century. In: Saunders, S. R., and Herring, A. (eds.): *Grave Reflections: Portraying the Past Through Cemetery Studies.* Toronto: Canadian Scholars' Press, pp. 199–222.

Molleson, T., and Cox, M. 1993. *The Middling Sort. The Spitalfields Project,* vol. 2. *The Anthropology.* CBA Research Report 86. York: York Council for British Archaeology.

Molnar, S. 1971. Human tooth wear, tooth function and cultural variability. *Am J Phys Anthropol* 34:175–190.

Moore, K. L. 1974. *The Developing Human: Clinically Oriented Embryology.* Philadelphia: W. B. Saunders.

Moorrees, C. F. A. 1957. *The Aleut Dentition: A Correlative Study of Dental Characteristics in an Eskimoid People.* Cambridge, MA: Harvard University Press.

Moorrees, C. F. A., Fanning, E. A., and Hunt, E. E., Jr. 1963a. Age variation of formation stages for ten permanent teeth. *J Dent Res* 42:1490–1502.

Moorrees, C. F. A., Fanning, E. A., and Hunt, E. E., Jr. 1963b. Formation and resorption of three deciduous teeth in children. *Am J Phys Anthropol* 21:205–213.

Morgan, T. H. 1903. *Evolution and Adaptation.* New York: Macmillan.

Morgan, T. H. 1916. *A Critique of the Theory of Evolution.* Princeton: University of Princeton Press.

Morse, D. 1978. *Ancient Disease in the Midwest.* Springfield: Illinois State Museum.

Moyers, R. E. 1959. Le stade moyen de calcification est indique pour chaque dent selon les 10 stades de calcification de la table de Nolla. *Rev Odontostomatol (Bordeaux)* 9:1424–1433.

Murray, K. A., and Murray, S. A. 1989. Computer software for hypoplasia analysis. *Am J Phys Anthropol* 78:277–278.

Nery, E. B., Kraus, B. S., and Croup, M. 1970. Timing and topography of early human tooth development. *Arch Oral Biol* 15:1315–1326.

Netter, F. H., and Crelin, E. S. 1987. *The CIBA Collection of Medical Illustrations,* vol. 8. *Musculoskeletal System,* part I. *Anatomy, Physiology and Metabolic Disorders,* sect. II. *Embryology.* Summit, NJ: CIBA-GEIGY Corp.

Nomina Anatomica, 5th ed. 1983. Baltimore: Williams & Wilkins.

Ogden, J. A. 1980. Chondro-osseous development and growth. In: Urist, M. R. (ed.): *Fundamental and Clinical Bone Physiology.* Philadelphia: J. B. Lippincott, pp. 108–171.

Olivier, G. 1960. *Pratique Anthropologique.* Paris: Vigot.

Ooë, T. 1956. On the development of position of the tooth germs in the human deciduous front teeth. *Okajimas Folia Anat Jpn* 28:317–340.

Ooë, T. 1957. On the early development of human dental lamina. *Okajimas Folia Anat Jpn* 30:197–211.

Ooë, T. 1965. A study of the ontogenetic origin of human permanent tooth germs. *Okajimas Folia Anat Jpn 40*:429–437.

Ooë, T. 1969. Epithelial anlagen of human third dentition and their migrations in the mandible and maxilla. *Okajimas Folia Anat Jpn 46*:243–251.

Ooë, T. 1971. Three instances of supernumerary tooth germs observed with serial sections of human foetal jaws. *Z Anat Entwickl Geschungfors 135*:202–209.

Ooë, T. 1979. Development of the human first and second permanent molar, with special reference to the distal portion of the dental lamina. *Anat Embryol 155*:221–240.

Ortner, D. J. 2003a. Infectious diseases: Introduction, biology, osteomyelitis, periositis, brucellosis, glanders, and septic arthritis. In: Ortner, D. J. (ed.): *Identification of Pathological Conditions in Human Skeletal Remains*. New York: Academic Press, pp. 179–226.

Ortner, D. J. 2003b. Infectious diseases: Mycotic, viral, and multicelled parasitic diseases of the human skeleton. In: Ortner, D. J. (ed.): *Identification of Pathological Conditions in Human Skeletal Remains*. New York: Academic Press, pp. 325–341.

Ortner, D. J. 2003c. Methods used in the analysis of skeletal lesions. In: Ortner, D. J. (ed.): *Identification of Pathological Conditions in Human Skeletal Remains*. New York: Academic Press, pp. 45–64.

Ortner, D. J. 2003d. Osteoarthritis and diffuse idiopathic skeletal hyperostosis. In: Ortner, D. J. (ed.): *Identification of Pathological Conditions in Human Skeletal Remains*. New York: Academic Press, pp. 545–560.

Ortner, D. J. 2003e. Theoretical issues in paleopathology. In: Ortner, D. J. (ed.): *Identification of Pathological Conditions in Human Skeletal Remains*. New York: Academic Press, pp. 109–118.

Ortner, D. J., and Putschar, W. G. J. 1981. *Identification of Pathological Conditions in Human Skeletal Remains*. Washington, DC: Smithsonian Institution Press.

Osborn, J. W. 1970. New approach to Zahnreihen. *Nature 225*:343–346.

Osborn, J. W. 1971. The ontogeny of tooth succession in *Lacerta vivipara* Jacquin (1787). *Proc R Soc Lond B Biol Sci 179*:261–289.

Osborn, J. W. 1973. The evolution of dentitions. *Am Sci 61*:548–559.

Osborn, J. W. 1978. Morphogenetic gradients: fields versus clones. In: Butler, P. M., and Joysey, K. A. (eds.): *Development, Function, and Evolution of Teeth*. New York: Academic Press, pp. 171–201.

Ossenberg, N. S. 1969. Discontinuous morphological variation in the human cranium [PhD thesis]. Toronto: University of Toronto.

Ossenberg, N. S. 1970. The influence of artificial cranial deformation on discontinuous morphological traits. *Am J Phys Anthropol 33*:357–372.

Ossenberg, N. S. 1976. Within and between race distances in population studies based on discrete traits of the human skull. *Am J Phys Anthropol 45*:701–716.

Parfitt, A. M. 1979. Quantum concept of bone remodeling and turnover: Implications for the pathogenesis of osteoporosis. *Calcif Tissue Int 28*:1–5.

Parfitt, A. M. 1983. The physiologic and clinical significance of bone histomorphometric data. In: Recker, R. R. (ed.): *Bone Histomorphometry: Techniques and Interpretation*. Boca Raton, FL: CRC Press, pp. 143–223.

Pfeiffer, S. 1998. Variability in osteon size in recent human populations. *Am J Phys Anthropol 106*:219–227.

Pfeiffer, S., Lazenby, R., and Chiang, J. 1995. Cortical remodeling data are affected by sampling location. *Am J Phys Anthropol 96*:89–92.

Phenice, T. W. 1969. A newly developed visual method of sexing the os pubis. *Am J Phys Anthropol 30*:297–301.

Pietrusewsky, M. 2000. Metric analysis of skeletal remains: Methods and applications. In: Katzenberg, M. A., and Saunders, S. R. (eds.): *Biological Anthropology of the Human Skeleton*. New York: Wiley-Liss, pp. 375–415.

Putschar, W. G. J. 1966. Problems in the pathology and palaeopathology of bone. In: Jarcho, S. (ed.): *Human Palaeopathology*. New Haven, CT: Yale University Press, pp. 57–63.

Redfield, A. 1970. A new aid to aging immature skeletons: Development of the occipital bone. *Am J Phys Anthropol 33*:207–220.

Reed, J. C. 2006. The utility of cladistic analysis of nonmetric traits for biodistance analysis. [PhD thesis]. Pittsburgh: University of Pittsburgh.

Resnick, D., and Niwayama, G. 1988. *Diagnosis of Bone and Joint Disorders*. Philadelphia: W. B. Saunders.

Robling, A. G., and Stout, S. D. 2000. Histomorphometry of human cortical bone: Applications to age estimation. In: Katzenberg, M. A., and Saunders, S. R. (eds.): *Biological Anthropology of the Human Skeleton*. New York: Wiley-Liss, pp. 187–205.

Romero, J. 1970. Dental mutilation, trephination, and cranial deformation. In: Stewart, T. D. (ed.): *Handbook of Middle American Indians*, vol. 9. Austin: University of Texas Press, pp. 50–67.

Rothschild, B. M., and Martin, L. D. 1993. *Paleopathology: Disease in the Fossil Record*. Boca Raton, FL: CRC Press.

Russell, K. F., Simpson, S. W., Genovese, J., Kinkel, M. D., Meindl, R. S., and Lovejoy, C. O. 1993. Independent test of the fourth rib aging technique. *Am J Phys Anthropol* 92:53–62.

Sandberg, M. M. 1991. Matrix in cartilage and bone development: Current views on the function and regulation of major organic components. *Ann Med* 23:207–217.

Saunders, S. R. 1978. *The development and distribution of discontinuous morphological variation of the human infracranial skeleton*, Paper 81. Ottowa: National Museums of Canada, Archaeological Survey of Canada.

Saunders, S. R. 1989. Nonmetric skeletal variation. In: İşcan, M. Y., and Kennedy, K. A. R. (eds.): *Reconstruction of Life from the Skeleton*. New York: Alan R. Liss, pp. 95–108.

Saunders, S. R. 2000. Subadult skeletons and growth-related studies. In: Katzenberg, M. A., and Saunders, S. R. (eds.): *Biological Anthropology of the Human Skeleton*. New York: Wiley-Liss, pp. 135–161.

Scheuer, L., and Black, S. 2000. *Developmental Juvenile Osteology*. New York: Academic Press.

Schoeninger, M. J., and Moore, K. 1992. Bone stable isotope studies in archaeology. *J Arch Sci* 6:247–296.

Schranz, D. 1959. Kritik der Auswertung der Altersbestimmungsmerkmale van Zähnen und Knochen. *Deutsche Z Gesellschaft Gerichtllungskeit Med* 48:562–575.

Schultz, A. H. 1936. Characters common to higher primates and characters specific for man. *Q Rev Biol* 11:259–283, 425–455.

Schutkowski, H. 1993. Sex determination of infant and juvenile skeletons: I. Morphognostic features. *Am J Phys Anthropol* 90:199–205.

Schwartz, J. H. 1974. The human remains from Kition and Hala Sultan Tekke: A cultural interpretation. In: Karageorghis, V. (ed.): *Excavations at Kition*, vol. I. *The Tombs*. Nicosia: Cyprus Department of Antiquities, pp. 151–162.

Schwartz, J. H. 1980. Morphological approach to heterodonty and homology. In: Kurtén, B. (ed.): *Teeth: Form, Function, and Evolution*. New York: Columbia University Press, pp. 123–144.

Schwartz, J. H. 1982. Dentofacial growth and development in *Homo sapiens*: Evidence from perinatal individuals from Punic Carthage. *Anat Anzeiger Jena* 152:1–26.

Schwartz, J. H. 1983. Premaxillary–maxillary suture asymmetry in a juvenile *Gorilla*: Implications for understanding dentofacial growth and development. *Folia Primatol* 40:69–82.

Schwartz, J. H. 1984. Supernumerary teeth in anthropoid primates and models of tooth development. *Arch Oral Biol* 29:833–842.

Schwartz, J. H. 1986. Primate systematics and a classification of the order. In: Swindler, D., and Erwin, J. (eds.): *Comparative Primate Biology*, vol. 1. *Systematics, Evolution and Anatomy*. New York: Alan R. Liss, pp. 1–41.

Schwartz, J. H. 1987. *The Red Ape: Orang-utans and Human Origins*. Boston: Houghton Mifflin.

Schwartz, J. H. 1988. History, morphology, paleontology, and evolution. In: Schwartz, J. H. (ed.): *Orang-utan Biology*. New York: Oxford University Press, pp. 69–85.

Schwartz, J. H. 1989. The Tophet and "sacrifice" at Phoenician Carthage: An osteologist's perspective. *Terra* 28:16–25.

Schwartz, J. H. 1995. *Skeleton Keys: An Introduction to Human Skeletal Morphology, Development, and Analysis*. New York: Oxford University Press.

Schwartz, J. H. 1997. *Lufengpithecus* and hominoid phylogeny: Problems in delineating and evaluating phylogenetically relevant characters. In: Begun, D., Ward, C., and Rose, M. (eds.): *Miocene Hominoid Fossils: Functional and Phylogenetic Implications*. New York: Plenum Press, pp. 363–388.

Schwartz, J. H. 1999a. Can we really identify species, fossil or living? *Anthropologie XXXVII*:221–230.

Schwartz, J. H. 1999b. *Sudden Origins: Fossils, Genes, and the Emergence of Species*. New York: John Wiley.

Schwartz, J. H. 2004. Barking up the wrong ape—australopiths and the quest for chimpanzee characters in hominid fossils. *Coll Antropol* 28(Suppl. 2):87–101.

Schwartz, J. H. 2005a. Molecular systematics and evolution. In: Meyer, R. A. (ed.): *Encyclopedia of Molecular Cell Biology and Molecular Medicine*. Weinheim: Wiley-VCH Verlag, pp. 515–540.

Schwartz, J. H. 2005b. *The Red Ape: Orangutans and Human Origins*, rev. ed. Boulder, CO: Westview Press.

Schwartz, J. H. 2006. Decisions, decisions: Why Thomas Hunt Morgan was not the "father" of evo-devo. *Philos Sci*.

Schwartz, J. H., and Brauer, J. 1990. The Ipiutak dentition: Implications for interpreting sinodonty and sundadonty [abstract]. *Am J Phys Anthropol* 81:292.

Schwartz, J. H., and Tattersall, I. 2000. The human chin what is it and who has it? *J Hum Evol* 38:367–409.

Schwartz, J. H., and Tattersall, I. 2002. *The Human Fossil Record*, vol. 1. *Terminology and Craniodental Morphology of Genus* Homo (*Europe*). New York: Wiley-Liss.

Schwartz, J. H., and Tattersall, I. 2003. *The Human Fossil Record*, vol. 2. *Craniodental Morphology of Genus* Homo (*Africa and Asia*). New York: Wiley-Liss.

Schwartz, J. H., and Tattersall, I. 2005. *The Human Fossil Record*, vol. 4. *Craniodental Morphology of Australopithecus, Paranthropus, and* Orrorin. New York: Wiley-Liss.

Schwartz, J. H., Tattersall, I., and Laitman, J. L. 1999. New thoughts on Neanderthal behavior: Evidence from nasal morphology. In: Ullrich, H. (ed.): *Hominid Evolution: Lifestyles and Survival Strategies*. Schwelm: Edition Archaea, pp. 166–186.

Schwartz, J. H., and Yamada, T. 1998. Carpal anatomy and primate relationships. *Anthropol Sci 106*(Suppl.):47–65.

Scott, G. R. 1972. An analysis of population and family data on Carabelli's trait and shovel-shaped incisors [abstract]. *Am J Phys Anthropol 37*:449.

Scott, G. R., and Dahlberg, A. A. 1980. Microdifferentiation in tooth crown morphology among Indians of the American Southwest. In: Kurtén, B. (ed.): *Teeth: Form, Function, and Evolution*. New York: Columbia University Press, pp. 259–291.

Singer, R. 1953. Estimation of age from cranial suture closure. *J Forensic Med 1*:52–59.

Skinner, M., and Goodman, A. H. 1992. Anthropological uses of developmental defects of enamel. In: Saunders, S. R., and Katzenberg, M. A. (eds.): *Skeletal Biology of Past Peoples: Research Methods*. New York: Wiley-Liss, pp. 153–174.

Smith, B. H. 1991. Standards of human tooth formation and dental age assessment. In: Kelley, M. A., and Larsen, C. S. (eds.): *Advances in Dental Anthropology*. New York: Wiley-Liss, pp. 143–168.

Smith, P., and Avishai, G. 2004. The use of dental criteria for estimating postnatal survival in skeletal remains of infants. *J Archaeol Sci 32*:83–89.

Spalding, K. L., Buchholz, B. A., Bergman, L.-E., Druid, H., and Frisén, J. 2005. Age written in teeth by nuclear tests. *Nature 437*:333–334.

Sponheimer, M., and Lee-Thorpe, J. A. 1999. Isotopic evidence for the diet of an early hominid, *Australopithecus africanus*. *Science 283*:368–370.

Stewart, T. D. 1934. Sequence of epiphyseal union, third molar eruption and suture closure in Eskimos and American Indians. *Am J Phys Anthropol 19*:433–452.

Stewart, T. D. 1958. The rate of development of vertebral osteoarthritis in American whites and its significance in skeletal age identification. *Leech 28*:114–151.

Stewart, T. D. 1979. *Essential of Forensic Anthropology: Especially as Developed in the United States*. Springfield, IL: Charles C. Thomas.

Stewart, T. D., and McCormick, W. F. 1983. The gender predictive value of sternal length. *Am J Forensic Med Pathol 4*:217–220.

St. Hoyme, L. E., and İşcan, M. Y. 1989. Determination of sex and race: Accuracy and assumptions. In: İşcan, M. Y., and Kennedy, K. A. R. (eds.): *Reconstruction of Life from the Skeleton*. New York: Alan R. Liss, pp. 53–93.

Stout, S. D. 1989. Histomorphometric analysis of human skeletal remains. In: İşcan, M. Y., and Kennedy, K. A. R. (eds.): *Reconstruction of Life from the Skeleton*. New York: Alan R. Liss, pp. 41–52.

Stout, S. D., and Paine, R. R. 1992. Histological age estimation using rib and clavicle. *Am J Phys Anthropol 87*:111–115.

Stout, S. D., and Paine, R. R. 1994. Bone remodeling rates: A test of an algorithm for estimating missing osteons. *Am J Phys Anthropol 93*:123–129.

Stout, S. D., Porro, M. A., and Perotti, B. 1996. A test and correction of the clavicle method of Stout and Paine for histological age estimation of skeletal remains. *Am J Phys Anthropol 100*:111–115.

Stout, S. D., and Teitelbaum, S. L. 1976. Histological analysis of undecalcified thin sections of archeological bone. *Am J Phys Anthropol 44*:263–267.

Stringer, C. B., Hublin, J. J., and Vandermeersch, B. 1984. The origin of anatomically modern humans in western Europe. In: Smith, F. H., and Spencer, F. (eds.): *The Origins of Modern Humans: A World Survey of the Fossil Evidence*. New York: Alan R. Liss, pp. 51–135.

Stuart-Macadam, P. 1989. Porotic hyperostosis: Relationship between orbital and vault lesions. *Am J Phys Anthropol 80*:187–193.

Suchey, J. M., and Katz, D. 1986. Skeletal age standards derived from an extensive multiracial sample of modern Americans [abstract]. *Am J Phys Anthropol 69*:269.

Swindler, D. R. 1976. *Dentition of Living Primates*. New York: Academic Press.

Swindler, D. R., and Olshan, A. F. 1988. Comparative and evolutionary aspects of the permanent dentition. In: Schwartz, J. H. (ed.): *Orang-utan Biology*. New York: Oxford University Press, pp. 271–282.

Tattersall, I., and Schwartz, J. H. 1991. Phylogeny and nomenclature in the "*Lemur*-group" of Malagasy strepsirhine primates. *Anthropol Papers Am Mus Nat Hist 69*:1–18.

Ten Cate, A. R. 1989. Physiological tooth movement: Eruption and shedding. In: Ten Cate, A. R. (ed.): *Oral Histology: Development, Structure, and Function*. St. Louis: C. V. Mosby, pp. 275–298.

Ten Cate, A. R., and Mills, C. 1972. The development of the periodontium: The origin of alveolar bone. *Anat Rec 173*:69.

Ten Cate, A. R., and Osborn, J. W. 1976. *Advanced Dental Histology*, 3rd ed. Bristol: John Wright & Sons.

Thompson, D. D. 1978. Age-related changes in osteon remodelling and bone mineralization [PhD thesis]. Storrs, CT: University of Connecticut.

Tobias, P. V. T. 1967. *Olduvai Gorge*, vol. 2. *The Cranium and Maxillary Dentition of* Australopithecus (Zinjanthropus) boisei. Cambridge: Cambridge University Press.

Todd, T. W. 1920. Age changes in the pubic bone: I. The white male pubis. *Am J Phys Anthropol* 3:285–334.

Todd, T. W. 1930. Age changes in the pubic bone: VIII. Roentgenographic differentiation. *Am J Phys Anthropol* 14:255–271.

Todd, T. W., and Lyon, D. W., Jr. 1924. Endocranial suture closure: Its progress and age relationship. Part I. Adult males of white stock. *Am J Phys Anthropol* 7:324–384.

Todd, T. W., and Lyon, D. W., Jr. 1925a. Cranial suture closure: Its progress and age relationship. Part II. Ectocranial closure of adult males of white stock. *Am J Phys Anthropol* 8:23–43.

Todd, T. W., and Lyon, D. W., Jr. 1925b. Cranial suture closure: Its progress and age relationship. Part III. Endocranial suture closure of adult males of Negro stock. *Am J Phys Anthropol* 8:47–71.

Todd, T. W., and Lyon, D. W., Jr. 1925c. Cranial suture closure. Its progress and age relationship. Part IV. Ectocranial suture closure of adult males of Negro stock. *Am J Phys Anthropol* 8:149–168.

Tonge, C. H. 1976. Morphogenesis and development of teeth. In: Cohen, B., and Kramer, I. R. H. (eds.): *Scientific Foundations of Dentistry*. London: William Heinemann Medical Books, pp. 325–334.

Trotter, M. 1970. Estimation of stature from intact long limb bones. In: Stewart, T. D. (ed.): *Personal Identification in Mass Disasters*. Washington, DC: National Museum of Natural History.

Trotter, M., and Gleser, G. C. 1977. Corrigenda to "Estimation of stature from long limb bones of American whites and Negroes," *American Journal of Physical Anthropology* (1952). *Am J Phys Anthropol* 47:355–356.

Tsukahara, J., and Hall, B. K. 1994. Transmembrane signaling in bone cell differentiation. In: Hall, B. K. (ed.): *Bone*, vol. XII. *Mechanisms of Bone Development and Growth*. Ann Arbor: CRC Press, pp. 109–135.

Tucker, A., and Sharpe, P. 2004. The cutting-edge of mammalian development: How the embryo makes teeth. *Nat Rev Genet* 5:499–508.

Turner, C. G. II 1984. Advances in the dental search for Native American origins. *Acta Anthropol* 8:23–78.

Turner, W. 1886. The index of the pelvic brim as a basis of classification. *J Anat* 20:125–143.

Ubelaker, D. H. 1999. *Human Skeletal Remains: Excavation, Analysis, Interpretation*, 3rd ed. Washington, DC: Taraxacum.

Vallois, H. V. 1960. Vital statistics in prehistoric populations as determined from archaeological data. In: Heizer, R. F., and Cook, S. F. (eds.): *The Application of Quantitative Methods in Archaeology*. Chicago: Quadrangle Books, pp. 186–204.

Vallois, H. V. 1965. Anthropometric techniques. *Curr Anthropol* 6:127–143.

Van Vark, G. N., and Schaafsma, W. 1992. Advances in the quantitative analysis of skeletal morphology. In: Saunders, S. R., and Katzenberg, M. A. (eds.): *Skeletal Biology of Past Peoples: Research Methods*. New York: Wiley-Liss, pp. 225–257.

Weaver, D. S. 1979. Application of the likelihood ratio test to age estimation using the infant and child temporal bone. *Am J Phys Anthropol* 50:263–269.

Weisl, H. 1954. The articular surfaces of the sacroiliac joint and their relation to movements of the sacrum. *Acta Anat* 22:1–14.

Whittaker, D. K., and Richards, D. 1978. Scanning electron microscopy of the neonatal line in human enamel. *Arch Oral Biol* 23:45–50.

Wilder, H. H., and Wentworth, B. 1918. *Personal Identification: Methods for the Identification of Individuals, Living or Dead*. Boston: Gorham.

Winkler, L. A., Schwartz, J. H., and Swindler, D. R. 1991. Aspects of dental development in the orangutan prior to eruption of the permanent dentition. *Am J Phys Anthropol* 86:255–271.

Wolpert, L., Beddington, R., Jesseu, T., Lawrence, P., Meyerowitz, E., and Smith, J. 2002. *Principles of Development*. New York: Oxford University Press.

Wood, J. W., Milner, G. R., Harpending, H. C., and Weiss, K. M. 1992. The osteological paradox: Problems in inferring prehistoric health from skeletal samples. *Curr Anthropol* 33:343–370.

Index